Liquid INTELLIGENCE

W. W. NORTON & COMPANY NEW YORK LONDON

LIQUID INTELLIGENCE

THE ART AND SCIENCE
OF THE PERFECT COCKTAIL

DAVE ARNOLD

PHOTOGRAPHY BY TRAVIS HUGGETT

Page 267: Milk punch recipe by Benjamin Franklin, 11 October 1763.
Original manuscript from the Bowdoin and Temple Papers in the Winthrop
Family Papers. Massachusetts Historical Society.

For information about permission to reproduce selections from this book,
write to Permissions, W. W. Norton & Company, Inc.,
500 Fifth Avenue, New York, NY 10110

For information about special discounts for bulk purchases, please contact
W. W. Norton Special Sales at specialsales@wwnorton.com or 800-233-4830

Manufacturing by RR Donnelley, Shenzhen
Book design by Marysarah Quinn
Production manager: Anna Oler

Library of Congress Cataloging-in-Publication Data

Arnold, Dave, 1971– author.
Liquid intelligence : the art and science of the perfect cocktail /
Dave Arnold ; photography by Travis Huggett. — First edition.
    pages cm
 Includes bibliographical references and index.
 ISBN 978-0-393-08903-5 (hardcover)
1. Cocktails. 2. Measuring instruments. I. Title.
 TX951.A675 2014
 641.87'4—dc23
                                        2014022332

W. W. Norton & Company, Inc.
500 Fifth Avenue, New York, N.Y. 10110
www.wwnorton.com

W. W. Norton & Company Ltd.
Castle House, 75/76 Wells Street, London W1T 3QT

1 2 3 4 5 6 7 8 9 0

FOR MY WIFE, JENNIFER, AND MY SONS, BOOKER AND DAX

CONTENTS

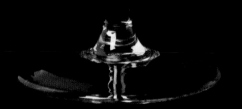

MANHATTAN

ACKNOWLEDGMENTS

Thanks to my wife, Jennifer, who spent an inordinate amount of time beating my rambling, incoherent first drafts into the language we call English, then revising and polishing the finished book. I have loved her clarity of thought and distaste for nonsense since we first met. Without her I could not write any sort of book anyone would want to read.

Thanks to my editor at W. W. Norton, Maria Guarnaschelli, who invited me to lunch one day three years ago and decided I was going to write a book. I often hear from author friends that old-school editors don't exist. I now know that they are wrong. I am extremely fortunate to have had such a passionate, involved, and caring advocate to shepherd this book into existence.

Thanks to:
Nastassia Lopez, for keeping the wheels on the bus.
My partners, the inimitable David Chang and Drew Salmon, for believing in Booker and Dax and in me. I am forever grateful.
Travis Huggett, for such a fine job on the photography.
Mitchell Kohles, for doing a lot of heavy lifting at W. W. Norton.
Harold McGee, for the constant inspiration and for his thoughtful review of drafts of this book.
James Carpenter, for going through the manuscript with his laser eye for detail.
John McGee, Paul Adams, Ariel Johnson, and Don Lee for their input.

And thanks to the people who have inspired me and supported me on my decidedly non-linear career path, even when my endeavors seemed far-fetched:
My mom and Gerard, my dad, and the whole Addonizio/Arnold clan; Maile, Ridge, and the rest of the Carpenters.
Early, Ludwick, Sweeney, and Strauss, who helped foot the bill.
Jeffrey Steingarten and Dana Cowin, early supporters who helped me find my place in the food world.

Michael Batterberry, who got me my first real job in the business.

Wylie Dufresne, who started me on the road to using new technologies in the kitchen.

Nils Noren, with whom I've done some of my best work—Skoal!

Johnny Iuzzini, from whom I learned a lot when we were knocking around together.

The chefs I worked with at the FCI, and the students in my FCI intern program (with a special shout-out to Mindy Nguyen.)

The talented crew I've worked with at Booker and Dax: Piper Kristensen, Tristan Willey, Robby Nelson, Maura McGuigan, and all the bartenders, bar-backs, servers and hosts—you too Ssam crew!

Finally, thank you to all my friends in the cocktail community—you motivate me every day.

LIQUID INTELLIGENCE

CHARTRUTH

INTRODUCTION

Cocktails are problems in need of solutions. How can I achieve a particular taste, texture, or look? How can I make the drink in front of me better? Taking cocktails seriously, as with all worthy inquiries, puts you on a lifelong journey. The more you know, the more questions you raise. The better a practitioner you become, the more you see the faults in your technique. Perfection is the goal, but perfection is, mercifully, unattainable. I have spent seven years and thousands of dollars on the problem of the perfect gin and tonic; I still have work to do. How boring if I were finished—if I were satisfied. Learning, studying, practicing—and drinking with friends: that is what this book is about. The premise: no cocktail detail is uninteresting, none unworthy of study.

A little dose of science will do you good. **THINK LIKE A SCIENTIST AND YOU WILL MAKE BETTER DRINKS**. You don't need to be a scientist, or even understand much science, to use the scientific method to your advantage. Control variables, observe, and test your results; that's pretty much it. This book shows you how to make your drinks more consistent, how to make them consistently better, and how to develop delicious new recipes without taking random shots in the dark.

Sometimes on our journey, in hot pursuit of a particular flavor or idea, I use methods that are preposterous and equipment that is unattainable for most readers. You'll see what it means to run an idea into the ground. I also hope you'll be entertained. I don't expect most people to tackle these more involved drinks, but you'll get enough information to give them a go if you are willing and able. I'll hold nothing back and keep no secrets, which means there will be as many mishaps as successes (mistakes are often the origins of my best ideas). Last, I promise that there will be plenty of techniques, flavors, and drink ideas you can use, even if all you have is a set of cocktail shakers and some ice. I'm out to change the way you look at drinks, no matter what kind of drinks you make.

This is not a book on molecular mixology (a term I detest). The connotations of *molecular* are all bad: gimmicks for gimmicks' sake, drinks that don't taste very good, science gone wrong. My guidelines are simple:

- Use new techniques and technologies only when they make the drink taste better.
- Strive to make an amazing drink with fewer rather than more ingredients.
- Don't expect a guest to know how you made a drink in order to enjoy it.
- Gauge success by whether your guest orders another, not by whether he or she thinks the drink is "interesting."
- Build and follow your palate.

This book is divided into four parts. The first part deals with preliminaries—equipment and ingredients—that pave the way for the rest. The second part is a careful study of how classic cocktails work: the basics of shaker, mixing glass, ice, and liquor. The third part is an overview of newer techniques and ideas and how they relate to classic cocktails. The last part is a series of recipes, mini-journeys based on a particular idea. At the end you will find an annotated bibliography of cocktail books, science books, and cookbooks, plus journal articles that I find interesting and germane to our subject.

WHAT I AM THINKING ABOUT PRETTY MUCH ALL THE TIME

I approach cocktails like everything I care about in life: persistently and from the ground up. Often I perceive an irksome problem in an existing cocktail, or become entranced with an idea or flavor, and my journey begins. I ask myself what I want to achieve, and then I beat down every path to get there. I want to see what is possible and what I'm capable of. In the initial phases of working through a problem, I don't much care if what I'm doing is reasonable. I prefer to go to absurd lengths to gain minute increments of improvement. I am okay with spending a week preparing a drink that's only marginally better than the one that took me five minutes. I'm interested in the margins. That's where I learn about the drink, about myself, and about the world. Sounds grandiose, but I mean it.

I am not unhappy, but I am never satisfied. There's always a better way. Constantly questioning yourself—especially your basic tenets and practices—makes you a better person behind the bar, in front of the stove, or in whatever field you choose. I love it when my dearly held beliefs are proved wrong. It means I'm alive and still learning.

I hate compromising, and I hate cutting corners, but sometimes I have to. You need to keep hating compromise at every turn while knowing how to compromise with minimum impact when necessary. Always be focused on the critical path to quality, from raw ingredients to the cup. I am often surprised by how much work someone will put into making ingredients for a drink, only to destroy all that work at the last moment. Remember, a drink can be ruined at any stage of its creation. Your responsibility for vigilance as a drink maker doesn't end until the drink is finished—and your responsibility as an alcoholic-drink maker doesn't end until the imbiber is safe and sound at home.

PART 1

PRELIMINARIES

Measurement, Units, Equipment

Having access to cool equipment has helped me develop ways of achieving good results *without* the equipment. In this section we'll look at the equipment I use at home and at my bar, Booker and Dax. Almost no one—not even well-heeled professional bartenders—will want or need all the equipment on this list. In the technique-based sections of the book I'll give you workarounds for the bigger-ticket and hard-to-find items as often as I can. At the end of this section you'll find shopping lists organized by budget and interest.

A note on measurement, before we launch into a discussion of tools.

HOW AND WHY YOU SHOULD MEASURE DRINKS

Drinks should be measured by volume. I am a big believer in cooking by weight, but I mix drinks by volume, and so should you. Pouring out small volumes is much faster than weighing a bunch of small ingredients. Furthermore, the densities of cocktail ingredients vary wildly, from about .94 grams per milliliter for straight booze to 1.33 grams per milliliter for maple syrup. For the bartender, the weight of the finished beverage isn't important, but the volume is. The volume determines how close the top of the finished drink will be to the rim of the glass. This liquid line is called the **wash line**, and maintaining a proper wash line is essential to good bartending. In a professional setting, it is essential that your drinks be consistent. Having standard wash lines for each drink you prepare gives you an instant visual check that everything is okay. If your wash line is wrong, something is wrong with the drink. Consistent wash lines are also important to your guests' well-being. Two people get the same drink, but one drink sits higher in the glass: do you like the person with the taller pour more, or are your techniques just a bit shaky?

Advocates of the free-pour don't measure their drinks with measuring tools. Some free-pourers gauge how much they have poured by looking at the liquid levels on the side of glass mixing cups. These bartenders recognize through practice what liquid increments look like in a standard mixing glass. Other free-pourers use speed-pour bottle tops, which produce a steady stream of liquor. Speed-pour mavens judge how much they pour by counting off the length of time they pour: so many counts equals so many ounces. These bartenders practice for hours and hours to attain a consistent counting technique.

Why do free-pourers eschew measuring cups? There are four main schools of thought: the Lazy, the Speedy, the Artist, and the Monk. The Lazy just don't care if they are accurate; enough said about them. The Speedy believe that free-pouring is accurate enough and saves valuable time behind the bar—a couple seconds saved on each drink adds up to a lot of time when customers are six deep at a busy bar. The Lazy and the Speedy free-pourers won't achieve accurate and repeatable results. Free-pour techniques are particularly inappropriate for the home bartender, who doesn't pour out dozens of drinks each night, hasn't practiced a lot, and should be spending the time to get each drink right. Free-pour Artists believe that measuring cups make them look unskilled, unpracticed, and lacking in finesse. I disagree. Masterful jigger-work is a pleasure to behold, and being accurate doesn't make you robotic. No free-pourer can be as consistent over as many different drinks in as many different conditions as someone who measures. The most intriguing argument in favor of free-pouring comes from the Monk, who believes that drinks should not be constrained to easily remembered recipes given in quarter-ounce increments. After all, why should we believe that exact quarter-ounce increments provide the ideal proportions for a drink? Monks pour by feel, tasting as they go, establishing the correct proportions for each drink individually by intuition and by how they size up their guests' tastes.

I like free-pour Monks, but in daily practice it is much better to have a standard recipe that you can remember and consistently follow than to worry about the constraints of the measuring system. Fixed recipes can still allow

The "wash line," where I'm pointing, is where the top of a drink hits on the glass.

IMPORTANT: UNITS IN THIS BOOK

In this book I always equate 1 ounce with 30 milliliters. My jiggers are sized this way as well. If you have a jigger set that uses a different-sized "ounce," your measurements will be consistently slightly different from mine—not usually a problem. Here's my decoder ring:

- Volume is (fluid) **ounces**, **milliliters** (ml), and **liters** (l); but in this book, a fluid ounce is 30 milliliters. Some bar-specific volume measurements in this book:

  - **1 bar spoon** = 4 milliliters, a bit more than $1/8$ ounce

  - **1 dash** = 0.8 milliliters, or 36 dashes to the ounce

  - **1 drop** = .05 milliliters, or 20 drops to the milliliter and 600 drops to the ounce

- Weight is always in **grams** (g) and **kilograms** (kg), not ounces and pounds. The persnickety will note that, unlike the ounce, the gram isn't a unit of weight; it is a unit of mass. Weight is a measure of force. Force, not mass, is actually what a scale weighs. The unit of mass in imperial units is the awesomely evocative, seldom-used slug. Really our metric scales should weigh in newtons (kg x m/s^2)—but that's just dumb.

- Pressure is in **psi** (pounds per square inch) and **bars** (atmospheric pressure is roughly 14.5 psi or 1 bar).

- Temperature is in both **Celsius** (C) and **Fahrenheit** (F), although for cocktail work I vastly prefer Celsius. Fahrenheit's proper place is in weather forecasts, baking, and frying.

- Energy and heat are in **calories**. Sorry, science world, but calories make intuitive sense when you are heating and cooling water a lot of the time. The standard unit, the joule, does not. (One calorie equals 4.2 joules.)

for nuance while taking advantage of standard measuring equipment and terms. While it is impossible to remember that a recipe should contain .833 ounces of lime juice (one-third of the way between ¾ ounce and 1 ounce), it's easy to remember "a fat ¾." Similarly, two-thirds of the way from ¾ to 1 ounce can be called out as a "short ounce." Some recipes that call out even smaller quantities have smaller but easily remembered increments: we use the bar spoon, the dash, and the drop. In answer to the Monk's main premise—that measuring doesn't guarantee a perfectly balanced drink—the good measuring bartender tastes each and every drink he or she makes. The pros use a drinking straw to extract a small sample of each drink to ensure that the measured proportions have produced the required effect.

Notice that in the above discussion my units are in *ounces* and not in milliliters. American cocktail recipes are always written in ounces. Before my metric friends get in a tizzy, let me say that in a cocktail recipe the word *ounces* is really just another way to say "parts." Cocktail recipes are all about ratios, and ratio measurements are in parts: 2 parts booze plus ½ part syrup and ¾ part juice and the like. In this book, 1 part equals 1 ounce equals 30 milliliters. The size of an actual "ounce" depends on whom you ask. Even without considering the various now-obsolete international *ounce* definitions, the United States *by itself* has a bewildering array of differently sized "ounces" that all hover around 1 fluid ounce equaling 30 milliliters. It turns out that the 30 milliliter ounce is an extremely convenient part size for making cocktails. So when I am making individual drinks, I usually speak in ounces. When I'm making large batches, I use a calculator or spreadsheet to convert my ounce-based recipes to any size I need, and I calculate in milliliters at 30 milliliters to the ounce.

Remember! In this book, 1 ounce of liquid equals 30 milliliters.

EVERYDAY EQUIPMENT FOR MEASURING DRINKS

Individual drinks are usually measured with jiggers. The jigger set I like to use consists of two vessels, each a double cone. They follow the convention of the 30 milliliter "ounce." The larger of the two jiggers, which I use for measuring anything between 1 and 2 ounces, has a 2-ounce cone with markings inside for 1½ ounces and a 1-ounce cone with markings inside for ¾ ounce. The smaller jigger has a ¾-ounce cone with a ½-ounce internal line and a ½-ounce cone with a ¼-ounce internal line. These two jiggers are all you need for most measurements. If you are measuring to an internal line in the jigger, pour all the way to the line. If you are measuring to the top of the jigger, pour *all the way to the top*. People tend to underpour when they are measuring a full jigger. Tip: have your jigger over your mixing vessel as you pour to prevent yourself from spilling precious, precious booze.

Given a choice between tall, skinny jiggers and short, squat jiggers, I always choose the tall. They are far more accurate. A pour that's a millimeter higher or lower in a wide jigger constitutes a much larger error than it does in a skinny jigger. The same principle applies to the next piece of equipment I recommend you acquire: a graduated cylinder.

ABOVE: Three jiggers and a graduated cylinder. Surprisingly, all three jiggers have the same volume. Because it is narrower, the tall one in back is much more accurate. The jigger in front is more accurate than the one on the left because its sides are straight near the top. The graduated cylinder is a paragon of accuracy.

Even with a tall, accurate jigger you must pour consistently to the top. The jigger on the left is holding 2 ounces (60 ml). The one on the right is short by a full ⅛ ounce (3.75 ml).

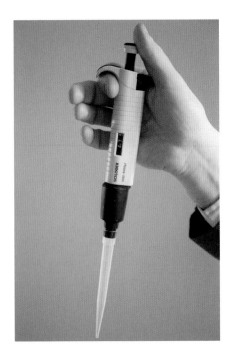

This micropipette can measure and dispense any volume between 1.00 ml and 5.00 ml.

Graduated cylinders are tall, skinny, straight columns with milliliter volume measurements on the sides. They range from tiny 10-milliliter (about ⅓ ounce) guys to 4-liter behemoths. Graduates are very accurate. Measuring cups and beakers are sloppy by comparison. I use them for recipe development and for batching large quantities of cocktails accurately. The ones I use most are 50 ml, a little under 2 ounces, which is a good jigger replacement for fine-tuning a highly technical recipe or adding very concentrated flavors to large batches; 250 ml, a little more than 8 ounces, which is perfect for mixing up to four cocktails at a time or measuring the smaller-volume ingredients in larger batches; and 1 liter, a little more than 1 quart, for when I'm batching by the bottle. Clear unbreakable plastic versions are available at modest prices; glass ones cost more.

Most people won't need one, but I really love my micropipette. Micropipettes allow you to measure a small volume of liquid very quickly and very accurately. Mine is adjustable to measure any liquid amount between 1 ml and 5 ml and is accurate to .01 ml. Using one takes only seconds, and unlike digital scales, micropipettes require no batteries. We use one every day at Booker and Dax for juice clarification. I also use it for recipe testing on very concentrated flavors, such as concentrated phosphoric acid or quinine sulfate solution. I use a micropipette to help figure out the proper recipe by adding small, well-defined quantities bit by bit.

EQUIPMENT FOR MAKING DRINKS

MIXING CUPS FOR STIRRED DRINKS

Mixing cups are a reflection of the bartender's personal style and can have a large effect on the finished cocktail. The most traditional mixing cup, a standard pint glass, has many advantages: it is cheap and fairly rugged, you can see through it, and it can be used for shaking. Many professional bartenders use pint glasses because they already have them on hand for serving beer. The main arguments against the pint glass are threefold: (1) they are made of glass (more on that in a minute); (2) they aren't supersexy; and (3) they have a rather narrow base, so they're easy to tip during a vigorous stirring bout.

Many fancy alternatives can combat the last two issues. The two most popular are the cut-glass crystal Japanese mixing cup, shaped like a beaker, and large, stemmed mixing glasses, which look a bit like squat, overgrown wineglasses. Many of my bartenders at Booker and Dax use the cut-crystal mixing glasses. They look great and have stable wide bases and useful pour spouts. They are also quite expensive, and only one careless drop away from the garbage bin.

A third option is a beaker from a scientific supply house, which makes a good, if goofy, mixing cup with a wide base and helpful, fairly accurate volume measurements on the side. These measurements are especially handy for the occasional home bartender, who doesn't have the time or inclination to learn to equate volume and pour-height-in-the-glass by rote.

Fancy mixing cups are an investment. If you are making drinks in front of guests, nice gear enhances their experience and gives a better overall impression of you as a bartender, and can therefore increase the guests' enjoyment of the drink. If you don't plan to mix in front of your guests, don't bother spending the extra cash.

My preferred mixing vessel is affordable and unbreakable: an 18-ounce metal shaker commonly referred to as a tin, even though they are almost all made of stainless steel. I favor metal because it has a much lower specific heat than glass. It takes less energy to cool or heat a gram of stainless steel than a gram of glass. Most glass mixing cups are thicker, and weigh much more than the average 18-ounce metal tin, so you can see that glass mixing cups represent a significant thermal mass—one that will affect the temperature and dilution of your drink. If you were to stir two drinks, one in a chilled glass mixing cup and one in a room-temperature glass mixing cup, there would be a noticeable difference between the two. To make sure that your drinks are consistent (and consistency should be one of your main bar goals), you must either *always* use chilled mixing glasses, or *never* use prechilled mixing glasses, or be some kind of savant who can autocorrect between the two. Metal tins require so little energy to cool down or heat up that they have very little effect on your cocktail.

For stirring drinks I prefer the metal tin. The lovely glass vessels have a large thermal mass, which can throw off your chilling unless you consistently prechill them.

SHAKERS (LEFT TO RIGHT): Three-piece cobbler shaker; Boston shaker with glass on metal; Boston shaker with metal on metal (my preference); two-piece Parisian shaker.

PARISIAN SHAKER

COBBLER SHAKER

MIDDLE LEFT: This Parisian shaker is pretty cool looking, but is not as versatile or inexpensive as a standard set of metal shaking tins. You build your drink in the small, top section, add ice to it, and then tap the larger cup on top. As you shake, the cup will chill, contract, and seal around the top.

BOTTOM LEFT: The cobbler shaker is good for people with small hands and people who like fancy-looking shakers. They are not very versatile and I dislike the integrated strainer.

RIGHT: To break a set of tins, smack the larger tin right where the seam between the two tins starts to turn into a gap.

COCKTAIL SHAKERS FOR SHAKEN DRINKS

Anyone even remotely interested in cocktails should own a set or two of cocktail shakers. The primary shaker criterion: it must withstand violent agitation without spilling liquids all over you and your guests. The two main styles of cocktail shaker are the three-piece and the two-piece. A third style, the Parisian shaker, looks really cool but is rather rare.

Three-piece shakers, or cobbler shakers, consist of a mixing cup, a top with an integral strainer, and a tiny cup-shaped cap sealer. You build the drink in the cup, add ice, put the sealed top onto the cup, shake, and then remove the top and strain into your glass through the integral strainer. These shakers come in a wide range of sizes, from single-drink shakers to large, multiliter party shakers. We don't use three-piece shakers at Booker and Dax, and I don't know many professional American bartenders who favor them. They don't strain as quickly as a two-piece setup, don't offer control over straining (more on that later), are prone to jamming after shaking, and are infuriatingly incompatible from set to set. You haven't had fun shaking drinks until you've had to shake a party's worth of cocktails with a random batch of nonstandard 3-piece shakers. Poorly made three-piece shakers—and most of them are—also tend to leak. On the upside, cobbler shakers, especially the small ones, are easier for bartenders with small hands to manipulate with skill and aplomb. Some bartenders who follow particular Japanese schools of bar artistry believe that the shape of the cobbler shaker improves the structure of the ice crystals produced while shaking; I think this is hokum. Nearly all professional three-piece shakers are made of metal, but there are exceptions. A bartender at one of the most renowned bars in Tokyo served me a cocktail from a neon-pink plastic cobbler shaker. The reason, a student of the Japanese school told me, is that the plastic, being soft, creates fewer and different ice crystals than a metal shaker would. I have yet to test this hypothesis, but I'm skeptical.

Two-piece shakers, or Boston shakers, are the most popular choice for professional bartenders in the United States. I use them at home and at work. They consist of two cups, one of which fits inside the other to form a seal. It is difficult to believe that a good seal will form the first time you use a Boston shaker—you fully expect to spray cocktail all over the room—but believe me, two-piece shakers seal reliably day in and night out. I'll save the physics for later. Metal-on-glass uses a large 28-ounce metal tin and a standard U.S. pint glass. You mix your cocktail in the glass part, then put the metal tin over the glass and shake. I don't like the metal-on-glass setup because the pint glass might break, all that glass has a large thermal mass, and it's hard to shake a metal-on-glass setup one-handed unless you have orangutan hands. Some people like the mixing glass because it lets them see how much liquid they have added to the cocktail. I don't care about this, because I measure my cocktails. I prefer metal-on-metal shakers, specifically the two-piece type known as "a set of tins"—a 28-ounce tin and an 18-ounce tin. The 18-ounce tin is often called a "cheater" tin by the pros because in a pinch it can be used as a strainer. These suckers are indestructible, have a low thermal mass, are fairly easy to handle, and look and sound good while shaking. A set of tins has a large internal volume, so while making one drink at a time is the norm, making two drinks, even three, in one shaker is no sweat. You build the drink in the smaller tin, add ice to the rim, place the larger tin over the top, tap to initiate the seal, and shake. After shaking, you break (take apart) the tins. Good bartenders break their tins with much flair and an audible commanding crack that makes my mouth water as if I were one of Pavlov's dogs. I myself am still working toward tin-breaking mastery. This is a skill you will never regret acquiring.

Until fairly recently, two-piece metal-on-metal tins weren't easy to find as matched pairs. You needed to scrounge around kitchen and bar supply shops to find two tins that worked well together. Now it is easy to find tins that go together, seal well, and are stiff enough to give a nice break. Poor-quality tins with noodlelike flexibility are notoriously difficult to break; they just slide back and forth against each other.

STRAINERS

You need three different strainers for bar work: julep, hawthorn, and tea.

Julep Strainer The julep strainer is an oval with rather large holes intended for straining stirred drinks. The large holes allow fast straining, and because the whole strainer fits inside the mixing cup, the pouring lip of the mixing cup is unobstructed.

Hawthorn Strainer The hawthorn strainer has a spring around its edge that lets it fit into a metal shaking tin and into many mixing glasses. In general, that spring makes the hawthorn better at straining out unwanted bits, such as chunks of mint and small ice pieces, than a julep strainer. Almost all hawthorn strainer springs, however, are too

STRAINERS (CLOCKWISE FROM TOP):
Tea strainer; julep strainer; hawthorn strainer.

large to capture everything that the caring bartender worries about, so many bartenders use the hawthorn in tandem with a fine tea strainer. I use a Cocktail Kingdom hawthorn with a very fine spring that obviates that problem.

Some hawthorn strainers are made so that the flow of cocktail can be broken into two streams, enabling the bartender to pour two drinks at once, like a barista pouring two espresso shots into different demitasse cups.

Hawthorn strainers are more difficult to use than julep strainers because they sit on the outside of the cup and are prone to drips and spills, but in skilled hands they provide much more control than a julep strainer. The bartender uses his or her index finger to slide the strainer up and down, changing the width of the pouring gap between the hawthorn strainer and the cup. This action is called "adjusting the gate." Pouring with a closed gate holds back ice crystals. Pouring with an open gate allows more ice crystals to float on top of your drink. I have been a party to many heated arguments about the merits and demerits of ice crystals on shaken drinks. For many years crystals were a strict no-no, a sign of poor craftsmanship. Today the tide has turned, and many people, including me, proudly proclaim their love of a beautiful, shimmering layer of crunchy crystals. The argument against the crystals is that they melt quickly to form unwanted excess dilution at the top of the drink. I say, Drink faster! Shaken drinks deteriorate immediately after they are shaken. Like cherry blossoms, they are dead before you know it. They should be made in small portions and consumed quickly.

ABOVE: This new hawthorn strainer has a very tightly wound spring for capturing small ice crystals that would make it through a standard hawthorn. Why have springs at all instead of mesh? Because the spring conforms to the shape of your mixing tin. BELOW: With practice you can use a hawthorn to pour into two glasses simultaneously. More of a gimmick for slow nights than a useful skill—but fun.

Tea Strainer The third strainer you'll need at the bar is a tea strainer, or any small fine-meshed strainer, to filter large particulates out of your drink. In the anti-ice-crystal days the bartender would use the tea strainer to insure that no ice particles entered the drink. At Booker and Dax, we use tea strainers to keep big herb particles out of our nitro-muddled drinks (see the section on nitro-muddling, page 165).

Don't pinch here! Spoon shaft must be free to twirl

Push forward with your ring finger.

The ice will keep the bowl of the spoon pinned against the glass and cause it to spin naturally

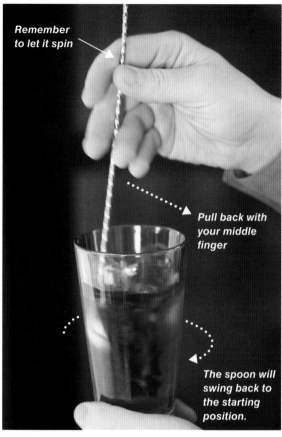

Remember to let it spin

Pull back with your middle finger

The spoon will swing back to the starting position.

THE TRICK TO STIRRING: simply push the bar spoon back and forth with your lower fingers while the top of the spoon rotates in the pocket between your thumb and pointer finger. Your hand never swirls around. Focus on always pushing the spoon against the inside wall of the glass—if the back of the spoon is in contact with the glass, the spoon will turn when you push.

BAR SPOONS

Bar spoons are long, thin spoons that usually have a twisty shaft. Before I knew better, I thought they were dumb and unnecessary. I know now that these spoons make elegant stirring possible, and they make stirring results more repeatable. Nothing looks clumsier than a hamfisted stir. Believe me—I'm not very good at stirring and am often stirring next to masters. Great stirring technique appears effortless, and it is efficient and, above all, repeatable. Proper stirring requires precise repeatability. Really great stir fiends can accurately stir two drinks at a time. Some ninjas can do four. This sort of high stirring artistry cannot be performed with a run-of-the-mill teaspoon. You need a well-designed, well-balanced bar spoon with a shaft that suits your style. In addition to stirring, you can use your bar spoon to measure. The bar spoons I use are 4 ml (a little more than 1/8 ounce). Measure yours.

Another tip: use your bar spoon to fish the cherry or olive garnish out of that irritating jar. Don't use your fingers!

BAR MATS

Bar mats are one of those simple, life-changing things that have been completely overlooked by the home market. They are cheap and awesome. Made of thousands of tiny, 1/4-inch high rubber nubbins, they capture spills and are nonslip, preventing further spills. I don't care how good a bartender you are, you will spill stuff. When you spill onto a countertop, you look like a slob and your counter becomes slippery. When you spill into a bar mat, it looks like nothing has happened at all. The bar mat stays nonslip, no matter what. At the end of a night (or after a particularly bad spill), carefully pick up the mat, dump it into a sink, and rinse it off. Good as new. Bar mats also make primo drying mats for cups and dishware. Buy a bar mat.

PICTURED ON THE BAR MAT: 1) Bad Ass Muddler (accept no other), *2)* a Kuhn-Rikon "Y" peeler (also accept no other—if you don't like this tool it's because you aren't using it properly), *3)* dasher tops, *4)* eye dropper, and *5)* a paring knife.

MISCELLANY

If you are like me, you will acquire a wealth of little bar tools and gimcracks. Most of them are of limited use, but some are cool enough to search out.

Muddlers: Get a good muddler to smash ingredients in the bottom of a mixing cup. Most muddlers are crap—they don't have a big enough crushing area, so they just push ingredients around. One muddler stands out: the Bad Ass Muddler from Cocktail Kingdom. This large, solid cylinder lives up to its name. My second favorite muddler is a simple, straight-sided rolling pin. If you are going to muddle with liquid nitrogen, avoid cheap plastic or rubberized muddlers; the liquid nitrogen makes them brittle and shatter-prone (the Bad Ass Muddler is made of plastic but is unaffected by liquid nitrogen).

Misters: Little spray misters are occasionally useful for applying an aroma to a glass before pouring. Using one is more accurate and less wasteful than using the traditional rinse. Misters are also useful for applying aromatic oils or other ingredients to the top of a cocktail. I don't really use the misting technique, but many of my friends do.

Knives: You will want a decent paring knife to keep handy with your bar kit. Store it with a knife guard so that it doesn't get dull when tossed in among other bar wares. Invest in a small cutting board as well.

Droppers: Get a couple of glass bottles with screw-on eyedroppers. I use them to dose all kinds of liquids into cocktails in small amounts: saltwater, bitters, etc. If you like fancy-looking things, get some bitters bottles to add dashes bigger than drops but smaller than the contents of bar spoons. Be aware that different dasher tops have different dash sizes. At the bar we use a half-dasher and a full dasher. At home I just save old Tabasco and Angostura bottles, wash off the labels, and refill. I'm a frugal man.

Ice Storage: At the bar, our ice is stored in ice wells: large, insulated containers with integral drains. We use metal scoops or shaking tins to pick up the ice. At home, you'll need an ice bucket and a scoop or tongs. It's a bad idea to handle ice with your hands in front of guests unless they are close family. Unfortunately, I have never seen an ice bucket that I like. They should have a drain tap, like a water cooler, and a plastic rack to keep the ice from swimming in the meltwater. They should also look good.

Capper: If you are going to make bottled drinks, get yourself a hand-held bottle capper. They are cheap and available at any home-brew shop.

EQUIPMENT FOR MEASURING, TESTING, AND MAKING INGREDIENTS

WEIGHT-MEASURING EQUIPMENT

Even though drinks should be measured by volume, it's useful to have scales (two, actually) for advanced bar work. Sugars for simple syrup, for instance, should be weighed. Hydrocolloids and other powdered ingredients should also be weighed. There is no reason to weigh anything in any unit other than the gram, and that is all you will see in this book. Water is the one liquid ingredient that can be easily converted from volume- to weight-based measurements—1 gram = 1 milliliter (at standard temperature)—and so I sometimes weigh my water (see the section on Calculating Dilution, page 123). Get two digital scales: one that can read in tenth-of-a-gram increments and one that can read in one-gram increments. For the smaller units, ask for a drug scale. If you just ask for a scale that can read by ¹/₁₀ gram, people always make the mistake of getting a scale that isn't accurate enough. If you ask for a drug scale, everyone knows you mean business when it comes to accuracy. Drug scales can be purchased very inexpensively. If you start playing with hydrocolloids and other new ingredients, or need to measure herbs for tinctures and the like, this scale will serve you well.

You will also want a scale that measures larger weights in 1-gram increments. Get one that will measure at least 5000 grams. You might wonder why you can't get a scale that has a large capacity and also measures by tenth-of-a-gram increments. The answer is you can, if you have a lot of cash and patience—they cost a few thousand dollars and they are slow. Scales that measure large weights have large platforms. Large platforms are affected by wind currents. Even minor wind currents in a kitchen can make a large-platform scale fluctuate by a couple of tenths of a gram. Thus the two scales.

OTHER ANALYTICAL EQUIPMENT

Refractometers are instruments that measure refractive index, or how much light is bent as it enters a transparent substance. Because different concentrations of different substances dissolved in water bend light by a different and specific amount, refractometers can tell you the concentration of water-based solutions, such as the percent of sugar in syrups and fruit juices, alcohol levels in distilled beverages, and salt concentration in brines. Do you need one at home? Not really. But we use it every day at the bar, because I prize consistency.

Even though refractometers technically measure refractive index, that isn't what we care about—we care about how refractive index correlates to the concentration of whatever we're interested in: sugar, ethanol, salt, propylene glycol, whatever. So refractometers usually have scales that read in terms of those solutions. You can buy a refractometer for sugar, one for salt, and so on. The most common scale, and the most useful for the bar, is Brix, which is the measure of sucrose (table sugar) content by weight. Simple syrup with a Brix of 50 contains 50 grams of sugar in every 100-gram sample. That particular ratio is standard 1:1 simple syrup. Our rich simple syrup has a 2:1 sugar-to-water

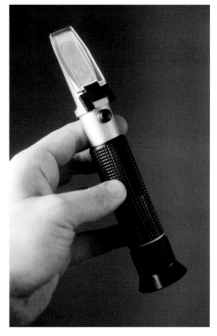

Hand-held refractometers like this one are pretty inexpensive. Be sure to get one that works in the range you need.

ratio and therefore a Brix of 66. Brix refractometers are a relatively cheap and easy way to standardize the syrups that you use. Want to make sure your honey syrup has the same sugar content as your simple syrup? Use a refractometer. Want to know how much sugar is in your fruit juice? Refractometer.

If you can spring for it, I recommend that you get yourself an electronic refractometer that can measure between 0 and 85 Brix. They are fast to use, even in a dark bar, and are accurate over their whole range. Traditional manual refractometers, which can be bought inexpensively, work fine. The trick with the nonelectronic jobs is making sure the Brix range they measure is sufficient. The most common range is 0–32 Brix, which is fine for any fruit juice work you do but won't work for syrups. You can buy a 0–80 or 0–90 manual unit, but then it is hard to read accurate numbers, because the scale on a 0–90 is the same size as the scale on a 0–32, making the 0–32 almost three times more accurate. Even so, if your primary use is at the bar, get a 0–80 Brix unit (make sure it is a Brix scale and not an alcohol scale). The other issue with manual refractometers is you need a light source to read them. It sucks trying to hold a flashlight up to the end of a refractometer stuck in your eye behind a dark bar during the press of a busy service. If you are on the fence about buying one, using a Brix refractometer is also helpful outside the bar, to standardize recipes for sorbet and the like when you are using ingredients like fruit with an unknown amount of sugar (these applications want a 0–32 Brix unit).

Caveat: it is extremely easy to misuse a refractometer. Brix scales assume that the only substances in your sample are sugar and water. For most products, like fruit juices, that assumption is okay because they don't contain a lot of salt, ethanol, or other ingredients that dork with refractive index. However, using a refractometer to measure liquids that are a mixture of sugar and ethanol *will not work*. Both alcohol concentration and sugar concentration affect refractive index; there is no way to know how much of each is present. Anything you measure with a refractometer must be considered a mixture of water and *one other substance only*.

Thermometers will be useful in the imbibing experiments that we will explore later in this book. Any relatively fast-acting digital thermometer that can reliably measure temperatures down to −20° Celsius (−4F°)

is fine. I have a fancy eight-channel, data-logging, computer-connected, thermocouple thermometer. It's probably overkill for normal home use, but it's not very expensive, so you may want to try one.

pH meters are rarely useful for cocktail work, but people in the trade often ask me about them. A pH meter can measure how acidic an ingredient is but cannot predict how sour an ingredient will taste. See the Ingredients section, page 50, for more detail on this phenomenon.

JUICERS FOR LIMES AND LEMONS

Juicing small citrus is important to me. I joke that I don't respect people who can't juice quickly—but I'm not really joking. Forget hand reamers; they suck. Upright lever-pull citrus presses are good for grapefruits that won't fit in smaller hand presses, but they are slow. The best small-citrus press is the lowly swing-away hand press.

Many years ago I was taught the secrets of the hand press by my San Francisco bartender friend Ryan Fitzgerald. He is still faster than me, and I hate him for it. First, have all your citrus washed, cut, and in an easy-to-reach pile next to you on the counter. Right in front of that pile, have a

*JUICING LIMES LIKE A MACHINE: Set up your station as shown so you can quickly grab fresh limes, with the juice-catcher and the spent-peel-catcher next to each other. **1)** Hold the juicer as shown with the lime face-down. **2)** Violently smash the hell out of the lime and then **3)** immediately throw open and release the handle. The force from the handle popping open will eject the spent peel into the waste bin while you are reaching for the next lime and placing it in the juicer. Repeat again and again and again . . .*

wide bowl to catch the juice. Right next to the juice bowl, have something to catch the spent citrus. To juice, hold the press open in your weak hand. Quickly grab a citrus half with your strong hand and place it *cut face down* into the cup of the press. In one forceful and strong motion, use your strong hand to close the press and crush out all the juice in one violent spray into the bowl. Now flick the handle back sharply as your weak hand aims the cup of the press toward the spent citrus bowl. As the handle flies back and comes to a halt, the jerk that you created should pop the spent citrus into the waste container *without your touching it*. This is very important, because your strong hand should already be reaching for the next fruit. Done properly, you should be able to achieve juicing speeds in excess of 300 ml per minute. The trick to the technique is choosing the right press. Bad presses are too deep. Deep-cupped presses seem like they'd be good, but what you need for real speed-juicing is a shallow cup that can properly eject spents. Good action on the handle is also a must. Finally, the handle of the press should open only about 120 degrees; a press that opens all the way to 180 degrees or more is wasting movement and time.

Why not use an electric juicer, you may ask. My favorite juicer used to be the Sunkist-style electric juicer. It is fast. With a two-handed grab-press-toss juicing technique, I can easily achieve 800 ml per minute or more. Juice flows from my hands like a mighty river on the Sunkist, with spent lime halves raining matrixlike into trashcans. The yield is 25 percent higher than with the hand press. The problem? The juice doesn't taste as good. In blind taste tests, people universally preferred the juice from a hand press over that obtained from the Sunkist, probably because the spinning reamer scrapes bitterness out of the white pith, or albedo, of the fruit. If you persist in using the Sunkist, do yourself a favor: rip out the goofy strainers. After 3 or 4 quarts of juice they clog up and are a pain to clean.

If unlike me, you have infinite money and space, you can use the Zumex automatic juicer favored by the godfather of juice himself, Don Lee, one of the masterminds of the cocktail world. Pour a box of washed but uncut citrus in, juice comes out. I could stare at that thing all day. Don uses it to make lime and lemon juice for the thousands of drinks served daily at the cocktail industry's yearly bacchanalia, Tales of the Cocktail, in New Orleans.

No matter how you press your citrus, you should stain it through a fine chinois or tea strainer before use. Large pulpy bits on the side of a cocktail glass look terrible.

JUICERS FOR LARGER CITRUS

For larger citrus fruits such as oranges and grapefruits, the vertical lever-pulled press, OrangeX or equivalent, is the best choice. It makes short work of large fruit. Get a heavy-duty one, because cheap ones break. It will also work on pomegranates.

JUICERS FOR OTHER FRUITS AND VEGETABLES

To juice hard fruits like apples, or veggies like carrots, I use a Champion Juicer. It's a workhorse that can take a severe beating and keep on juicing. It once helped me rapidly juice six cases of apples in a row without pausing. Its housing got so hot that it boiled the water I tried to cool it with—and kept right on juicing. I threw wet towels around the Champion so I could keep working, and it kept working right alongside me, sending steam out of the towels the whole time. I eventually melted the magnets on the safety interlock (put there so you can't grate your hand off), but the motor kept on chugging. This baby will juice just about anything except wheatgrass and sugarcane. It works great on things you might not expect, like horseradish and ginger.

If I were running a fresh juice business, I'd invest in a Nutrifaster juicer, the ones that look like chrome spaceships from the 1960s. Nutrifasters are amazing. They require no operator force to make juice (you have to push pretty hard on a Champion to go fast), but they cost ten times more than a Champion and take up twice the space.

BLENDERS

I use only Vita-Prep high-speed blenders. They have balls and a very intuitive interface: two paddle switches and a knob. Everyone loves the two paddle switches and a knob. I don't know anyone who bought a Vita-Prep and regretted it. Vitamix is the home version of the Vita-Prep—it's the same basic machine with a better warranty and a lower price tag. Home folks should go for the Vitamix. If you're a pro, you should spring for the Vita-Prep—you'll void the warranty on the Vitamix if you use it in a professional setting. Though I love them, Vita-Preps aren't perfect. Their pitcher chokes down toward the blade so thick products spin up and away from the blade, creating an airlock. This forces you to use the plunger that is sold with the unit to keep the blending going. Guess what always gets lost in a restaurant or bar? The dang plunger. The speed control knob, which I love, is implemented with a cheap potentiometer, which I hate. Over the years the potentiometer goes wonky and causes the speed of the blender to jump around wildly. It's funny to get sprayed with the contents of a blender unexpectedly, right? I do not recommend the BarBoss, the unit the company makes specifically for bars. It doesn't have a real speed control, just a timer. Good for a drone making smoothies, bad for everyone else.

The other megablender on the market, the Blendtec, has the power to go toe to toe with the Vita-Prep and has a well-shaped pitcher that doesn't require a plunger (the lids on the Blendtecs I have used are wretched, however—they leak). You can see the Blendtec in the various "will it blend" videos that pepper the Internet. I cannot recommend these blenders, however, because their controls are bad for thinking people. Are you a smoothie robot who wants to make consistent smoothies without paying any attention at all? Get a Blendtec. I have pleaded with Blendtec to make a unit with controls built for cooks and bartenders with brains, to no avail.

If you don't want to shell out hundreds of dollars for a blender, fear not. Cheaper blenders are worthwhile tools and will make most of the recipes in this book, but until I can find a cheap blender that can suck up a pitcher of liquid nitrogen without choking or blend a pound of bacon to a beautiful smooth paste, I will never go back.

STRAINERS FOR INGREDIENTS

If you make juices or infusions, you are going to need to strain them. The strainers I use, in order from coarsest to finest: china caps—large holes, fast-draining; fine chinois—fine mesh, slow; clean muslin cloth (not cheesecloth); and coffee filters. Don't use a strainer finer than you need, because fine filters take longer. When you need a superfine strainer, like a coffee filter, strain your product through a coarser strainer first, or the fine one will choke up immediately. For juices or syrups I typically put a china cap inside a chinois and pour the juice through both at the same time, saving myself a step. I then decide if I need the muslin. Only as a last resort do I bust out the coffee filters, because they take so dang long.

STRAINING IMPLEMENTS FOR YOUR PREP WORK: *I use **1)** a coarse china cap and **2)** a fine chinois, sometimes in combination. These tools are very useful but not necessary—you can also use a standard kitchen strainer. I use **3)** coffee filters constantly, but don't enjoy them because they always clog. When I am straining a large amount of product I often go through five or six, one after another. A **4)** straining bag is a useful intermediate between a kitchen strainer and a coffee filter.*

There are very nice, overpriced, straining bags on the market called superbags that come in a range of mesh sizes from fine to very, very fine. They are useful as an alternative or adjunct to ordinary strainers and filters.

CENTRIFUGES

Years ago, when I started telling people to buy a centrifuge, they just laughed. Now more and more chefs and bartenders are using them for a very simple reason: they save time and money. I can use the centrifuge to convert 2.5 kilos of fresh strawberries into two liters of

A 3-liter, 4000 g, benchtop, swinging-bucket centrifuge.

clear, pure strawberry juice in 20 minutes flat without adding any heat. That is a game-changing ability. Centrifuges aren't really home-friendly— *yet.* The ones I use are rather large, and can cost quite a bit. Centrifuges use centrifugal force to separate ingredients based on density. They can spin the pulp out of juices, spin the solids out of nut milks, spin the oil out of nut pastes, and squeeze the fluid out of almost anything you can blend. The heart of a centrifuge is its rotor, the thing that spins around. Most rotors are one of two styles: fixed and swinging bucket. Fixed rotors hold sample tubes rigidly and spin them about. Mixtures spun in these tubes form a solid pellet on the bottom side of the tube. Swinging-bucket rotors are what they sound like—swinging buckets on the end of spinning arms. The solids in a mixture inside a swinging-bucket rotor get smashed onto the bottom of the buckets.

Centrifuges vary wildly in capability, cost, and size. The centrifuge I use at Booker and Dax is a 3-liter benchtop centrifuge with a swinging-bucket rotor holding four 750 ml buckets. It has a refrigerator to keep my products cool (spinning rotors generate heat from friction, so refrigeration is nice but not necessary) and can produce four thousand times the force of gravity at a speed of 4000 rpm. The one at the bar costs $8000 new, but we bought a refurbished unit for $3000. In my test lab I use an identical one that I bought on eBay for $200, but I had to fix it up a bit, and it could break again at any moment. The 3-liter benchtop hits the sweet spot for kitchen-friendly centrifuges. Smaller units don't have the capacity to make them worthwhile. Slower units don't produce enough force to make them useful in a busy kitchen. Larger units are bigger, more expensive, and more dangerous without offering results that are much better. I have spent years perfecting

The buckets for my centrifuge hold 750ml each. Notice the balance they are sitting on. Buckets that sit across from each other in the centrifuge must weigh the same amount or all hell breaks loose.

This tiny centrifuge can be had for under $200. It works. It helps you make the same awesome cocktail stuff as the big one, just in very small quantities.

recipes that work at 4000 g's or lower so that you don't have to go out and buy that 48,000-g model. The largest, fastest centrifuges produce so much force that if something goes wrong with them, they can rip themselves apart like a bomb.

You should never buy a used high-speed or superspeed centrifuge unless you know what you are doing. These units have rotors that must be inspected and retired after a certain number of spins, and a used rotor rarely comes with a guarantee of how it has been used or treated. Most rotors are made of aluminum, which experiences fatigue after spinning, starting, and stopping year in and year out. That fatigue can cause the rotor to crack and then suddenly fly apart without warning. I once had a freestanding spinning rotor centrifuge that was totally unprotected, the SS1 superspeed centrifuge from Sorvall. Made in the 1950s when it was still okay to kill a lab tech every once and a while, that sucker could do 20,000 g's and was little more than an aluminum rotor mounted on a motor with some legs bolted to it. Spinning that fifty-year-old aluminum rotor was the dumbest thing I've ever done in a kitchen, which is saying a lot. We called it the dangerfuge and immediately retired it to my bookshelf. Even on slower centrifuges, you should never use a rotor or buckets that are damaged or corroded in any way. An older motor or frame isn't a safety issue on a centrifuge, but old rotors and buckets are. Consider getting a rotor that is guaranteed safe.

The second safety issue with used centrifuges is you don't know what's been in them. Assume the worst: prions, Ebola, whatever. The swinging-bucket types that I get are more often than not retired from labs doing bloodwork. I bleach the hell out of them when they show up, then pressure-sterilize the buckets in a home canning rig, then bleach the whole thing again. I call this procedure bleaching the rabies out.

For ambitious home enthusiasts, there is a worthwhile centrifuge unit selling now for under $200. It handles only 120 milliliters (a little over 4 ounces) at a time, and spins at only 1,300 times the force of gravity, but it can give you a realistic picture of what is possible with a centrifuge, weighs only 10 pounds, is the size of a toaster, and is safe.

LIQUID NITROGEN

I love liquid nitrogen (LN or LN$_2$). Liquid nitrogen is liquefied nitrogen gas, which makes up three-quarters of the air we breathe. N$_2$ is in no way toxic. It is a chemical, just like the H$_2$O we drink is a chemical—which isn't to say you shouldn't be cautious when you use it. There are safety rules which must be followed at all times.

At −196° Celsius, liquid nitrogen can give you a nasty cold-burn. If you ingest it, the results can be catastrophic. Never serve or allow someone else to serve a drink that contains any cryogenic material, ever. Forget serving drinks with clouds of liquid nitrogen vapor emanating from them. While I'm at it, don't serve drinks with chunks of dry ice in them either. Ingested cryogens cause permanent damage. A young woman in England lost most of her stomach and was put in critical condition because a bartender thought it would be cool to serve a drink with some LN rolling around on top of it. It wasn't. In practice, it is easy to prevent customers from coming into contact with LN. You must always be vigilant.

When liquid nitrogen hits your hand it instantly vaporizes, forming a protective insulating vapor shield that prevents you from freezing. This is called the Leidenfrost effect.

Liquid nitrogen in the eyes can blind you. You must be extremely careful to prevent situations in which liquid nitrogen might get in someone's eyes. Never pour it over anyone's head, for instance. Some people recommend using gloves to handle liquid nitrogen. I don't. The only time I have been seriously burned with liquid nitrogen was when the cold made my glove so brittle that it cracked and allowed the liquid nitrogen to pour in. I couldn't shake the glove off fast enough to prevent a burn. You can, however, dip your hand directly into liquid nitrogen without getting frostbite. A layer of nitrogen vapor immediately forms around your hand, and that vapor temporarily protects you from the extreme cold. This phenomenon is called the Leidenfrost effect. You can observe it if you dump liquid nitrogen on the floor: it will roll around in little beads that skitter across the floor, almost frictionless, protected by a thin layer of nitrogen vapor, as when you throw a drop of water into a pan so hot that the water skitters around like a small marble rather than spreading out and boiling away. Remember, the Leidenfrost effect only helps you if a vapor can form between your hand and the nitrogen. Grab a superchilled metal cup and you are SOL.

More safety: just because liquid nitrogen isn't toxic doesn't mean it can't asphyxiate you. Large amounts of vaporizing nitrogen in a small space can displace oxygen—the oxygen you need to survive. Perversely, your body doesn't react negatively to oxygen deprivation. The panic you feel when you can't breathe isn't from lack of oxygen, it's from excess carbon dioxide (CO_2) in your blood. Trying to breath in a pure nitrogen environment doesn't cause an excess of CO_2 in your blood, so you feel great—loopy, in fact. Without extensive training (which some pilots go through), it is very difficult to self-diagnose a lack of oxygen. Breathing in pure nitrogen is much, much worse than breathing nothing at all. Your lungs aren't a one-way system for putting oxygen in your blood. They only function properly when the air contains more oxygen than your blood does. In a pure nitrogen environment, your blood contains more oxygen than the "air" in your lungs, and the oxygen is literally sucked out of your blood. Only a few breaths will take you out. The rule in industry is, don't try to rescue someone trapped in a nitrogen environment—he is already dead. For the good news, no cook or bartender has ever suffocated himself or herself with liquid nitrogen—yet. Don't travel with large amounts of liquid nitrogen in an elevator. Ever. An elevator is an enclosed space where, if your LN storage vessel failed catastrophically, you'd be trapped. Don't carry liquid nitrogen in a car with you. If you get in an accident and are knocked unconscious, you could be asphyxiated.

Yet more safety: liquid nitrogen should never be kept in a closed container. Ever. As liquid nitrogen boils in a room-temperature environment, it expands by almost seven-hundred times. Tremendous pressure will build up inside a sealed vessel as the gas expands—thousands of psi. Usually the vessel you have chosen won't withstand the pressure and will explode. A young cook in Germany was permanently maimed and almost killed this way in 2009, in a tragic accident that occurred because of a simple mistake.

So why use liquid nitrogen? It is mesmerizing, fantastic stuff. It can chill glasses almost instantly. It can chill and freeze herbs, fruits, drinks, and other products without contaminating or diluting them in any way. As I've said, I love liquid nitrogen, and I don't know anyone who uses it who doesn't. I like it so much that I worry myself. One further caveat: in general, liquid nitrogen isn't good for chilling single drinks; it is way too easy to overchill a drink using LN. A mildly overchilled drink might not taste good, but that scoop of frozen booze you might try to serve someone could be cold enough to burn the tongue. I've been served overchilled LN booze sorbets that have ruined my taste buds for the night.

Everyone I know gets liquid nitrogen delivered from a local welding supply shop. You store it in a piece of equipment called a dewar, an insulated vessel built to hold cryogenic fluids for a long time with minimal loss. Standard dewar sizes are 5, 10, 25, 35, 50, 160, 180, and 240 liters. At Booker and Dax we have a 160-liter dewar that our supplier refills every week. Larger amounts of liquid nitrogen are much more economical than smaller amounts. It costs me only $120 to fill my 160 but over $80 to fill a 35-liter dewar. The large dewars are also economical to rent; a small monthly fee and a hefty up-front deposit are all that is required to start using liquid nitrogen. A properly functioning 160-liter dewar will hold liquid nitrogen a long time before all the liquid evaporates away. If you have any skills with a torch at all, avoid the costly take-off hose the company will try to sell you; make your own from commonly available copper parts. One more note: large LN dewars are kept under a small amount of pressure, typically 22 psi. This pressure provides the force for removing the liquid from the dewar. To maintain that 22 psi, the dewar has a relief valve that occasionally opens up and relieves excess pressure, making a hissing sound. The hissing really freaks people out. I reassure people by smiling and saying, "If it didn't hiss, we'd all blow up."

We also have smaller 5- and 10-liter dewars that we use to shuttle LN around the bar during service. They are pretty pricey, a couple hundred bucks each. What we actually use to pour LN into drinks and glasses are vacuum-insulated coffee carafes and camp thermoses—unsealed, of course. Avoid carafes with a lot of plastic near the pour spout—they won't last long.

To dispense liquid nitrogen out of your tanks, don't bother with the expensive hoses the LN folks try to sell you. You can make your own takeoffs from readily available copper pipe and copper fittings sweated together with lead-free solder. The doo-dad at the end is a sintered bronze muffler. It costs less than $9. If you call that same thing a cryogenic phase separator, it costs $135.

DRY ICE

Dry ice, the other cryogenic cooking material, made of solidified carbon dioxide, is not nearly as useful as liquid nitrogen. It seems attractive. While dry ice is considerably warmer than liquid nitrogen, at −78.5°C, it has considerably more chilling power per pound. Dry ice is also easier to buy and has fewer safety concerns than LN does, but it loses out because it is a solid. You cannot immerse foods in dry ice. Dry ice mixed into liquids doesn't disappear as quickly as LN does. You can't chill glasses very well with chunks of dry ice. Also, carbon dioxide gas is soluble in water. If you aren't careful, that drink you chill with a chunk of dry ice might get a little carbonated. I primarily use dry ice to maintain the temperature of large batches of drinks at events (see the Alternative Chilling section, page 140).

iSi CREAM WHIPPERS

I love my whippers. They have three main uses at a bar: making foams (which I don't do), infusing nitrogen rapidly, and, in a pinch, carbonating. The best whippers I've used are made by the iSi company. Cheaper whippers often leak, and some really bad ones lack the safety features that iSi builds in.

Essentially, whippers are metal pressure vessels that enable you to pressurize liquids with a gas. The gas comes in small, 7.5-gram cartridges. You can buy either carbon dioxide cartridges or nitrous oxide (N_2O) cartridges. The bubbles from CO_2 make things taste carbonated, like seltzer, while the bubbles from N_2O are a bit sweet and not at all prickly. N_2O is also known as laughing gas, so some people use it as a drug, in which case the cartridges are known as whippets. N_2O is the gas I use most often at the bar because I use it for infusions, to which I don't want to add residual carbonation.

The main downside of whippers is the high cost of the cartridges. Even though they cost less than a dollar apiece, I often use two or three at a time, and costs add up. In fact, companies like iSi aren't really making much money on the whipper; they want you to buy the cartridges. A last note on cartridges: you may not bring them on airplanes, even as checked luggage, because they are considered pressure vessels. This fact has always amused me, because almost every seat in an airplane has a life vest under it powered by—guess what? An iSi compressed gas cartridge.

CARBON DIOXIDE GAS AND PARAPHERNALIA

Carbon dioxide is the gas used to carbonate drinks. Ten years ago there were two main options available for carbonation: soda siphons (which work poorly) and commercial carbonation rigs (for making soda in bars). Around 2005 I became aware of the carbonator cap made by the Liquid Bread company. Liquid Bread was started by home-brew nuts who wanted a way to take samples of home-brewed beer to competitions and their buddies without losing carbonation. They developed a plastic cap that fits on ordinary soda bottles and easily hooks up to a CO_2 line via a cheap ball-lock connector obtainable from any home-brew shop. I started using that cap to carbonate cocktails, and it changed my world. There are now tons of options for carbonating at homes and at bars, and I have used many of them. None beats the carbonator cap. Other systems look better and use nicer-looking containers, but I've never had bubbles that I really liked out of any other system. As you'll see later in the Carbonation section, I'm a bit of a bubble nut.

The whole carbonator-cap system is pretty inexpensive. All you need, besides the cap and connector, are a length of gas hose, a regulator, and a 5- or 20-pound CO_2 tank. The 5-pounder is small and easy to lug around and, depending on your technique, will carbonate between 75 and 375 liters (20–100 gallons) of liquid. A 20-pound tank isn't easy to lug around, but it will fit in a standard home under-counter cabinet. A word of caution: you should use a chain or strap to prevent your tanks from falling over. CO_2 tanks are easily refilled at any welding shop. In fact, you can buy your tanks at the local welding shop, but it is often cheaper to buy them online; they are shipped empty. When you buy a regulator, make sure you buy a pressure regulator and not a flow regulator, because the latter won't work. Also make sure you get a regulator that can produce at least 60, and preferably 100 psi (the higher pressure will be handy if you take up soda kegging as a hobby). Lower-pressure beer regulators will not work.

ICE TOOLS

I am a fan of shaking with large ice cubes. You can purchase silicone molds that make six 2-inch-by-2-inch square ice cubes that work great for shaking. Those molds will not produce the crystal-clear-presentation ice I use for my rocks drinks; for those you'll need a rectangular Igloo cooler or other insulated container that fits in your freezer. To work with ice, I use two styles of ice pick: a multipronged pick and a single-pointed pick. Both are of very high quality. Stay away from cheap ice picks; they will bend and frustrate you. I use an inexpensive flat-bladed slicing knife in conjunction with my ice picks to break large ice into smaller pieces in a controlled fashion. You should have some way to make crushed ice as well. The fancy way is to get a sturdy canvas bag called a Lewis bag and a wooden ice mallet to crush your ice right before you need it. The canvas soaks up melting water so your

ICE TOOLS (CLOCKWISE FROM TOP):
1) Ice mallet, 2) multi-prong ice chipper,
and 3) ice pick on top of 4) a Lewis bag.
Use the mallet to knock on a knife when
cutting large blocks of ice, or to smash
ice cubes that you put in your Lewis
bag. It's better to crush ice in a Lewis
bag than a plastic bag because the
Lewis bag is absorbent and will yield dry
crushed ice. Crushing in a cloth napkin
also works, but the bag prevents ice from
flying around the room. The ice pick at
the bottom and ice chipper on the right
are high-quality tools—avoid low-quality
substitutes.

crushed ice is relatively dry. I also use an old-school Metrokane crank-style ice crusher to make what I call pebble ice, which is a bit larger-grained than crushed ice from a Lewis bag.

For shaved-ice drinks, which I enjoy immensely, I use a hand-cranked Hatsuyuki ice shaver—a thing of beauty. Swan makes a similar model. For pros, I recommend getting one of these. I like just looking at the cast-iron lines of my Hatsuyuki. It makes no noise, something I prize behind the bar. The noise from shaking a drink: inviting; the noise from whirring electric machinery: not so much. These shavers provide very fine control over the texture of the shaved ice. They shave blocks of ice, not small cubes. The cheapest way to make the blocks is to freeze small plastic soup tubs full of water.

For the budget-conscious, some remarkably cheap professional-size electric ice shavers are on the market now and perform admirably well, but they look kind of junky. It'd be worthwhile to get one of these for a shaved-ice party at home, as long as you have room to store the shaver afterward. You wouldn't want one on your counter. On the very inexpensive end, you can purchase hand ice shavers that look like little block planes. They are hard to adjust properly and are usually of poor quality. In my experience, you can get them to work, but they are a pain in the butt. Apparently, I'm just incompetent at using them, however, because every octogenarian street-corner shaved-ice peddler in my Lower East Side neighborhood seems to have no problem with them. If you really have patience, there is always the Snoopy Sno-Cone Machine and its modern equivalents.

REFRIGERATION

At the bar I use very accurate Randell FX fridge/freezers to chill my carbonated drinks and bottled cocktails. The FX can maintain any temperature I want between $-4°F$ ($-20°C$) and $50°F$ ($10°C$) within $2°F$. I'm very particular about the temperature of my drinks. Without the FX, it would be difficult to maintain the quality of my prebatched drinks. I have one set to $18°F$ ($-8°C$) which I like for carbonated drinks, and one set to $22°F$ ($-5.5°C$), which is the temperature I like for my stirred-drink-style bottled cocktails. Regular refrigerators are too warm for either of these applications, and freezers are too cold. I can't stress enough how important accurate drink temperature control is to the functioning of the bar.

RED-HOT POKER

I build red-hot pokers to ignite and heat drinks at my bar. For why, and how to do it yourself, see the Red-Hot Pokers section, page 177.

VACUUM MACHINE

Vacuum machines are designed to seal foods in vacuum bags for preservation and *sous vide* cooking. For cocktails, I use them to infuse flavors into fruits and vegetables. Good machines are pricey—well over a thousand dollars—although you can play around with vacuum infusion for much less money.

ROTARY EVAPORATOR

The rotary evaporator, or rotovap as it's called in the biz, is a piece of laboratory equipment that enables you to distill in a vacuum instead of at atmospheric pressure. This is a good thing, for several reasons.

In distillation you boil a mixture of ingredients—typically water, alcohol, flavors, and (unavoidably) impurities—and convert a portion of that mixture into vapor. Everything that *can* boil in the liquid will be present in the resulting vapor to some degree, but the concentration of substances with lower boiling points, like alcohol and aromatics, will be higher in the vapor than it was in the liquid. The alcohol and flavor-enriched vapor then feeds into an area called the condenser, where it cools down and condenses back into a liquid.

In *atmospheric pressure distillation*, this process occurs at elevated temperatures in the presence of oxygen. In *vacuum distillation*, boiling occurs at lower temperatures, even at room temperature or lower, because reducing pressure reduces the boiling point. Vacuum distillation, therefore, is very gentle, because it happens at low temperatures and in a reduced-oxygen environment that prevents ingredients from oxidizing.

Another nifty feature of the rotovap is its rotating distillation flask. The rotation creates a tremendous fresh liquid surface area, which enhances distillation and promotes gentle, even heating of the mixture. Strangely, even if you are distilling at room temperature, you need to add heat. If you

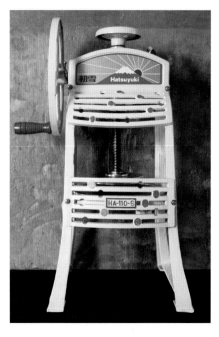

Hatsuyuki manual ice shaver–a thing of beauty and silence.

ROTARY EVAPORATOR: *You place liquid in the distillation flask (colored red) where it is warmed by a water bath, which is shown unfilled in this illustration for clarity. The distillation flask rotates to promote even heating and distillation. The vapor travels from the distillation flask to the condenser area (colored blue), where it is cooled and either condenses back to a liquid or is frozen solid on the condenser, depending on conditions. This particular condenser is cooled with liquid nitrogen. Anything that remains liquid after condensing drips into the receiving flask (colored green.) The entire process takes place at low temperatures because the system is kept under vacuum by a vacuum pump (colored yellow.)*

don't add heat, the mixture will cool down as it distills because of evaporative cooling. If you had a good enough vacuum, you could even freeze the mixture this way.

Why Use One: The Good: Rotary evaporation can make distillates of fresh products that taste fresher and cleaner than you'd think possible. Properly used, rotovaps can recover nearly all the flavors of the original mixture. Unlike atmospheric stills, they can split flavors without changing them or losing them. I've distilled mixtures, recombined the leftovers with the distillate, and then blind-tasted against the undistilled liquid: people cannot tell them apart. The rotary evaporator is like a flavor scalpel. Use it artfully, and it allows you to manipulate flavors like no other piece of equipment.

The rotovap has taught me to think about flavors in new ways. My favorite uses highlight how your brain integrates complex flavor inputs. A distillation of red habanero peppers smells like it will kill you with spice but isn't spicy at all, because capsaicin, the chemical that causes the burn, does not distill. Cacao distillates taste like pure chocolate but lack the inherent bitterness of unsweetened chocolate because the bitter principles don't distill. I've made distillations of herbs and broken the flavors down into dozens of fractions that I can recombine however I like. Flavor scalpel. I love my rotovap.

Sometimes you don't want the distillate, you want the leftovers. Imagine the freshness and punch of concentrated strawberry syrup that has never been heated. Simply remove the water from fresh clarified strawberry juice using the rotary evaporator. Delicious. Port wine reductions done at room temperature? Ridiculous (don't forget to drink the port "brandy").

The Bad: Unfortunately, rotovaps have some problems that will keep them from landing in most homes anytime soon. First, they are expensive: fully decked out, they cost well over ten grand. The cheaper ones are horrible, often leaky and therefore useless. Second, rotovaps are very fragile because of all the fancy glassware they contain. When you break a piece of glass in a rotovap—and you will—it'll set you back a couple hundred bucks. Third, there is a steep learning curve to achieving good results. You'll get okay results quickly enough, but a novice operator will never produce products as good as those made by someone who's been flying the rotovap for years.

The Ugly: Last, laws prevent the distillation of alcohol in bars here in the United States. Consequently, many rotovap owners are constrained to distilling only water-based mixtures—no alcohol. Standard rotovap setups are woefully inadequate to produce decent results without ethanol. Water just doesn't hold on to flavors the way ethanol does, and the majority of the delicate aromas the rotovap is such a genius at capturing are lost. I have worked long and hard to produce good water-based distillates. You must use a condenser filled with liquid nitrogen that freezes all the flavor compounds solid so they cannot escape. After you finish distilling, you melt your frozen flavor directly into high-proof ethanol. A pain. The upshot of this gloomy graph is that while I would like to write a whole chapter on the rotovap—it is worthy—the world hasn't caught up to it yet.

My rotovap flying along.

These shopping lists will help you navigate the maze of cocktail equipment. They are organized by desires and needs. Don't want to make carbonated drinks? Skip the bubbles section. Want to try carbonation but don't want to buy the rig on my list? I'll give other options later. The only mandatory shopping list is the first one. After you have your basic setup, you can add equipment from other lists as you wish. Sources for most of these items can be found in the section on page 378. Photographs of many of the items appear on page 49.

Remember not to get discouraged by the daunting nature of some of the equipment. In the techniques section of the book, I always try to give you a way to test cool techniques without breaking the bank.

HEY, I JUST WANNA MAKE SOME GOOD DRINKS

1. Two sets of shaking tins
2. Good jigger set
3. a) Hawthorn strainer, b) julep strainer, and c) tea strainer
4. Muddler
5. Paring knife
6. Y-peeler
7. Bar spoon

I WANNA ACT LIKE A FANCY PRO WITHOUT BREAKING THE BANK

Add:
8. Bar mat
9. Good ice pick
10. Dasher tops for bitters
11. Flat-bladed slicing knife (for ice) (not pictured)
12. Ice bucket and scoop (not pictured)
13. 2-inch ice-cube molds
14. Small rectangular Igloo cooler (not pictured)
15. Glass bottles with eyedroppers
16. Hand citrus press
17. Lewis bag or other ice crusher (not pictured)

THIS IS GOOD FOR MY KITCHEN TOO

Add:
18. High-speed blender, like a Vita-Prep
19. iSi whipped cream siphon
20. If you plan on making a lot of juice, a Champion juicer or equivalent

I'M GONNA MAKE SOME RECIPES THAT REQUIRE ACCURACY

Add:
21. Drug scale: 250 grams by the $1/10$ gram
22. Kitchen scale: 5 kilos by the gram
23. Decent digital thermometer
24. Money permitting, 50 ml, 250 ml, and 1000 ml plastic graduated cylinders

BUBBLES

Add:
25. 5- or 20-pound CO_2 tank
26. Regulator, hose, ball-lock connector
27. Three Liquid Bread carbonator caps

I'M AN EXPERIMENTER OR A STICKLER FOR ACCURACY

Add, in this order:
28. Refractometer
29. Micropipette

I'M CRAZY

Add:
30. Professional ice shaver
31. Red-hot poker

I'M NOT BANKRUPT YET: BIG-TICKET TECH I'M GONNA GET ONE AT A TIME, IF EVER

Add in this order:
32. Liquid nitrogen
33. Vacuum machine
34. Centrifuge (add this first if you are buying for a professional bar)
35. Rotary evaporator

ABOVE LEFT: BASIC BAR KIT: *If you have everything in this picture you can execute any classic cocktail with style and grace.*
ABOVE RIGHT: GOOD TO HAVE (LEFT TO RIGHT): *Vita-Prep (Vitamix for home use)—if you can afford it, the only blender you should buy; ½-liter iSi cream whipper; Champion juicer.*

IF YOU WANT TO ACT LIKE A FANCY PRO: Get a bar-mat to capture unsightly drips and spills; a set of two-inch ice-cube molds to make big ice for shaking (you can use these cubes for rocks drinks as well); ice picks to work with larger ice cubes and chunks; dasher tops to make your own bitters bottles; an eye dropper for saline solution or tinctures; and a hand citrus press—because squeezing citrus between your hands or reaming with a fork is goofy.

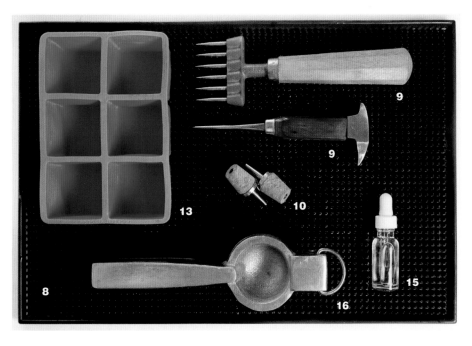

Ingredients

The most important ingredient in a cocktail is the liquor, but I'm not going to spend a long time on liquors here. There are innumerable books on the subject of spirits, and I list some of my favorites in the bibliography. Please buy good spirits, make sure you always have high-quality vermouth on hand (store it in small containers in the fridge), and never run out of Angostura bitters. Otherwise you are on your own, because I prefer to discuss ingredients that don't receive enough scrutiny elsewhere—the sweeteners, acids, and salt that we use to augment the liquors.

SWEETENERS

Honey (right) is far too viscous to pour and incorporate into cocktails. Honey syrup (left) is the way to go.

Almost every cocktail contains something sweet, whether vermouth, liqueur, juice, or sugar. The sweetness in these ingredients comes from a small group of basic sugars, the most important of which are sucrose (table sugar), glucose, and fructose. Fructose and glucose are simple sugars, while sucrose is composed of one molecule of glucose bonded to one molecule of fructose. You will commonly hear the oversimplification that fructose is 1.7 times sweeter than sucrose (table sugar, remember), while glucose is only 0.6 times as sweet. Luckily, very few sweeteners contain a preponderance of glucose or fructose. Relatively equal mixtures of glucose and fructose, such as you find in honey and most fruit juices, behave pretty much like sucrose does for cocktail purposes, so most of the time these ingredients can be used interchangeably. Agave syrup is an exception: over 70 percent of the sugar in agave nectar is fructose, so it does not act at all like table sugar. Read on to learn how sugars behave in cocktails and syrups.

SWEETNESS AND TEMPERATURE

The colder your drink is, the less sweet it tastes. Thus drinks served very cold—shaken cocktails—typically have more added sweetener than ones served warmer—stirred cocktails . . . and those shaken drinks seem to get sweeter if they are allowed to warm up. We've all had the experience of tasting a drink and deeming it well balanced, only to taste it again a few minutes later and find it too sweet—one reason to drink quickly, and in moderation, of course.

SUGAR AND CONCENTRATION

A wrinkle: it's very difficult to taste and judge sweeteners when they are in their concentrated ingredient form, like simple syrup and liqueurs. While our taste response to sugars is fairly linear up to about 20 percent concentration, which includes the 4 percent to 12 percent concentrations typically found in finished drinks, very weak and very high sugar concentrations don't behave linearly. Once you jack sugar content above 20 percent, your perception starts getting warped, and at 40 percent your palate is pretty much worthless. Simple syrup is 50 percent sugar, and many liqueurs have 200 to 260 grams of sugar per liter (20–26 percent). These sweeteners must be tasted at the dilution and temperature at which they will be served to be properly judged.

SUGAR AT THE BAR

Most of the time we use liquid sugars (syrups) at the bar, because granulated sugars don't dissolve quickly enough. These syrups need to measure and pour well in a jigger and quickly and easily disperse with other cocktail ingredients. They need to be fairly liquid but can't contain so much water that they spoil easily or overdilute your drink. Two sugar levels hit the sweet spot and are thus commonly used in bar recipes: 50 percent and 66 percent sugar by weight. Anything below 50 percent spoils too quickly and overdilutes. Anything over 66 percent is hard to pour quickly and may crystallize in the fridge.

Simple Syrup Simple syrup is the default sweetener for cocktails, and simple it is: just sugar and water. Simple comes in two varieties, 1:1 (regular simple) and 2:1 (rich simple). Regular simple is 1 part sugar to 1 part water *by weight*. Rich simple has 2 parts sugar to 1 part water *by weight*. Perhaps counterintuitively, rich simple is not twice as sweet as regular simple. Every fluid ounce of rich simple provides the same sweetening power as 1.5 ounces of regular simple. At the bar and in this book I use

FRUCTOSE

The sweetness of fructose comes on faster and more intensely than sucrose but fades faster, too—hit and run—something to pay attention to when mixing cocktails with agave nectar. The weirdest thing about fructose, however, is that it maintains its sweetness as it gets cold. Cold fructose is much sweeter than cold sucrose. Conversely, heating fructose makes it seem less sweet than sucrose. The upshot: drinks made with agave nectar might be balanced at room temperature but too sweet when chilled and not sweet enough when heated. Why? Because fructose can exist in a number of different configurations with radically different sweetnesses. The amount of each configuration present depends on temperature. In colder cocktail conditions, the sweet forms predominate. In hot drinks, the less sweet configurations are more prevalent.

1:1, regular simple, almost exclusively. It pours better and mixes quicker than 2:1. It is also more tolerant of bad jiggering than rich simple is.

Regular simple is easy to make. Put equal weights of sugar and water in a blender and blend on high till the sugar is dissolved. If you have time, let the syrup settle for a few minutes to get rid of air bubbles. To make rich simple, or regular simple without a blender, heat the ingredients on a stove until the syrup turns clear (indicating total dissolution) and then let the syrup cool. The disadvantages of stovetop simple are (1) you can't use the syrup right away because it's too hot, and (2) some water evaporates, throwing off your recipe. If you don't have a blender, stove, scale, or time, use superfine sugar. The crystals in superfine sugar are small enough to dissolve without a blender, and superfine comes in conveniently, though expensive, preweighed 1-pound (454-gram) boxes. Just add one box of superfine sugar to a pint (2 cups or 454 milliliters) of water, shake in a covered container for a minute, and you're off to the races, no weighing necessary.

Many bartenders measure their sugar by volume, a practice I discourage. Granulated sugar and water do not have the same density. Domino-brand granulated sugar has a density of .84 gram per milliliter straight out of the bag, while room-temperature water has a density of 1 gram per milliliter—a 16 percent difference. If you tap on your measuring cup repeatedly to compact the sugar, you can achieve a density very close to that of water, but few people go through that process, and it is more difficult than weighing.

Both regular and rich simple can be left unrefrigerated for several hours at a time, but eventually they will develop moldy, floaty occlusions. Store yours in the fridge.

Brown Sugar, Demerara Sugar, and Cane Syrup: Brown sugar is made by adding molasses to refined white sugar. Demerara is crystallized sugar that was never refined into white sugar. Cane syrup is unrefined sugar that has been concentrated but not crystallized. All these sweeteners have more or less of a rich molasses note. When using brown sugar or Demerara, make a 1:1 syrup. Cane syrup has no standardized sweetness but is almost always sweeter than 1:1 syrup.

Honey: The taste of honey varies dramatically depending on the species of flower the bees visited during pollenation. While it's fun to experiment with different varieties, most bartenders settle on fairly neutral honeys, such as clover. I have tried many times to make a good drink with buckwheat honey, which is superdark and funky-tasting, but have had no luck.

Honey is roughly 82 percent sugar and very, very thick. It is hard to use at the bar unless you make it into a thinner syrup. To make a honey syrup that can be substituted for simple syrup in any recipe, add 64 grams of water to every 100 grams of honey (note that the honey must be weighed; it is much denser than water). Unlike simple syrup, honey has some protein in it. Those proteins will increase the foaminess of shaken drinks, especially those with acidity in them as well.

Maple Syrup: Maple syrup is a fantastic cocktail sweetener. It is roughly 67 percent sugar by weight, comparable to rich simple syrup. Every fluid ounce of maple syrup is as sweet as 1½ fluid ounces of simple syrup. Flipped around, to replace 1 ounce of simple syrup, use ⅔ ounce maple syrup; to replace ¾ ounce simple, use ½ ounce maple (remember, these conversions are by volume!). Because maple syrup is expensive and I want the longest possible shelf life, I never water it down to typical simple-syrup levels. Maple syrup doesn't need to be refrigerated on a short-term basis, and it is never dangerous to leave it at room temperature. A disgusting mold can grow and spoil your syrup overnight, so keep it in the fridge for long-term storage, or boil it periodically. That mold tastes nasty—really nasty. Always smell the syrup before you pour it into your drink. I once ruined a hundred dollars' worth of batched cocktail by adding moldy maple syrup.

Agave nectar is composed mainly of fructose, with a little bit of glucose mixed in. Usually it is around 75 percent sugar by weight—between maple syrup and honey. The flavor changes from brand to brand. Fructose sweetness hits fast and decays fast, so use agave when you don't want the sweetness to linger. When using agave nectar straight, use about 60 percent of what you would use for simple syrup by volume. To make an agave nectar that you can substitute for simple syrup, add 50 grams of water to every 100 grams of agave nectar (note that the agave must be weighed; it is much denser than water). Agave nectar goes well in margaritas, but not because tequila is made from agave; that's a coincidence. Agave does well with lemon-based drinks because the acidity of lemon is also fast attack, fast decay (see the Acids section, page 58).

Quinine Simple: Quinine is the intensely bitter bark that gives bitterness to tonic water. Use this simple syrup for making tonic water, or any time you want the characteristic bitterness of tonic. (See the recipe on page 367)

EMULSIFIED SYRUPS, BUTTER, AND ORGEATS

Fats don't want to meld into drinks on their own. Oil and water, as everyone knows, don't mix. But you can force them into drinks by making emulsions, typically in the form of a sweetened syrup. To make an emulsion you need an emulsifier, whose job it is to make oil and water live side-by-side. The emulsifier I use has the horrible name Ticaloid 210S and is made by TIC Gums, which, unlike most industrial food companies, will sell to normal humans and not just large corporations. Ticaloid 210S is a mixture of gum arabic, a fantastic emulsifier derived from tree sap, and xanthan gum. Gum arabic is great for cocktail applications because its emulsions don't break when they are suddenly diluted, and they are immune to temperature changes, acidity, and alcohol. The xanthan is a stabilizer, which protects the emulsion from separating. If you can't get Ticaloid 210S, you can substitute a mixture of powdered gum arabic and xanthan gum in the ratio of nine to one.

In 2009 I made my first Ticaloid syrup, a butter syrup that I used to make a cold buttered rum. I love this stuff. This syrup has a lot of butter in it, so you have to use more of it than you would regular simple syrup—1.5 times as much.

Butter Syrup

INGREDIENTS

200 grams water

10 allspice berries, crushed

3 grams TIC Gums Pretested Ticaloid 210S

150 grams melted butter

200 grams granulated sugar

PROCEDURE

Heat the water and infuse the allspice berries for 5 minutes at a simmer, then strain out the allspice. Hydrate the Ticaloid 210S in the allspice-infused water with a hand blender. Add the melted butter and blend until smooth. Add the sugar and blend until smooth. This syrup can be stored at the bar until needed. It will separate over time but can be stirred back together by hand.

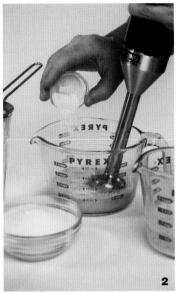

MAKING BUTTER SYRUP: 1) *Strain allspice out of hot water;* **2)** *Blend in Ticaloid 210S emulsifier;* **3)** *Emulsify in melted butter;* **4)** *Blend in and dissolve sugar.*

Cold Buttered Rum

MAKES ONE 5³/₅-OUNCE (168-ML) DRINK AT 16.4% ALCOHOL BY VOLUME, 8.6 G/100 ML SUGAR, .54% ACID

INGREDIENTS

2 ounces (60 ml) spiced rum, such as Sailor Jerry

Fat ounce (1⅛ ounces, 33.75 ml) Butter Syrup

½ ounce (15 ml) freshly strained lime juice

PROCEDURE

Combine the ingredients, shake with ice, and strain into a chilled old-fashioned glass.

COLD BUTTERED RUM: *Notice the butter syrup doesn't break after it has been diluted into a drink—the magic of gum arabic.*

Soon after I developed the butter syrup, I started using this same technique with other oils—pumpkin seed, olive, and so on. My biggest successes came with nut oils, specifically pecan, which makes a truly fantastic syrup, especially when you add some nut solids into the mix. I realized I could use the Ticaloid to make my own orgeats. Technically, orgeat is an almond simple syrup mixed with a bit of rosewater, but I use the word to mean any nut-flavored simple syrup made with nut milk and stabilized with Ticaloid (I don't add rosewater). I've made pecan orgeats, peanut orgeats, pistachio orgeats. Any nut will work. Here is the procedure.

Any Nut Orgeat

FIRST MAKE NUT MILK

600 grams very hot water (660 if you don't have a centrifuge)

200 grams nut of your choice

If the nuts aren't salted you can add some salt if you like

PROCEDURE

Blend the water and nuts together in a Vita-Prep high-speed blender. Either strain and press the nut milk through a fine strainer or spin the mixture in a centrifuge at 4000 times the force of gravity for 15 minutes (see the Clarification section, page 235). If you use the centrifuge, remove the fat and liquid portions from the top and save, discarding the solids at the bottom—or save them to make cookies. Add salt if you like.

Straining nut milk can be tedious. At the bar I use a centrifuge, but at home I use a very fine strainer. A nut milk bag is your other straining option.

THEN MAKE THE ORGEAT

FOR EVERY 500 GRAMS NUT MILK:

1.75 grams Ticaloid 210S

0.2 gram xanthan gum

500 grams granulated sugar

PROCEDURE

In a high-speed blender, combine the nut milk, Ticaloid, and xanthan. When they are combined, add the sugar and blend until combined.

Pecan orgeat with a pecan bourbon sour in the background. If your nut milk contains too many nut-solids the syrup might break when you shake the drink. When this screwup happens, a quick spin with a stick blender will fix it.

MEASURING SWEETNESS AT THE BAR

It is very helpful to have a quantitative idea of the sugar content of common drink ingredients. At the bar I use a refractometer to measure sugar content in Brix. Brix is the percent sucrose (table sugar) in a solution by weight. One hundred grams of a 10-Brix solution will contain 10 grams of sugar. The problem is, refractometers don't actually measure sucrose directly; they measure how much light passing through a solution is bent. Ingredients besides sugar, like alcohol, also affect how light is bent and throw off your readings, so the cardinal no-no with a Brix refractometer is trying to measure the sugar level of alcoholic beverages.

Remember that Brix denotes sugar concentration by *weight*, not by volume. A 50-Brix syrup definitely does *not* contain 500 grams of sugar in a *liter* of syrup. A 50-Brix syrup (like regular simple syrup) has a density of 1.23 grams per milliliter. A liter of simple syrup, therefore, weighs 1,230 grams and contains 615 grams of sugar. Since cocktails are normally measured by volume, it is best to think of simple syrup as containing 615 grams of sugar per liter. Rich simple (66 Brix) has a density of 1.33 grams per milliliter. A liter of it weighs 1,330 grams and contains 887 grams of sugar, or 887 grams per liter, a little less than 50 percent sweeter than regular simple.

Refractometers are most useful when measuring fresh fruit juices, which can vary widely in sugar content. No two batches of fruit are exactly the same. The sugar level of a particular fruit and its juice can change radically from day to day. If blueberry juice is 11 Brix today and 15 Brix tomorrow, the drinks made with those juices will obviously be different. At home you can change the balance of a recipe by taste based on the juice you've got, but in a bar that practice is impractical. You can't be sure those corrections will be made properly day in and day out by each person. So at the bar we chose a Brix level a couple points higher than the level the juice typically arrives at, and we correct each batch to that sweetness every time we make juice. Some juices we correct with sugar, some with honey, and some with cane syrup—whatever sweetener we think will show off the best qualities of the juice. You don't want to add too much sugar to a juice; you're not looking to add sweetness, you're just looking to standardize. When making juices at home, you don't need to use a refractometer to correct your juices, but the idea of Brix correction still comes in handy. Low-Brix versions of fruit juices tend to taste poor: flat, weak, and one-dimensional. Sometimes when you Brix-correct it by adding sugar, it tastes every bit as full and delicious as its high-Brix cousin. It doesn't take a lot of sugar to accomplish this miracle.

ACIDS

It is the rare cocktail that contains no acid. Sometimes the acidity is hidden in the form of vermouth or other wine-based ingredients, but it is almost always there. Almost all the common acidic cocktail ingredients listed below are available in pure form from home-brew shops.

TASTING ACIDS

Acids are molecules that produce free hydrogen ions in solution. More hydrogen ions mean more acidity. When scientists measure acidity, they measure pH, which directly correlates to the number of free hydrogen ions. Your tongue doesn't work like a pH meter, which means that pH meters are useless for cocktail work. Instead of sensing how acidic something is, you much more closely sense *how many acid molecules* are present. Chemically, this measure is referred to as *titratable acidity*. This fact makes it very easy to convert back and forth between different types of acids, because most organic acids have roughly similar weights. You can switch a gram of citric acid for a gram of malic or a gram of tartaric. It will taste different, but the acidity itself won't be too far off. In the rest of this book, therefore, I specify acidity as percent acidity per unit volume: 1 percent acidity means that a liter of juice contains 10 grams of acid.

LIME JUICE FRESHNESS

I have run several taste tests of lime juice at different ages both in cocktails and in limeade. In these tests, the batches are twenty-four, eight, five, three, and two hours old, and fresh. Not surprisingly, the day-old juice always loses. Surprisingly, the lime juice that is several hours old usually wins over lime juice that is superfresh. My subjects for these tests were predominantly professional American bartenders. Most good American bartenders use lime juice that is juiced at the beginning of the shift and therefore is several hours old during service. My cocktail compatriot Don Lee ran this same test with a bunch of European bartenders, who typically juice their limes à la minute. They tended to choose fresh lime juice over any of the other batches. Best, therefore, may just be what you are used to.

LEMONS AND LIMES

Lemon juice and lime juice are the two most common acids at the bar. Both of them are roughly 6 percent acid. The acidity of lemons is almost pure citric acid, while the acidity of lime juice is roughly 4 percent citric and 2 percent malic with a tiny bit of succinic acid. Succinic acid tastes terrible on its own—bitter, metallic, bloody. But in tiny amounts it really improves the flavor of lime juice. It is hard to get; you have to pay exorbitant rates for it at chemical supply houses.

Because lemon and lime juices are similar in acidity, they are roughly interchangeable in a recipe from a quantity standpoint, though lime's malic acid content means acidity lingers longer than lemon's. Unlike juices from grapefruit, orange, and apple, which can be kept for a couple days or even longer, lemon and lime juices must be used the day they are made. Lime is the most fragile, start-

COMMON ACIDS IN COCKTAILS

CITRIC ACID

Citric acid is the primary acid in lemon juice. On its own it tastes like lemon. Citric acid is clean, hits hard and fast, then fades rather quickly.

MALIC ACID

Malic acid is the primary acid of apples. On its own it tastes like green-apple candy. Its flavor lingers longer than that of citric acid.

TARTARIC ACID

Tartaric acid is the primary acid of grapes. On its own it tastes like sour-grape candy.

ACETIC ACID

Acetic acid is vinegar, the only common food acid that is aromatic. It is used as a minor acid, especially in bitters and savory cocktails.

LACTIC ACID

Lactic acid comes from fermentation. On its own it is reminiscent of sauerkraut, pickles, cheese, and salami. Its agreeable use in cocktails might surprise you.

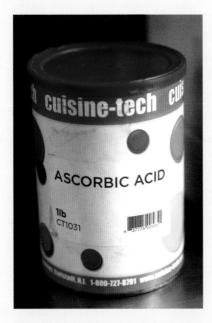

Ascorbic acid is the acid that stops fruits and juices from turning brown. It does not taste very acidic.

PHOSPHORIC ACID

Phosphoric acid is the only non-organic acid in this group. It is extremely strong and dry. It is the characteristic acidulant (along with citric) of colas, and it was extremely popular in the soda fountain days. It is not available from home-brew shops. I don't use it much.

ASCORBIC ACID

Ascorbic acid is vitamin C. On its own it does not have much flavor or impart much acidity. It is used primarily as an antioxidant: it stops juices and fruit from turning brown. It is often confused with citric acid.

COMBINATIONS OF ACIDS

Different acids have different flavors. Mixtures of acids taste different from single acids in surprising ways. Citric acid tastes like lemon and malic acid tastes like apple, but a mixture of the two tastes like lime. Tartaric acid tastes like grape and lactic acid tastes like sauerkraut, but a mixture of the two is the characteristic acid of champagne.

ing to change the moment it is juiced. I like lime juice best after it has rested for a couple hours.

The manner in which you juice a lemon or lime makes a difference. See the Juicers for Limes and Lemons section, page 33.

At 6 percent acidity, lemon and lime juice are fairly concentrated—a good thing for cocktails. Typical sour cocktails will require only ¾ ounce (22.5 ml) of lemon or lime juice to tart them up. Most other juices are not acidic enough, so I often make acid blends that mimic the strength of lime juice to augment my cocktail work. Some examples:

Lime Acid

Lime acid is what it says it is: a stand-in for lime juice. I would never use it instead of fruit, but it can be used to bolster fruit with a bit of acidity. For real authenticity, add the succinic, though it can be omitted.

INGREDIENTS

94 grams filtered water

4 grams citric acid

2 grams malic acid

0.04 gram succinic acid

PROCEDURE

Combine all the ingredients. Stir until dissolved.

Lime Acid Orange

Orange juice is typically 0.8 percent citric acid—not acidic enough to use in a proper sour. You can purchase sour oranges, which are delicious, but I often find myself with a glut of regular oranges whose peels I have used as a cocktail garnish. I correct their juice with acids to give it the same acid profile as lime juice. Be careful when juicing oranges; some of them, including many navel varieties, will turn bitter after the juice sits around for a while.

INGREDIENTS

1 liter freshly squeezed orange juice

32 grams citric acid

20 grams malic acid

PROCEDURE

Combine all the ingredients. Stir until dissolved.

The Dr. J (page 271) is made with lime-acid orange juice and tastes like an orange Julius.

Champagne Acid

The primary acids in grapes are tartaric and malic. But those aren't the primary acids in most champagne, which goes through a process called malolactic fermentation, in which malic acid is converted to lactic acid. White wines and champagnes (like Krug) that don't go through malolactic fermentation have a characteristic green-apple note to their acidity, whereas those that have gone through malolactic do not. So my standard champagne acid is a 1:1 mixture of tartaric and lactic acids. This acid mix tends to crystallize a bit when it sits around; don't worry about it, just shake it up. This acid blend is surprisingly versatile. I use it in carbonated drinks or any time I want to add a bit of that champagne feeling to a cocktail. I rarely use it as the sole acidity in a drink.

INGREDIENTS

94 grams warm water

3 grams tartaric acid

3 grams lactic acid (use the powder form)

PROCEDURE

Combine all the ingredients. Still until dissolved.

SALT

Salt is the secret ingredient in almost all my cocktails. Any cocktail that includes fruit, chocolate, or coffee benefits from a pinch of salt. I rarely want a drink to taste salty; the salt should be subthreshold. The next time you make a cocktail, divide it into two glasses and add a pinch of salt to one glass but not the other. Taste the difference—you will never forget the salt again. At home you can get away with adding a pinch. At the bar we have to be more precise, so we use a saline solution: 20 grams of salt in 80 milliliters of water (20 percent solution). A drop or two of this is all it takes to make a cocktail pop.

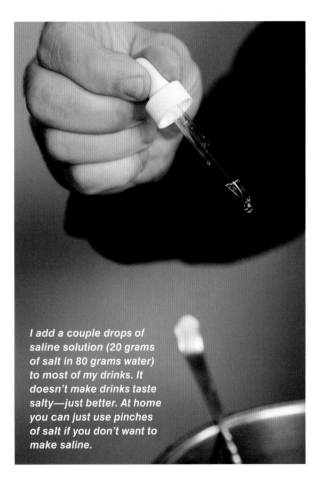

I add a couple drops of saline solution (20 grams of salt in 80 grams water) to most of my drinks. It doesn't make drinks taste salty—just better. At home you can just use pinches of salt if you don't want to make saline.

PART 2

TRADITIONAL COCKTAILS

Our investigation of cocktails begins where it must: with the basics. When I say traditional cocktails, I don't mean classic cocktails. I mean cocktails whose production requires nothing more than ice, booze, mixers, and a modicum of equipment—shakers, mixing glasses, spoons, and strainers. Generations of bartenders have developed thousands of delectable cocktails with these simple building blocks.

Section 1 deals with the science, production, and use of ice, the way ice interacts with booze, and the Fundamental Law of Cocktails.

Section 2 deals with shaken and stirred, built and blended drinks, with a final word on the underlying structure of all cocktail recipes.

Ice, Ice with Booze,
and the Fundamental Law

Ice by Itself

Ice is simply frozen water. It doesn't seem like there would be that much to say about it. Just put water in your freezer and freeze it. But in fact, making ice can be quite complicated, and modern bartenders spend a lot of their time trying to recreate a specific type of pure, clear ice that was the norm in the days before mechanical refrigeration. Here is the story of ice, the science behind it, and why you should care.

CLEAR ICE AND CLOUDY ICE, LAKE ICE AND FREEZER ICE

Before mechanical refrigeration, people harvested winter ice from lakes and rivers and stockpiled it in large icehouses for use throughout the year. By the mid 1800s ice dealers were shipping lake and river ice from northern locales to points around the globe, including the tropics. The golden age of the iced cocktail began. (For more on the subject of the early ice business, read *The Frozen Water Trade*; see Further Reading, page 379).

How could ice sit through sweltering heat on a non-air-conditioned boat for many weeks and survive well enough to make the trip worthwhile? The answer is in a relationship that's critical to cocktail making, and we'll return to it again and again: the relationship between surface area and volume. The amount of ice that melts in a given length of time is proportional to how much heat gets transferred into the ice. The amount of heat transferred is in turn directly proportional to the surface area of ice exposed to the environment. As something gets larger, its surface area increases, but not nearly as fast as its volume does. Tripling the size of a cube increases the surface area by a factor of 9 (surface area goes by the square : $3^2 = 9$) and the volume by a factor of 27 (volume goes by the third power: $3^3 = 27$).

Together, the 27 small ice cubes on the bottom have the exact same weight and volume as the single cube on the top, but they have three times the surface area, and therefore three times the surface water.

Huge volumes of ice, therefore, melt much, much more slowly than small volumes of ice. This fact made intercontinental ice shipments possible, and this same fact is fundamental to figuring out how cocktails work.

You might think that ice harvested from lakes and rivers would be inferior to ice made from purified water in modern freezers. Not so. It is crystal clear, while ice from your freezer typically is not. Cloudy ice is every bit as good at chilling as clear ice is, but clear ice looks a whole lot nicer in your cocktail. Clear ice is also easier to carve into whatever shapes you want (big, beautiful 2¼-inch cubes are my favorite). Cloudy ice shatters when cut and when shaken. Many bartenders believe that the tiny shards created by cloudy ice shattering in shaking tins water down their drinks excessively, a statement that is both true and false, as we will see when we deal with the science of shaking. Regardless, clear ice is alluring, and pretty much every bartender wants it. Nothing beats the look of an old-fashioned served over a water-clear hand-cut ice cube. If you don't care about beautiful clear ice, go ahead and use cloudy ice-cube-tray ice, just please read the section on Making Good Everyday Ice on page 73. Your cocktails will taste just fine. But if you like good things, read on. The next section tells the strange story of how ice forms and why lake ice is clear. If all you want to know is how to make your own clear ice, skip to the section on Making Clear Ice in Your Freezer, page 68.

HOW ICE FORMS

To understand how to make clear ice, you must understand how it forms. Lake ice is clear because it is formed layer by layer from the top down. The crystals first form on the surface of the water and then grow downward, getting thicker and thicker. But why does the ice form on the top of the lake, and why does it matter?

Most everything gets denser and shrinks as it cools down. Water doesn't. Liquid water is actually densest at around 4°C. Chilling water below 4°C will actually cause it to expand. This property is super-duper rare, and is fittingly called the "anomalous expansion of water." This anomalous expansion is lucky, because it means that the densest water—the water that will sink to the bottom of a lake in winter—isn't the water that is about to freeze. The water that is about to freeze—the stuff at 0°C—will float at the top. If water didn't behave this way, no aquatic life could survive the winter in cold climes. It would be frozen out.

Even weirder, while almost everything contracts when it solidifies, water expands by about 9 percent as it freezes. Ice floats because liquid water molecules, which are free to move around, can pack more densely than the rigidly aligned water molecules in ice. Force from expanding ice is tremendous, easily shattering rocks and water pipes in winter and bottles of beer accidentally left in your freezer. The anomalous expansion of water and the freeze expansion of ice explain why ice forms at the top and grows down. Because of anomalous expansion, the top of the water is coldest, so it freezes first, and because of the freeze expansion of ice, the ice floats on top.

But there is more to clear ice than that. Why doesn't ice form as masses of tiny crystals, like snow, instead of in large clear sheets? The answer has to do with a phenomenon known as supercooling.

The freezing point of water is 0° Celsius, but water won't *start* to freeze at 0° unless ice is already present. Water needs to be chilled below 0° to form ice crystals—it needs to be supercooled. Remember our discussion of surface area and volume: very small crystals have a very large surface-area-to-volume ratio. Large surface areas encourage melting. Even at the freezing point, tiny crystals want to melt. For ice crystals to grow at 0°, they need something to grow on—either an existing ice crystal or something of a similar size and shape, like a speck of dust. Without a place for crystals to grow, the water will continue to cool below 0° without freezing—a process called supercooling. As water supercools, it becomes easier for new crystals to form. Eventually, in a process called nucleation, a batch of crystals will form in the supercold water. After nucleation occurs, those initial crystals will start to grow and the water will heat back up to 0° again. Why? Water *heats up* as ice is formed because freezing ice gives off heat—a counterintuitive fact. Water freezing to ice gives off heat because it is going from a higher-energy state—a liquid—to a lower-energy state—a solid.

You might think that after the initial nucleation, ice crystals could continue to form anywhere in a freezing lake. They don't. After initial nucleation—once the water has supercooled and ice crystals have formed—those crystals grow and the surrounding water heats back up to 0°. Because the growing ice crystals keep the temperature of the nearby water at or near 0°, new crystals can't form. Forming new crystals would require more supercooling.

As those crystals near the top of the lake grow, they grow fairly slowly, and they grow clear. Unpurified water contains all kinds of dissolved and suspended crud: gas, salt, minerals, bacteria, dust. Crud. But as water freezes onto an existing ice crystal, it sheds and pushes out impurities such as trapped air, dust, dirt, minerals, and other contaminants; those things just don't fit in ice's crystal lattice. Very rapid freezing, in contrast, tends to produce many nucleation sites with smaller crystals; these smaller, fast-forming crystals can grow around and trap impurities, making disruptions in the crystal lattice and producing gassy, cloudy ice.

Now we are in a position to understand why standard freezer ice is cloudy. Ice-cube trays freeze water relatively rapidly, leading to the inclusion of impurities between crystal boundaries and thus cloudy ice. Ice in a freezer also tends to form from all sides of its container and grow toward the center of the cube, with the center freezing last. The water trapped in the center of a cube has no place to release gas and other impurities, so cloudiness is guaranteed. Even worse, as the water freezes inside its ice prison, it builds up tremendous force by freeze expansion, eventually shattering the outer portions of the ice cube and forming those mountain peaks you often see in home ice. The solution: get your freezer to act like a lake, so it freezes ice slowly and from one direction only.

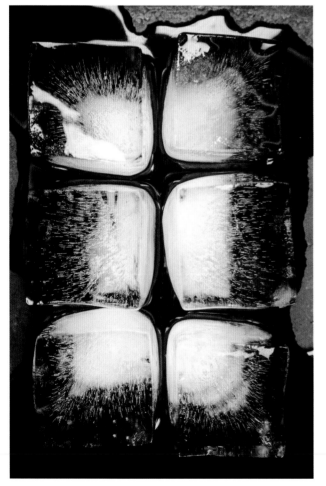

These ice cubes are arranged just as they were in their freezer tray. Notice the outsides—which froze first—are pretty clear. As the cubes continued to freeze, gas came out of solution and left trails of air bubbles. At some point, the ice formed a shell around the whole cube, trapping liquid water. When that last stuff froze there was no place for impurities to go and no place for the freezing water to expand, so you get bulging, fractures, and white haze.

MAKING CLEAR ICE IN YOUR FREEZER

Large blocks of clear ice are professionally produced for bartenders and ice sculptors in a machine called a Clinebell freezer, which acts like nature upside down. Clinebells freeze single blocks of ice weighing hundreds of pounds by freezing from the bottom only, constantly stirring the top of the water to prevent it from freezing over and scouring the surface of the forming ice to keep air bubbles out. As water freezes in a Clinebell, impurities are concentrated in the remaining liquid. You throw out the last of the water with all the impurities before it freezes. I once built my own version of a Clinebell and froze a 200-pound cylinder (25 gallons) of water. The gallon of water that I threw away at the end was dark brown and disgusting. Going in, the water appeared crystal clear. After concentrating, the impurities were quite apparent.

You don't need a Clinebell to make good ice. At home, freeze large blocks of ice in an insulated container with an open top, like a small Igloo cooler without the lid. The water in the cooler will start to freeze from the top, because the top is the only side that's not insulated. As freezing progresses, the water will keep freezing from the top, because heat from the freezing water is conducted more easily through the solid block of ice (a surprisingly good conductor of heat) than through the insulated sides. Because most of the water and ice is shielded with insulation, freezing is slow, which promotes the growth of large crystals that exclude gases and impurities, which get concentrated in the remaining liquid water.

Fill the cooler fairly high, but not so full that you spill water getting the thing into your freezer. Use hot water, which has fewer trapped gases than cold water. Don't put the hot water directly in your freezer; let it cool first, but don't pour it after it cools or you'll just trap more gases. (Hot water in a freezer will partially defrost your food, causing small ice crystals to melt. When your food refreezes the water will recrystallize on the remaining large crystals, making those crystals even larger and ruining the texture of your food.) A couple gallons of water will take several days to freeze in a standard freezer. Pull the cooler out of the freezer before it freezes completely, so you can drain off the impure residue. If you screw up and freeze all the water, no big deal: you can just cut the unclear stuff off the bottom. It is very difficult to judge how thick an ice cube is just by looking into the top of the cooler. I have been fooled into thinking that the water in my cooler was almost frozen solid, just to find out that I had a mere 2- or 3-inch-thick frozen slab.

If you try to make clear ice starting with cold water you'll get lots of air bubbles from trapped gasses.

When you remove the cooler from the freezer, don't immediately try to cut the ice. Let it sit and temper.

TEMPERING CLEAR BUT TEMPERAMENTAL ICE

All ice is 0°C or colder, but it can be much colder or barely colder, and if you're making cocktails, barely colder is better. Cold ice shatters and is no fun to work with, while ice that has warmed up to freezing temperature is a pleasure. You can tell the difference just by looking at it. Look at ice as it comes out of the freezer: it will take on a satiny appearance as moisture condenses on it and freezes. Large and exceptionally cold pieces of ice might accumulate a layer of ice crystals. The ice won't look wet. This is a visual indication that your ice is too cold to work with. As it warms up, however, you will see the ice change from dry and frosted to wet and clear. When the block is clear and glistening, it is sufficiently tempered—its tem-

HOW TO MAKE CRYSTAL CLEAR ICE AT HOME

To ensure bubble-free ice, pour hot water into cooler.*

For this size cooler (6.5 gallon), ice will be thick enough after 24–48 hours.

Invert cooler and allow ice to unmold itself—some water will spill out.

Shave excess ice shards off bottom of block.

---

* If you have small, gas-free ice in the cooler you can pour hot water directly over it in the freezer, without stirring, and making sure it all melts.

Finished block of bubble-free ice.

Score block on both sides with straight-bladed serrated knife.

Lightly tap along back of blade with mallet or rolling pin so ice cleaves into perfect columns.

Using same tapping motion, cut columns into cubes.

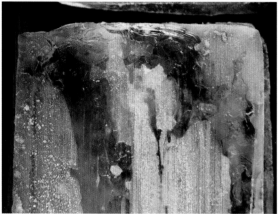

TOP: *I tried to cut this ice while it was too cold. It shattered like glass.*
BOTTOM: *This piece of ice is too cold to cut. It looks dry and frosty.*

perature has almost reached the freezing point, and it is ready to be cut or used in drinks.

Ice is a good conductor of heat, so it won't take long to temper most of the ice you'll be working with. Doubling the thickness of the ice block increases the tempering time by a factor of 4; a 2-inch-thick slab should temper in less than 15 minutes, a 4-inch block in less than an hour.

Don't try to speed up the tempering process. Ice straight out of the freezer is under tremendous stress as it warms. Stress develops because ice expands (as most substances do) as it warms up. The ice on the outside warms faster than that on the inside, causing the outside to expand faster than the inside. Ice left alone on a counter in air usually withstands this stress just fine, because air is a poor conductor of heat; but try to speed up the warming and the ice will crack, just like ice cracks when you throw it straight from the freezer into a glass of water.

CHOPPY CHOPPY: CUTTING CLEAR ICE

Once you have successfully made clear and tempered ice, you will be surprised to see how easy it is to work with. Woodworking tools will cut it with ease. Professional ice carvers use chainsaws, wood gouges, chisels, and electric grinders. In the bar I mainly use ice picks and a long, straight, inexpensive slicing or bread knife. Ice picks can be used to chip and sculpt and to score lines in ice to facilitate cutting, but really 99 percent of all your ice work can be done with just a bread knife. The knife can neatly divide a slab or block into perfect cubes and shape a cube into smaller diamonds or spheres. The trick is to lay the edge of that knife on the surface of the ice and gently rub it back and forth a bit. The highly conductive metal knife will tend to melt a small, thin crevice on the surface of the ice. This line concentrates stress in the ice block and provides a place for cracks to start—much like scoring a tile or piece of glass before you cut it. Place the ice on a nonslip surface (I use a bar mat), keep the blade in contact with the ice, and simply tap on the back of the blade with a mallet or other heavy object, and the block will cleave cleanly in two. Cutting ice this way is one of the easiest things you can learn to do, yet it always garners the admiration of onlookers who aren't in the know. For more complicated manipulations, see the ice books in the Further Reading section, page 379.

MAKING GOOD EVERYDAY ICE
(WHEN YOU DON'T NEED CLEAR ICE)

If your ice will appear in a finished cocktail or needs to be cut, you should use clear ice (remember, cloudy ice just shatters). When you are shaking or stirring drinks, you don't need presentation ice. But randomly made megacloudy ice is no good, as it will shatter unpredictably for almost no reason at all. Shattering ice creates an unpredictable slush of ice crystals when you shake—not what you want. Megacloudy ice is also really ugly—much uglier than moderately cloudy ice. So take a little extra trouble to make only moderately cloudy ice.

Megacloudy ice can be caused by quickly freezing contaminated or gassy water. If your water has a lot of impurities, you can invest in a filtration or reverse osmosis system. Getting rid of trapped gases and chlorine is easy: just use hot water. If your pipes have lead, you can heat cold water to drive off chlorine and other gases on the stove. Less dissolved gas equals less cloudy ice.

Avoid stacking ice trays on top of one another. The trays in the middle produce hopelessly cloudy, useless cubes, because they freeze almost equally from all sides, trapping the maximum amount of crud inside.

This ice is fully tempered and cuts with ease.

THE SHAPE AND SIZE OF ICE TO MAKE

For stirred cocktails not served on a rock, you can use any size or shape of ice. As we will see, you may have to alter your stirring technique to accommodate your ice, but any ice will work fine. For the very best texture for shaken drinks, however, you should freeze large cubes. Cubes 2 inches on a side are good, and very easy to make. Buy flexible ice-cube molds, but be choosy about which ones. My bartender friend Eben Freeman discovered years ago that some silicone molds impart an off taste to ice. The ones I purchase from Cocktail Kingdom are made from flexible polyurethane, and after thorough testing, I'm convinced they produce no off tastes. If you don't want to buy anything, you can freeze water in square metal cake pans. Some of the ice from such a pan will be megacloudy and useless, but a lot will be presentation-clear, and you can cut it away from the cloudy part after the ice is tempered.

Now that we have our ice, let's turn to ice with booze.

Ice with Booze

Ice at 0° Celsius can chill a cocktail *below* 0° Celsius. In fact, ice-chilled cocktails are routinely as cold as −6°C (21°F). Some people have a difficult time believing this important fact. They think their ice must have started below 0°C to make their cocktail that cold. Let's make a martini to verify this chilling phenomenon. We'll go through a bit more rigamarole than the drink requires in order to prove our point.

EXPERIMENT 1

EXPERIMENT 1

Practice Stirring a Martini

Water pitcher

Copious ice

Digital thermometer

2 ounces (60 ml) of your favorite gin (or vodka), at room temperature

Between ⅜ and ½ ounce (10–14 ml) Dolin dry vermouth (or your favorite), at room temperature

Metal shaking tin

Bar or kitchen towel

Martini glass or cocktail coupe waiting in the freezer

Strainer

1 or 3 olives on a toothpick, or a lemon twist

INGREDIENT NOTE: I am allowing for quite a bit of lee-way in the formulation of this cocktail. Half the fun of a martini is arguing over how it should be made. (And by the way, you should never apologize for your taste preferences. The only faux pas is not caring.) Because this is a stirring experiment, we will put the debate over shaking versus stirring aside—for now.

PROCEDURE

Fill the pitcher with ice and top up with water. Stir the water and ice, using your instant-read thermometer, until the temperature of the water is 0°C (32°F), which means your ice is also at 0°. Don't skip this step: it is crucial that you witness the fact that the ice is in fact at 0°C. Many people believe that the ability of ice to chill a cocktail below 0° comes from "extra coldness" that ice stores up in the freezer. If you are in doubt, continue stirring for a couple of minutes to verify that nothing is changing and everything in the pitcher is at 0°.

Put the gin and vermouth into the metal tin and measure the temperature with your thermometer. It should be room temperature, roughly 20°C. Scoop a large handful of ice out of the water pitcher—about 120 grams' worth (3 ounces), pat it with a towel to dry it a bit, shake off any surface water with your hands, then add it to the shaking tin. Start stirring with your digital thermometer and don't stop.

Within about 10 seconds, the drink should be at about 5°C. Keep stirring. Within about 30 seconds, the drink should be around 0°C. Most bartenders (including myself) would have stopped by now. Keep stirring. The temperature will drop below 0°. Keep stirring! The temperature will keep dropping! After a minute of constant stirring, your drink might be as cold as −4° Celsius (−25°F), depending on the size of your ice and how fast you are stirring. If you stir for up to 2 minutes, you might be able to get the drink down to −6.75°C. You'll notice that after 2 minutes or so the temperature stabilizes and doesn't get much lower. You've reached equilibrium, or close to it.

Pull your cocktail glass out of the freezer and strain the martini into it. Garnish with the olive(s) or lemon twist.

If I were making this drink for consumption instead of an experiment, I would never stir so long—the martini would be too diluted. If you prefer your martini shaken (which usually dilutes more than stirring does), you might find this 2-minute martini refreshing.

Now that you have proven that ice can chill a cocktail to a temperature below ice's own freezing point, you might want to know why this happens. How can ice make something colder than itself? (This gets pretty technical, so if you can't deal, skip ahead to Chilling and Diluting, page 80, and uncomplainingly accept some of my future claims at face value.)

INTERESTING COCKTAIL PHYSICS THAT YOU CAN IGNORE IF YOU DON'T CARE

You can go down a few different paths to understanding how ice chills below 0° (colligative properties and vapor pressure, for all you science types), but the best approach is to see the problem as a tug of war between *enthalpy* and *entropy*, two difficult and oft-misunderstood concepts at the heart of thermodynamics. Thermodynamics is the branch of science that explains why perpetual-motion machines can't exist and how the universe will eventually die. Deep stuff.

For our purposes, enthalpy can be thought of simply as heat energy, because we can think of a *change* in enthalpy as the measure of the heat absorbed or released during a reaction or process. (Science disclaimer: this statement is true only when our reactions take place at a constant pressure.) Entropy is a much wackier concept. It is most often explained as a measure of how disordered a system is, but entropy is much more than that. The following explanation might be a bit much for some readers to bear, but I promise that, apart from the word *entropy*, there isn't much jargon involved. Remember, we are trying to figure out why ice can chill an alcoholic drink below 0°.

HEAT AND ENTROPY TUG OF WAR: THINGS ARE LAZY BUT WANT TO BE FREE

The tug of war works like this: heat energy always wants to freeze your ice cube; entropy always wants to melt your ice cube. The relative power of these two contestants is determined by temperature. When you are making a cocktail, the tug of war always ends in a tie, but where it ends—the freezing temperature—can change.

Heat—the Lazy One: When I say heat, I don't mean temperature. Heat is not the same as temperature. To melt ice, you *add* heat but you *don't* change its temperature. Ice begins to melt at 0° and stays at 0° until it has completely melted, even though you are constantly adding heat energy to break water molecules free from their icy crystal prisons. Heat is a form of energy. Temperature is merely a measure of the average speed of the molecules within a substance. People often confuse these terms because to increase the molecules' speed—to increase temperature—you add heat.

When water freezes, it *gives off* heat. The heat that the ice gives off is what your freezer absorbs. Because ice gives off heat as it freezes, the internal energy of ice is lower than the internal energy of water at the same temperature. It is super-important to remember that point: freezing water gives off heat. Melting ice absorbs energy—it requires heat to melt. In general, all things being equal, reactions that give off heat and result in a lower internal energy are favored in nature, because in general, things want to go to a lower-energy state. Things are lazy. Making ice gives off heat, so *the change in heat favors water turning to ice.*

Entropy Pines to Be Free: Entropy is a different story. If you look at entropy as a measure of disorder, increasing entropy increases disorder. A fundamental tenet of thermodynamics is that the entropy of the universe is constantly increasing; hence the universe is constantly becoming more disordered (yay). A better way to define entropy is as a measure of how many different states something can be in. Scientists call these microstates. Things tend to maximize the number of available microstates and then commence to occupy those microstates in a random way. Things tend to increase in entropy. Things want to be free.

At any given temperature, there are more available positions, speeds, configurations—microstates—in a liquid than in a solid. Water molecules, for instance, are free to spin around and find new neighbors, while ice molecules are locked in a crystal. Being in a solid is more constraining than being in a liquid, so *changes in entropy favor ice melting into water.*

So Which Wins, Enthalpy or Entropy? It depends on temperature. Temperature, remember, is a measure of how fast, on average, the molecules in a substance are moving around. The higher the temperature is, the faster the molecules are moving. Faster molecules can achieve more disorder than slower ones, so the higher the temperature is,

SPECIFIC HEAT, HEAT OF FUSION, AND CALORIES

Different substances require different amounts of heat to raise and lower their temperature. The measure of this property is called **specific heat**. In calories, specific heat is the amount of energy required to raise the temperature of 1 gram of something 1° Celsius. Although the calorie is an outdated scientific unit (the joule is preferred), calories are useful for cooks and bartenders because they relate to temperatures and weights we can easily understand (note: calories in food are really kilocalories—1000 calories). For water, the specific heat is a very convenient 1 calorie per gram per degree. Ice has a lower specific heat than water: 0.5 calories per gram per degree. That means it requires only half the energy to heat or cool ice that it does to cool the same amount of water, and that unless ice is melting or freezing, it can supply only half the heating or cooling power that water can. Pure alcohol has a specific heat of 0.6 calories per gram per degree. The specific heats of water-alcohol mixtures (cocktails) are, unfortunately, nonlinear. Cocktails actually take more energy to heat or cool than either water or alcohol alone. Weird.

There's a second important heat-related property. Remember, to melt ice you must add heat. The amount of heat required to melt ice is called the **heat of fusion** (or the enthalpy of fusion). The heat of fusion works both ways. That is, it takes the same amount of heat to freeze something as it does to melt it. The heat of fusion of water is about 80 calories per gram. Thinking in calories lets you visualize how powerful ice really is. Eighty calories per gram means the heat required to melt 1 gram of ice is sufficient to heat 1 gram of water all the way from 0° to 80° C! More to the point, melting 1 gram of ice is sufficient to chill 4 grams of water from room temperature (20°C) all the way to 0°. We take ice for granted, but it is a miraculous substance. Gram for gram, liquid nitrogen, positively frigid at −196°C (−320°F), has only 15 percent more chilling power than ice does at a measly 0°C. A mere 15 percent! This startling fact explains why techie neophytes always underestimate how much liquid nitrogen they will need for a given project.

the more likely it is that entropy will win the tug of war and melt your ice. As the temperature goes down, the energy release from freezing tends to dominate, and water freezes. The freezing point of water (0°C) is the point at which the entropy gain from ice melting to water is exactly balanced by the amount of heat given off by water freezing into ice.

The surface of an ice cube in water is not static. Water molecules are constantly freezing onto and melting off the surface. If more molecules are sticking to the ice than leaving it, we say the ice is freezing. If more molecules are leaving than sticking we say the ice is melting. At the freezing point, water molecules are constantly freezing into ice and melting into water at the same rate—they are in equilibrium.

If you lower the temperature, the entropy gain from melting becomes puny and water freezes. If you raise the temperature, the entropy win from melting outstrips the enthalpy and the ice melts. Got it?

WHAT ABOUT MY MARTINI? WHAT HAPPENED WHEN I ADDED ALCOHOL?

Let's look at the point where you've stirred your martini and it has just reached 0°C. You have ice at 0° and a water-booze mixture also at 0°. When ice molecules melt into your cocktail, they absorb heat. The amount of heat absorbed is the same as if the ice were melting into pure water. The amount of heat absorbed from melting—the *heat* change—hasn't been altered by placing the ice in alcohol, because the ice is still pure ice. The *entropy* change associated with melting in alcohol is different, however. If a water molecule in our ice melts into the gin, the entropy gain is greater than it would be in the pure-water situation. Why? A mixture of water and alcohol is more disordered than a mixture of water alone. A scientist might say that there are more ways to arrange a group of water molecules and alcohol molecules uniquely than there are to arrange *the same number of identical* water molecules. More disorder, more available microstates. Entropy is winning again. When entropy wins, ice melts. So what happens? The ice starts to melt. What happens when ice melts? Cooling takes place. Melting ice absorbs heat and chills our drink below 0°. There is no external heat source to supply the heat needed to melt ice, so the heat is drawn from the system itself, and as a consequence, the entire system chills. The *drink and the ice itself* go below 0°.

As the ice melts and the gin gets more diluted, the size of the entropy win over the heat loss from melting decreases. Melting continues to happen until a new equilibrium is reached, when the entropy and heat become balanced again. That balance point is the new freezing temperature of our martini.

By the way, this same argument explains why salt added to ice can lower the temperature of the ice enough to freeze ice cream. Unfortunately, instead of getting the real explanation, most kids are just taught that the ice-and-salt trick works because "salt lowers the freezing point of water." Weak. Now that you are an adult, the truth can be told.

Chilling and Diluting

Every gram of ice melted provides 80 calories of chilling power. To put that power in perspective, an average 3.5-ounce (90-ml) daiquiri will melt between 55 and 65 grams of ice when you shake it for 10 seconds. That averages 2000 watts of chilling power . . . per drink. Shake four of those bad boys at once and you are blasting 8000 watts of chilling power.

All ice has the same 80 calories per gram of chilling power, regardless of how big it is or how fancy it is, but how that chilling power is delivered depends on the ice's size and shape. The difference between big ice cubes and small ones is their surface area. Smaller pieces of ice have more surface area for a given weight than larger pieces. These smaller pieces can therefore chill faster, which is good, but they also have more liquid water stuck to their surfaces, which is often bad. The surface of ice can also trap some of your cocktail so that it never makes it into your glass. Let's look at these three issues—surface area and chilling rate, surface area and trapped water, surface area and trapped cocktail—one at a time.

SURFACE AREA AND RATE OF CHILLING

Ice melts at its surface, so increasing the surface area increases the melting area and therefore the rate at which that ice can melt. Increasing the rate at which ice melts increases the rate at which it chills. But the surface area of the ice isn't the only factor. The surface area of the cocktail is important, too. A block of ice sitting still in a cocktail doesn't melt very fast. Stirring or shaking a cocktail brings fresh liquid into contact with the ice, essentially increasing the surface area of the cocktail and thus the rate at which it chills. The faster the drink moves, the faster the drink can chill.

As important as the rate at which the cocktail flows over the ice is *the rate at which melted water leaves the surface of the ice*. Rapid mixing of cold meltwater is a cocktail's main speed chiller. Plastic bags full of ice cubes and those blue gel packs you store in your freezer can't chill a drink nearly as fast as a good ol' ice cube, because their meltwater doesn't mix with the cocktail. Nor can ice cubes in plastic bags chill cocktails below 0°C—you don't get the entropy gain from melting into the alcohol. Even extremely cold things like blocks of steel stored in liquid nitrogen don't chill as fast as melting ice cubes.

Ice is fantastic.

SURFACE AREA AND SURFACE WATER

As ice approaches its melting point, it becomes jewel-like, with liquid water simmering on its surface. The more surface area your ice has, the more surface water it carries and the more water you'll add to your cocktail. Even though the liquid water stuck to the ice is at 0°C, it isn't an effective chiller, because it has already melted and relinquished most of its chilling power. Changing the amount of ice you use changes the surface area of the ice in contact with the cocktail and therefore the amount of water you are adding even before melting starts. In practice it is difficult to control the amount of ice you use, which means your drinks will have inconsistent amounts of dilution. This effect is magnified when you use ice with a high surface-area-to-volume ratio. When bartenders complain that small or thin varieties of ice (in bar parlance, "shitty" ice) overdilute their drinks, I suspect that they are reacting to the initial dilution they get from their ice's surface water.

To test this idea I made crushed ice with a very high surface-to-volume ratio (shitty ice) and then spun it in a centrifuge to get rid of the excess water. In the cocktails made with this ice, my final dilutions were the same as those made with larger ice cubes, when all the drinks were chilled to the same temperature.

The upshot: smaller ice with a large surface area relative to volume area can overdilute your drinks or make them inconsistent. To remedy this prob-

SUPERCHILLED ICE

A counterintuitive fact: colder ice chills more slowly than warmer ice. If you shake with ice that is well below the freezing point, it will actually chill more slowly than ice at 0°. The surface of very cold ice doesn't melt right away. Instead, the energy that the ice absorbs from the drink is used to heat up the ice cube to the freezing point. After the ice warms up and starts melting, the chilling rate increases. The supercold ice will eventually make your drink colder with less dilution than tempered ice will, but unless the superchilled ice is really cold, the difference will be small. Let's say you start with ice that is −1°C. Heating a gram of −1°C ice up to 0°C requires less than half a calorie—less than half the amount it takes to chill a gram of water 1°C and less than one 160th the amount it takes to melt 1 gram of ice. In other words, the amount of excess chilling energy stored in 160 grams of ice at −1°C would be provided by melting only one more gram of ice at 0°. Puny difference. Dramatically overchilled ice will dramatically reduce dilution. That isn't good. It is very rare that you want dramatically reduced dilution in a shaken drink.

In all likelihood you don't need to worry about your ice being too cold. Ice is a fairly good conductor of heat—about three and a half times better than water (as long as the water doesn't cheat by mixing and moving around). Ice's good conductivity plus the small amount of energy it takes to heat it up means that ice heats to the freezing point pretty quickly. If your ice has been out of the freezer for a while, it is probably very close to 0°C and you can ignore its actual temperature.

lem, shake off ice before you use it by putting a strainer over your shaking tin or mixing glass and throwing the water off the ice. I won't make you get rid of the water with a salad spinner, but I probably should. If you have access to large ice, make smaller, faster-chilling pieces by cracking the large pieces. The surface of freshly cracked ice doesn't have as much water on it as ice that has been cracked previously and allowed to sit.

SURFACE AREA AND HOLDBACK

The flip side of the water trapped on the surface of the ice before you chill is the cocktail stuck to your ice after you have chilled. I call this stolen bit of cocktail the holdback, and it can be quite substantial. More ice, smaller ice, and ice with nooks and crannies will increase holdback. To minimize holdback in your drink, make sure to give your chilling vessel a sharp snap with your hand after the last drip pours out of it. Right the vessel again and strain one last time. Based on tests I've done with lackadaisical pouring technique, stirring with crushed ice can hold back between 12 and 25 percent of your drink. Stirring with small ice-machine ice can hold back between 7 and 9 percent of your drink. Even with lousy pouring technique, stirring with block ice cut into large rectangles gives you holdback as low as 1 to 4 percent of your cocktail. The drain, snap, right, and drain again can knock all these numbers down to the 1 to 4 percent area, regardless of the ice you use. Proper cocktail pouring technique is important.

There is a widespread belief that ice should never be used twice. Hogwash. If you are stirring a drink and plan to serve it over ice, you should use the ice you stirred with, as long as it looks good. The ice you have stirred with is colder than fresh ice (if you have stirred the drink below 0°), and the used ice doesn't have water all over it, but cocktail instead.

MEASURING HOLDBACK

To quantify holdback, I measured the phenomenon in sugar-water-ice mixtures; I don't have the analytical equipment required to measure these ingredients plus alcohol. I made solutions that were 10, 20, and 40 percent sugar by weight, diluted 90-gram samples by stirring with crushed ice, ice-machine ice, and cut block ice, then measured the weight of the strained liquid on a scale and the final sugar concentration of the liquid with a refractometer. With that data I could calculate the amount of liquid left behind.

You would expect that holdback would primarily be a function of the surface area of the ice you use, just as dilution rate is, but holdback is more complicated. It is very strongly affected by the shape of the ice's surface as well as the surface area. Block ice with smooth sides has less surface area per gram than ice-machine ice does, and dilutes and chills more slowly than machine ice does, but it holds back far less cocktail than you'd predict from surface area alone.

Surprisingly, sugar content did not play a substantial repeatable role in the holdback of drinks.

CHILLING, EQUILIBRIUM, AND ULTIMATE TEMPERATURE

When we stirred our experimental martini, the temperature initially dropped rapidly. After a while, the temperature plateaued and further chilling was much slower. As the chilling slowed, so did the dilution, and after a couple minutes the drink wasn't changing much even though we were constantly stirring. The drink was approaching its equilibrium chilling point, the freezing point of the cocktail: the point at which the entropy gain from melting is balanced by the heat requirement of melting. You'll never really get to the equilibrium point, because your chilling techniques aren't rapid enough or efficient enough, but you can get close. Blended drinks come closest, getting really dang cold because their chilling is both extremely rapid and extremely efficient.

Before you reach equilibrium, either you stop chilling by removing the ice from your drink, or your chilling rate drops below the rate that heat enters the cocktail from the environment because your chilling isn't efficient enough. At this point your drink will get no colder. In fact, it will begin to warm up. How close you get to the theoretical freezing temperature of your drink depends on your chilling technique. The different techniques we use for traditional cocktails—stirring, shaking, building, and blending—all have inherent chilling parameters, and these parameters determine the structure of each drink style.

Before you add wet ice to your drink, snap the water off of it. This technique will prevent the ice's surface water from overdiluting your drink even if you don't have access to big ice cubes.

The Fundamental Law of Traditional Cocktails

Remember our definition of traditional cocktails: the only ingredients are liquor, mixers, and ice at 0°C (32°F). The only chilling comes from ice. We will assume that whatever mixing cups or shakers we use do not affect dilution and are isolated from the rest of the universe—no heat enters or leaves. This kind of assumption, while not strictly true, gives us useful results. With those givens, you have my **Fundamental Law of Traditional Cocktails**:

THERE IS NO CHILLING WITHOUT DILUTION, AND THERE IS NO DILUTION WITHOUT CHILLING.

The two are inextricably linked.

The only reason a cocktail gets diluted is that ice melts and turns into water. Conversely, the only way ice can melt is if it chills your drink. Stupidly simple, but the ramifications are deep. The law explains, for instance, why stirred drinks are warmer and less diluted than shaken drinks are, and why traditional cocktail ratios work so well.

MAKING A DRINK THE SCIENTIFIC WAY: THE MANHATTAN

Stirring a cocktail is a gentle proposition where the *only* things happening are chilling and dilution, as opposed to shaking, which also adds texture. The Fundamental Law of Cocktails states that chilling and diluting are inextricably linked. Here, as my son Booker says, is where you should prepare to be amazed. The upshot of that law: as long as two drinks with the same recipe reach the same temperature, they will be diluted the same amount. Any combination of ice size, stirring rate, and time that ends with the same temperature will make *exactly identical* drinks! When you are stirring, it doesn't matter how you get there—just that you know when you've arrived. I'll prove it. Let's make some Manhattans.

Because we are making drinks for science, you will need the same equipment you needed for the martini experiment, and you may want to enlist the help of a friend. Heck, why not just throw a cocktail party featuring Manhattans? You'll be making at least two drinks for the first experiment and three drinks for the second. You can substitute any Manhattan variant you wish—or any stirred drink recipe of your choosing, should you somehow get sick of Manhattans.

THE MANHATTAN

The Manhattan is my all-time favorite stirred drink. While I still enjoy a Manhattan made with bourbon, my standard Manhattan is made with Rittenhouse Rye. There is something about the mix of rye whiskey, sweet vermouth, and bitters that makes it always appropriate. Fifteen years ago, when rye wasn't as common as it is now, most Manhattans were made with bourbon instead of rye and used much less vermouth than is common today. It took the constant effort of the traditional cocktail cognoscenti to bring rye back to its rightful place of prominence at the bar.

Vermouth levels used to be lower partially because people were using bourbon instead of rye for the drink but mostly because people were using lousy vermouth, or worse—lousy old and oxidized vermouth. Common recipes in the early 1990s were three parts or four parts bourbon to one part vermouth. Nowadays a plethora of delicious vermouths is available, and many of us store them properly: refrigerated with a vacuum stopper. I use Carpano Antica Formula vermouth in this Manhattan, at a ratio of 2.25 parts rye to 1 part vermouth. Bitters are, of course, necessary to the drink, and here only Angostura will do—2 dashes to the drink. The standard garnish for the Manhattan is a brandied cherry. Unfortunately, I have to omit that step; I became deathly allergic to cherries when I turned thirty-one and am now confined to an orange twist.

EXPERIMENT 2

Different Ice, Different Stir, Same Manhattan

MAKES TWO 4¹/₃-OUNCE (129-ML) DRINKS AT 27% ALCOHOL BY VOLUME,
3.3 G/100 ML SUGAR, 0.12% ACID

INGREDIENTS FOR 2 DRINKS

4 ounces (120 ml) Rittenhouse rye (50% alcohol by volume)

A fat ¾ ounces (53 ml) Carpano Antica Formula Vermouth (16.5% alcohol by volume)

4 dashes Angostura bitters

2 piles of ice of two different sizes (The easiest way to do this is to gather a pile of your typical ice and then make another pile of that same ice broken in half.)

2 brandied cherries or orange twists

EQUIPMENT

2 metal shaking tins

2 hawthorn or julep strainers

2 digital thermometers

2 coupe glasses

PROCEDURE

Mix the rye, vermouth, and bitters together and divide the mix into two equal volumes. Mix it in one batch so the mix will be the same in both drinks. Add the pile of larger ice to one shaking tin and the small ice to the other. Using the strainers as covers for the tins, shake the excess water off the ice. Add the mixed cocktail to the tins and stir them with the digital thermometers (it is easier if you do one and a friend does the other). Stir the drinks until their digital thermometer reads −2°C and immediately strain the drinks into the coupe glasses and add the cherries or orange twists. Try to get all the cocktail out of the mixing tin. It doesn't matter how fast or how long each drink is stirred so long as the temperature is the same. Amazingly, the drinks will taste the same, look the same, and have the same wash line (the line in the glass where the drink hits). The two drinks will in essence be identical. The first time I ran this experiment I was astonished that it worked, even though I knew the physics of the situation guaranteed that it would.

Made according to these instructions, each Manhattan should pick up between 1¼ ounces (38 ml) and 1½ ounces (45 ml) of water and have a finished alcohol by volume between 27 and 26 percent.

This experiment gives you a good trick to use at home. Bartenders stir drinks night in and night out. They have perfected their technique like golfers perfect their swing. While you are learning, your technique might not be so spot-on. If you want your stirred drinks to be exactly the same every time, just stir them with a thermometer as you did in the experiment. Stop stirring whenever you hit your preferred temperature, and the dilution—and thus your drink—will be the same every time.

In photos 1–3, glass mixing cups are used for clarity. You should use metal. **1)** *Pour the same Manhattan over two different sizes of ice.* **2)** *Stir both with a digital thermometer.* **3)** *When the first drink reaches -2°C stop stirring it; continue stirring the other to -2°C.* **4)** *Results are identical.*

ONE MORE PROOF OF THE FUNDAMENTAL LAW

When bartenders are making a large round of drinks, they want to serve all the drinks at the same time. Some might be shaken, some might be stirred, but they all need to be served at once. The savvy mixer will not make each drink from start to finish, because the first set of drinks would be dying in the glass waiting for the later drinks to be completed. Instead, assembly-line techniques are the norm. Drinks are left in the mixing tin languishing with ice before stirring, or after stirring waiting to be strained. Luckily and surprisingly, this extra contact time with the ice doesn't hurt your drink as long as the time is reasonably short (a couple minutes) and the ice you are using is fairly large (not shaved). As I've mentioned, stationary ice is not very effective at chilling, so it doesn't dilute much and doesn't chill much. I know you probably don't believe me, so try for yourself. This time we will make three drinks; one that has the ice sitting in it before it is stirred ("sitting before"), one that has ice sitting in it after it is stirred ("sitting after"), and one that is stirred and poured right away ("normal").

EXPERIMENT 3

Languishing Ice Manhattans

INGREDIENTS FOR 3 DRINKS

6 ounces (180 ml) Rittenhouse rye

2¾ ounces (79 ml) Carpano Antica Formula vermouth

6 dashes Angostura bitters

3 piles of ice of the same size and type

3 brandied cherries or orange twists

EQUIPMENT

3 metal shaking tins

3 coupe glasses

Stopwatch

Mixing spoon

Hawthorn or julep strainer

PROCEDURE

Mix the rye, vermouth, and bitters together, then divide the mix into three equal volumes and add them to your shaking tins. Label the tins "sitting before," "sitting after," and "normal." Put the coupe glasses behind the tins. Once you start experimenting with three different samples, it is very easy to mix them up by accident, so don't get lazy here: label clearly.

Dump a pile of ice into the mix in the "sitting after" tin and a pile of ice into the mix in the "sitting before" tin. Start the stopwatch. Immediately stir the "sitting after" tin for 15 seconds. Try to use a consistent stir—you will need to replicate it two times. After you are done, wait 90 seconds. Your timer should read just over 1 minute 45 seconds. Dump the last pile of ice into the "normal" tin, stir it and the "before" tin for 15 seconds, and then strain all three drinks into their coupe glasses. Add the garnishes.

The three drinks should be almost identical. If you run the test only once, you may be able to convince yourself that you like one of the three more than the rest, but if you perform the test multiple times, as I have, you'll see that any differences between the drinks are random and nonreplicable.

This experiment still shocks me. Intuitively, it seems unreasonable that letting the ice sit in the drink for upward of 2 minutes won't ruin the drink, but it is true. There is marginally more dilution in the drinks that have ice sitting in them longer, but the difference is remarkably negligible.

Even with this knowledge—which should set me free from fretting over my rounds of stirred drinks—I still get physically uncomfortable when I see ice sitting in a stirred drink doing nothing. My heart tells me the drink is dying, even though my brain knows better. Getting rid of old prejudices can take time.

Although the instructions say to use metal tins, I'm using glass for clarity.

Start with 3 identical undiluted Manhattans labeled as shown.

Time = 0 seconds.
Add ice to the "Before" and "After" Manhattans and stir the "After" for 15 seconds.

Time = 15 seconds.
Stop stirring. You are going to let the "Before" drink sit with ice before you stir it and you are going to let the "After" drink sit with ice after you have stirred it.

Wait 90 seconds.

Time = 105 seconds.
Add ice to the "Normal" Manhattan and stir both it and the "Before" Manhattan simultaneously.

Stir for 15 seconds.

Time = 120 seconds.
Stop stirring and strain all three drinks. They should be identical, even though the "After" drink sat around for 105 seconds after it was stirred, the "Before" drink had ice added 105 seconds before stirring, and the "Normal" drink was stirred and served right away.

DAIQUIRI

Shaken and Stirred, Built and Blended

Shaken Drinks: The Daiquiri

The daiquiri is one of my all-time favorite drinks. Its reputation has been slandered by generations of careless people slinging Kool-Aid-tasting swill and calling it a daiquiri. The original is a thing of beauty. By the end of this chapter I hope you'll be a convert, if you're not already a believer. Let's look at shaking and ice before we get into the specifics of the drink.

Some cocktail neophytes have been led to believe that shaking is a kung fu–style-art whose heights can be scaled only after years of training. Not true. They might also believe, watching some bartenders shake their tins, that proper technique involves an aerobic workout. Also not true. And they may be under the impression that the best cocktails can be achieved only with the very best ice. (Mostly) not true.

Good news! Any reasonable shaking technique that lasts at least 10 seconds, using almost any kind of ice, can make a delicious and consistent shaken cocktail.

Cocktail shaking is violent. Banging ice rapidly around inside a shaking tin is the most turbulent, efficient, and effective manual chilling/diluting technique we drink makers use. Shaking is so efficient, in fact, that cocktails rapidly approach thermal equilibrium inside the shaker. Once equilibrium is reached, very little further chilling or dilution will take place, whether your ice is big or small, whether you continue to shake or don't. Shaken drinks get colder than and more diluted than the stirred Manhattans we tested earlier.

In addition to diluting and chilling, shaking adds texture to a drink in the form of tiny air bubbles. Sometimes you can see these bubbles in the form of foam on the top of

MY EARLY THINKING ON THE SCIENCE OF CHILLING COCKTAILS

In the early part of this century, the bar world hyperfocused on shaking technique and on the ice used in the tins. In general I am a proponent of such hyperfocus, because I believe that when you pay very close attention to what you are doing, your results are likely to improve, even when it turns out that your assumptions are off-base. At the dawn of our millennium, however, bar folk were making completely outlandish claims in favor of particular shaking techniques and particular ice cubes. In 2009, Eben Klemm, a bartender friend of mine with a mischievous bent, convinced me to cohost a debunking seminar at Tales of the Cocktail, the bar industry's annual booze-soaked convention in New Orleans. The preparation we did for that seminar, and the follow-up seminars we gave over the next two years, helped me form the tenets presented in this section.

your drink; sometimes you can only see the bubbles when you look under a microscope. But one thing is certain: without the air bubbles, you do not have a proper shaken cocktail. Because those air bubbles don't last long, the texture of a shaken cocktail is fleeting. A shaken cocktail is at its peak the moment it is strained and dies a little bit each moment it sits around waiting to be consumed—a strong argument against serving shaken drinks in large portions. Keep your guests happy and serve small shaken cocktails that they can drink in their prime.

CHOOSE THE RIGHT INGREDIENTS FOR THE RIGHT TEXTURE

You cannot make a proper shaken drink with booze and ice alone. Booze won't hold texture when shaken. The most common texturizers we use might surprise you: citrus juices—lemon and lime. You may not think of lemon and lime juices as foam promoters, but just squeeze some lime juice into a glass and then pour seltzer on top of it to witness the persistent foam it forms. In fact, if you shake a drink with clarified lime juice, the result will taste dead and flat next to its unclarified sibling, because clarification removes the juice's foam-enhancing characteristics. Most unclarified juices contain lots of plant cell-wall bits and other plant polysaccharides, like pectin, which add texture to a shaken drink. You don't need to start with goopy or pulpy juice—just unclarified. Strained citrus juice, for example, works as well as and looks oodles better than unstrained citrus juice. Some juices foam well even when clarified; cabbage juice and cucumber juice come to mind. Each time you take on a new juice, shake it to see how well it holds a foam. Or add a bit to a glass and do the seltzer test; if it foams up, you might have a good candidate for shaken cocktails.

I shook the cocktail on the left with normal lime juice and the cocktail on the right with clarified lime juice. Note the lack of texture on the right.

Milk and cream—and liquors that contain them—make good texturizers, especially in cocktails like the brandy Alexander that do not contain acid. I often use clarified milk (whey) as a texturizer in milk-washed booze (if that holds no meaning for you now, it will after you read the section on Washing, page 263). Whey is a versatile cocktail ingredient because it contains powerful foaming proteins but neither looks nor tastes milky. Unlike milk, it plays nicely with acidic ingredients. Honey syrup also contains proteins that promote foaming in acidic cocktails. But the old-school favorite for cocktail foaming, over any of these options, is egg white. Egg whites are a protein powerhouse and, used properly, can produce powerful and delicious foams on cocktails. To learn more, read the egg-white sidebar.

UPSHOT: Always include foam-enhancing ingredients in shaken drinks.

THE ICE YOU SHAKE WITH

Over the years I have run many tests shaking different cocktails with different kinds of ice. I've measured the dilution analytically and I've done side-by-side taste tests. The results are almost always the same: provided you follow a few simple rules, the kind of ice you use almost never affects dilution. Any ice, from the ¾-inch (20 mm) hollow cubes of hotel dispensers up to the 1¼-inch (30 mm) solid cubes from a Kold-Draft machine (this range includes almost all the ice you will come across), will dilute your drink the same amount, even though these cubes have radically different surface areas. The trick is to make sure you throw the surface water off the cubes before you shake: mix your cocktail in a small mixing tin, then put your ice into the larger tin, cap that tin with a strainer, and violently flick the tin downward to get rid of excess water. Put the ice into the smaller tin and shake.

Ice that is smaller or larger than the range given above will dilute differently. Shaking with crushed ice produces a very cold but way too diluted drink. Shaking with a single ice cube 2 inches (5 cm) square does not dilute nearly as much as shaking with the same weight in multiple smaller cubes, but it does produce a better-textured drink than the smaller cubes would—a fact that I denied for years before proving it to myself definitively. For years I scoffed at the numerous bartenders I heard waxing poetic on the virtues of shaking with one big cube. One year in front of a large audience I ran a test intended to prove that big cubes were all

EGG WHITES

There are a few things you need to know about egg whites before you use them in cocktails.

Egg whites will not make your cocktail taste weird and eggy.

Adding egg white to a cocktail will *not* kill the bacteria in the egg white. Depending on the proof of your cocktail, it can take days or weeks for any salmonella in a contaminated egg to die. If you are concerned about food safety or have a compromised immune system, use good, in-shell, pasteurized eggs. The pasteurized stuff sold in cartons is horrible. In my side-by-side egg-white-cocktail taste tests, fresh eggs win, followed closely by in-shell pasteurized eggs and trailed far, far behind by pasteurized egg whites in cartons. Please warn people before you serve them raw eggs—it's not a choice you should make for someone else.

Freshly cracked egg whites have no off aroma. Some egg whites can develop a strange and unpleasant aroma (reminds me of wet dog)10 or 15 minutes after being cracked. It is especially evident in cocktails and ruins their nose. Luckily, the aroma dissipates after a few hours. So you need to either crack your eggs just before you make a drink and don't let the drink sit around, or crack them several hours in advance and store them uncovered in the fridge until the odor dissipates. Nothing in between.

Some eggs never develop the dog aroma. I don't know why.

Many bartenders crack their egg whites in advance and store them in squeeze bottles—a good practice for busy bars.

To use egg white, first mix all your other cocktail ingredients together—booze, sugar, acid, juice—then add the egg white. You never want to go the other way around, because very high acid or alcohol levels will curdle the egg white before you shake it, and curdled egg white is nasty. The very upper limit for the alcohol by volume of a cocktail mixture before you add egg is 26 percent. Any higher and curdling is almost guaranteed. Next you must perform the dry shake, so called because you use no ice. Simply close your shaker and shake vigorously for 10 seconds or so to break up and disperse the egg white and prefoam your cocktail. Yes, this step is nec-

essary; I've done the side-by-side tests so you don't have to! Be careful when you dry shake—your cocktail shaker might try to push itself apart, because warm drinks don't suck the shaker shut like cold ones do. After you dry shake, open your tins, add ice, and shake again. Be sure to strain your egg drinks through a tea strainer. Although the strainer will kill some of your hard-won foam, you want to be sure to remove any curdled egg white and the chalaza, that unsavory little bump that suspends the egg yolk in the shell.

In addition to foaming, egg whites soften the flavors of oaky and tannic drinks, which is why they work so well in whiskey sours.

TESTING THE EFFECT OF EGG WHITES ON WHISKEY SOURS

Let's make two different whiskey sours, one with egg white and one without. We will add a bit of extra water to the eggless whiskey sour so the dilution of the two drinks will be fairly similar. Feel free to omit the extra water, though. The results of the test still stand.

EGGLESS WHISKEY SOUR

MAKES ONE 6¹/₅-OUNCE (185-ML) DRINK AT 16.2% ALCOHOL
BY VOLUME, 7.6 G/100 ML SUGAR, 0.57% ACID

INGREDIENTS

2 ounces (60 ml) bourbon or rye (50% alcohol
by volume)

Fat ½ ounce (17.5 ml) freshly strained lemon
juice

¾ ounce (22.5 ml) simple syrup

Short ¾ ounce (20 ml) filtered water

Pinch of salt

PROCEDURE

Combine the ingredients in a shaking tin and shake
with copious ice. Strain into a chilled coupe glass.
Now hurry up and make the next recipe.

WHISKEY SOUR WITH EGG WHITE

MAKES ONE 6³/₅-OUNCE (197-ML) DRINK AT 15.2% ALCOHOL
BY VOLUME, 7.1 G/100 ML SUGAR, 0.53% ACID

INGREDIENTS

2 ounces (60 ml) bourbon or rye (50% alcohol
by volume)

Fat ½ ounce (17.5 ml) freshly strained lemon juice

¾ ounce (22.5 ml) simple syrup

Pinch of salt

1 large egg white (1 ounce or 30 ml)

PROCEDURE

Combine all the ingredients except the egg white in
a shaking tin and make sure they are well combined.
Add the egg white to the mixture, cover the tin, and
shake violently for 8 to 10 seconds, making sure to
hold your tins together so you don't spill. Open the
tins, add copious ice, and shake again for 10 seconds. Immediately strain through a tea strainer into a
chilled coupe glass. Observe the glorious foam on this
drink. Marvel as it settles into a dense creamy head.

Once you have basked sufficiently in the greatness of
the foam and enjoyed its supremely smooth texture,
try to ignore it and just focus on the flavor difference
between the egg-free and egg-full cocktails. The
egg-white cocktail will taste less oaky and harsh.

CONTINUES

***HOW TO USE EGG WHITES: 1)** After you add the rest of your
ingredients to the mixing tin, add your egg white. **2)** Shake like
hell without ice. This is called the "dry shake." Careful, shaking
without ice doesn't seal your tins.*

3) *Open your tins; this is what you should see.* **4)** *Add ice and shake.* **5)** *Pour and behold. I'm using a hawthorn strainer with a fine spring. If you don't have one, pour through a tea strainer.* **BELOW:** *The progression of an egg white–containing whiskey sour after it is poured.*

show. I shook with different kinds of ice and dumped the drinks into graduated cylinders to measure the amount of foam the shaking had produced. To my surprise—and embarrassment—the large cube had a positive repeatable effect on foam quantity. I don't know *why* the big cube does a better job; it just does. I instantly changed my tune and insisted that all shaken drinks at my bar would henceforth be made with large cubes. But remember, that one big cube does not dilute as much as the smaller cubes do, which isn't good. The solution is to add a couple smaller ice cubes to your tin along with the big cube before shaking. The extra cubes don't seem to mess with the awesome texturizing effects of the big cube, and give all the extra dilution you need.

> **UPSHOT:** Unless you have 2-inch (5 cm) ice cubes on hand, it doesn't matter what kind of ice you use. Shake off the water. If you've got the big cubes, shake one big with two little.

YOUR SHAKING TECHNIQUE

From a technical standpoint, your shaking technique doesn't matter at all. It's possible to shake so languidly that you underdilute, but I've almost never seen this happen. On the flip side, my tests of a maniacal shake that I dubbed the "crazy monkey" reveal that going bonkers doesn't decrease final temperature or increase dilution. As long as you shake for between 8 and 12 seconds, your cocktails will be about the same, no matter what you do. If you shake for less than 8, you might be underdiluted. If you shake more than 12, almost nothing additional happens—you are just wasting time and energy. Similarly, in side-by-side tests with different bartenders using the same ingredients and the same ice, I was not able to detect any appreciable difference in the texture of drinks shaken with different styles. None of this stuff matters from a technical point of view. Shaking technique does matter quite considerably, however, in style points—and style is not to be ignored in the cocktail world.

I shook these cocktails with the ice shown. Look how gorgeous the drink on the left is. A big cube is better for shaking.

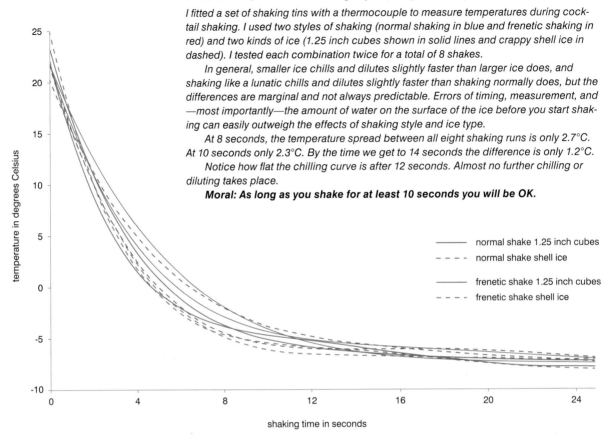

Different Ice, Different Shaking Styles, Very Little Impact

I fitted a set of shaking tins with a thermocouple to measure temperatures during cocktail shaking. I used two styles of shaking (normal shaking in blue and frenetic shaking in red) and two kinds of ice (1.25 inch cubes shown in solid lines and crappy shell ice in dashed). I tested each combination twice for a total of 8 shakes.

In general, smaller ice chills and dilutes slightly faster than larger ice does, and shaking like a lunatic chills and dilutes slightly faster than shaking normally does, but the differences are marginal and not always predictable. Errors of timing, measurement, and —most importantly—the amount of water on the surface of the ice before you start shaking can easily outweigh the effects of shaking style and ice type.

At 8 seconds, the temperature spread between all eight shaking runs is only 2.7°C. At 10 seconds only 2.3°C. By the time we get to 14 seconds the difference is only 1.2°C.

Notice how flat the chilling curve is after 12 seconds. Almost no further chilling or diluting takes place.

Moral: As long as you shake for at least 10 seconds you will be OK.

— normal shake 1.25 inch cubes

- - - normal shake shell ice

— frenetic shake 1.25 inch cubes

- - - frenetic shake shell ice

AVOID THE SPLASH

Adding ice to the drink in the small tin as I recommend may splash cocktail if you aren't careful. At home you can avoid the splash by using your fingers to put the ice in, but at the bar this won't do. Instead, I mix in the small tin, scoop ice into the big tin, and pour the mix over that ice. Unlike at home, I'm typically only mixing one or two drinks at a time in the shaker, so overfilling is not a big risk.

While I won't advise you on style—that you must develop for yourself—I will give you some advice on using your tins. Always measure your ingredients into the small tin, then add the ice to that small tin. Do this and you will never be embarrassed by filling a shaker set with more than it can hold and spilling cocktail everywhere. After the ice is in the small tin, place the large tin over it at a slight angle and give it a tap to set and seal the tins together. The angle will make it easier to break the tins apart later. As you shake, the tins will suck themselves together and will not come apart . . . but always shake with the small tin pointing toward you, just in case. If the seal fails and you get sprayed with cocktail, hilarity will ensue. If your guest gets sprayed, arguments might ensue. When it comes time to break the tins apart, push the small tin toward the gap between the shakers while you strike the large shaker at the beginning of the gap with the palm of your hand (this sounds harder than it is).

UPSHOT: Shake the way you like.

STRAINING AND SERVING YOUR DRINK

Even though your shaking technique isn't that important, your straining technique is. Shaking a drink produces lots of tiny ice crystals. If you use large, clear ice, these crystals tend to be pebblelike and somewhat regular. If you use ice that shatters or has lots of thin, small edges, you are likely to get crystals of widely varying sizes, with some shards thrown in. While they are in your shaker these crystals don't affect your dilution much, but if they make it into your drink, they melt relatively quickly and mess with your dilution. Many bartenders hate crystals on the top of their drinks. I happen to like them, in moderation. You can control how many crystals end up in your drink through your straining. Shaken drinks are strained with a hawthorn strainer, a flat plate with a spring attached to it. The bottom of the hawthorn is called the gate, and how you position the gate relative to the lip of your mixing tin determines how many crystals you let through. I advocate using a fully closed gate, which still lets some crystals through. Real crystal haters will pour their cocktail through a tea strainer after using the hawthorn. A good middle ground between the standard hawthorn and the hawthorn plus tea strainer is the fine-spring hawthorn strainer from Cocktail Kingdom.

UPSHOT: Pay attention to the gate!

WHY YOUR TINS GET SUCKED TOGETHER WHEN YOU SHAKE

Initially your tins are held together by the seal you make when you tap them together. This tapping motion forces out some of the air and leaves you with a partial vacuum. The real sucking, however, happens when you shake.

Remember that ice expands as it freezes, so as it melts, it contracts. Shaking a cocktail melts a lot of ice. The air inside your shaking tin, which starts at room temperature, is rapidly chilled during shaking and also contracts, as do the liquid contents of the drink. These three contractions cause a pretty strong partial vacuum inside your shaker tins, preventing them from flying apart while you shake.

Putting It All Together: Making Some Daiquiris

At long last, it's time to learn while shaking some daiquiris. The classic drink is a fairly low-alcohol, high-sugar cocktail, lying at one end of the shaken-drink spectrum. The Hemingway Daiquiri is a variant high in alcohol and low in sugar (Hemingway was both a diabetic and a booze hound) and thus sits at the opposite end of this spectrum. If you fix both of these drinks in your mind, you will be able to judge all other shaken sour drinks in between.

CLASSIC DAIQUIRI INGREDIENTS

MAKES 5$\frac{1}{3}$-OUNCE (159-ML) DRINK AT 15% ALCOHOL BY VOLUME, 8.9 G/100 ML SUGAR, 0.85% ACID

- 2 ounces (60 ml) light, clean rum (40% alcohol by volume)
- ¾ ounce (22.5 ml) simple syrup
- ¾ ounce (22.5 ml) freshly strained lime juice
- 2 drops saline solution or a pinch of salt

Before dilution this drink has a volume of 105 milliliters, is 22% alcohol by volume, and has 13.5 grams of sugar per 100 ml of fluid and 1.29% acid.

CONTINUES

HEMINGWAY DAIQUIRI INGREDIENTS

MAKES ONE 5⁴/₅-OUNCE (174-ML) DRINK AT 16.5% ALCOHOL
BY VOLUME, 4.2 G/100 ML SUGAR, 0.98% ACID

- 2 ounces (60 ml) light, clean rum (40% alcohol by volume)

- ¾ ounce (22.5 ml) freshly strained lime juice

- ½ ounce (15 ml) Luxardo Maraschino (32% alcohol by volume)

- ½ ounce (15 ml) freshly strained grapefruit juice

- 2 drops saline solution or a pinch of salt

Before dilution this drink has a volume of 112 milliliters, is 25.6% alcohol by volume, and has a minuscule 6.4 g/100 ml sugar and a whopping 1.52% acid.

PROCEDURE

Shake the ingredients for both drinks separately as described above for 10 seconds, using normal ice, and strain them into chilled coupe glasses. The Classic Daiquiri should have picked up roughly 55 milliliters of water and now be at 15% alcohol by volume with 8.9 grams per 100 ml of sugar and 0.84% acid. The Hemingway should pick up roughly 60 milliliters of water and have a final alcohol by volume of 16.7%, with 4.2 grams per 100 ml sugar and 0.99% acid. If you crunch the numbers, both drinks were diluted a similar amount on a percentage basis. I have run tests showing that in general, high-alcohol drinks dilute more on a percentage basis and get colder than low-alcohol drinks do, although a high-alcohol drink shaken side-by-side with a low-alcohol drink will always have higher alcohol by volume in the end. At the same time, I have run tests that show that high-sugar drinks dilute more than lower-sugar drinks do on a percentage basis. In the case of the Classic Daiquiri versus the Hemingway, these two dilution factors roughly cancel each other out.

Taste them. See what you think. I vastly prefer the Classic Daiquiri, even though it is a bit too sweet.

When I make it, I just short the simple syrup a bit—use a flat or short ¾ ounce of simple and a full ¾ ounce of lime.

For a second experiment, make both drinks again with the same specs. This time shake for only 5 or 6 seconds. The results should be totally different. Your shaking time was not long enough, and your drinks will be underdiluted. The Classic should be almost undrinkably sweet. The Hemingway will be too tart but acceptable. The lesson here: as sugar gets diluted, it attenuates in flavor more rapidly than acid, so high dilution favors adding more sugar, or less acid, for a particular flavor profile.

If you are up for a third experiment, make the two drinks again but shake for 20 seconds. They will be slightly more diluted than they were with the 10-second shake, but not as much as you'd expect—the last 10 seconds of shaking took place in the shallow part of the chilling curve and very little dilution took place. If there's any difference in taste, the Hemingway should taste a bit thin, as its meager sugar peters out even more.

On the right, a Hemingway daiquiri. Standard variety on the left.

Stirred Drinks: Manhattan versus Negroni

Stirring a drink may seem like a simple task. In fact, it is more difficult to make a consistent stirred cocktail than almost any other type of cocktail. Stirring is a relatively inefficient chilling technique. It takes a long period of constant stirring for your drink to approach its minimum possible temperature. In our martini experiment, it took over 2 minutes of stirring for the temperature and dilution to plateau—way longer than anyone should stir a drink. The main variables are the size of the ice cubes you use and how fast and long you stir. Smaller ice has a large surface area, and chills and dilutes quickly. Really fine shaved ice can chill and dilute extremely quickly, getting down to near-equilibrium in a few short seconds of flaccid stirring (see the section on Blended Drinks, page 114)—not ideal for most stirred drinks, which aren't designed to be highly diluted or overly cold. Conversely, a stirred drink made with giant 2-inch ice cubes dilutes and chills very slowly, usually leading to an overly warm, underdiluted drink. Some bartenders like this style, but I don't. If I want a low-dilution, merely cool drink, I'll order a built drink (see page 107). Stirred drinks should fall somewhere in between these and shaken drinks. The best ice to use for stirred drinks, therefore, is medium-sized and relatively dry: large cubes that you freshly crack into smaller cubes with the back of your bar spoon, or regular machine ice from which you shake the excess water. The size isn't vitally important, because you can adjust the dilution and temperature of the drink by how long and how fast you stir. If your ice is on the large side, you will stir longer or faster to achieve the same result you'll get with smaller ice. Got small ice? Stir less or stir slower.

In real life, you often can't choose what ice you'll be working with at any given time, and you should thus be prepared to adjust your stirring to match your ice. It isn't a good idea to change your technique; just adjust how long you stir. Practice stirring so that you can stir two drinks at a time with both hands stirring at the same rate. Try to make your stirring style consistent. If you are consistent, you can taste the first few cocktails you make in an evening and adjust how long you stir based on the ice you have.

I always prefer stirring in stainless steel shaking tins, because they have a very low thermal mass. It doesn't take very much ice melting to chill them down. They don't affect the temperature or dilution of your drink. Large glass stirring vessels can be gorgeous, but they have a huge thermal mass and can change the finished temperature of the drink by several degrees—enough to make the drinks seem too warm when served. You can get around this problem by prechilling all your stirring vessels with ice and water, but don't bother. Most of my bartenders want to use glass. I tell them that's fine, so long as they pledge to prechill every single mixing glass with ice and water every time they stir a drink, without exception. Metal tends to win out.

Many bartenders add a boatload of ice to their mixing vessels when they are stirring. But if ice doesn't touch the liquid, it does no good. Some excess ice can be helpful if you think the ice will settle a lot during stirring, but most bartenders use far more than is necessary. At best, the extra ice is useless; at worst, the extra water contained on its surface will overdilute the drink. Too little ice is also bad. Tiny amounts of ice don't have enough chilling power to complete the task and can't deliver what power they have rapidly enough. Ideally, you want the ice to contact the drink you are chilling from the bottom of the vessel all the way to the top of the liquid. To insure that this happens, you need to add more ice than is strictly necessary, because ice floats and you need a bit extra on top to weigh the mass of ice down into your drink. Any more ice than that is counterproductive, and a pain if you are making drinks at home, where your supply of ice is usually limited.

Remember, stirring drinks is a game of repeatability. Develop a stirring style and remain consistent and your drinks will remain consistent. The essence of stirring is pure dilution and chilling—no texture development, no aeration. Because stirred drinks aren't aerated, they can look amazingly clear. I love that clarity and don't want to add anything to my stirred cocktails to ruin it. Stirred drinks are relatively boozy (our Manhattan was 26 percent alcohol). The high booze level of the finished cocktails is why most stirred drinks tend to be very spirit-forward, not light and refreshing. Refreshing is hard to achieve at 26 percent alcohol by volume.

SERVING STIRRED DRINKS:
UP OR ON THE ROCK(S)? THE NEGRONI

Years ago when I saw someone order a Negroni, I could safely assume that he or she worked in the food industry. Chefs and bartenders have always loved them. Now everybody seems to know how good they are. The classic recipe, to which I adhere, is equal parts gin, Campari, and sweet vermouth. Modern bartenders often shift the proportion of gin upward, leaving the one-to-one correspondence between vermouth and Campari inviolate. This makes an acceptable Negroni. Some people swap Campari with Aperol. This too makes a decent Negroni. Some people add a splash of citrus juice or a bit of soda water to the classic drink—not my cup of tea, but it doesn't taste bad. The Negroni tastes good with normal dilution and tastes good with extra dilution. It is good carbonated. Even though the flavors of a Negroni are distinctive, rich, and complex, its basic structure can be beaten around and contorted into dozens of variants that all taste good and retain an essential Negroniness.

The eternal question: Do you want your Negroni up or on the rocks?

Even if you follow the classic Negroni recipe, which I wholeheartedly recommend, you still have one final decision to make: Do you serve it up or on the rocks?

There are two reasons to serve a drink up. One: up drinks are very elegant. Two: you might not want a stirred drink to pick up dilution as the ice melts. Martinis, for example, should always be served up, but they get warm if not drunk with celerity, and martinis don't taste good warm. There are two non-rocks solutions to the warming problem: drink faster or serve smaller portions. Responsibility dictates that I recommend the latter course. There was a regrettable trend years ago to make cocktails big simply because people associated small drinks with getting ripped off. I associate warm martinis with sadness.

There are four good reasons to serve a drink on the rocks: you want the drink to change over time as it waters down; the drink is otherwise likely to linger in the glass and get too warm to be enjoyable; you find the drink too alcoholic and prefer the extra bit of instant dilution that freshly tempered ice brings from the water clinging to its surface (though you could just stir more to solve this problem); or you like the way a drink looks on ice. Use those reasons as your guide and you will rarely go wrong. Negronis and other drinks that can tolerate a wide range of dilutions do just fine on the rocks. Many people can't tolerate a Negroni after it starts to warm up. If a drink can't tolerate watering, like a martini, serve smaller portions up, and don't let them sit around.

When I serve a drink on the rocks, I actually serve it on the rock—a single large cube, inefficient enough not to dilute my drink too quickly but efficient enough to keep it pleasantly cool.

HOW COLD IS TOO COLD? MANHATTANS VERSUS NEGRONIS

Temperature dramatically affects the way a drink tastes. Some drinks completely fall apart when they are chilled too much, including most drinks with aged and oaked spirits, like the Manhattan. Other drinks can be chilled quite deeply without losing their appeal but go out of whack when they get too warm, like the Negroni. In this last stirring experiment, you'll make two Negronis and two Manhattans at different initial temperatures and compare them as they warm up.

INGREDIENTS FOR 2 DRINKS OF EACH TYPE

MANHATTANS

MAKES TWO 4$\frac{1}{3}$–OUNCE (129-ML) DRINKS AT 27% ALCOHOL BY VOLUME, 3.4 G/100 ML SUGAR, 0.12% ACID

- 4 ounces (120 ml) Rittenhouse rye (50% alcohol by volume)
- Fat $\frac{3}{4}$ ounces (53 ml) fine sweet vermouth (16.5% alcohol by volume)
- 4 dashes Angostura bitters
- 2 brandied cherries or orange twists

NEGRONIS

MAKES TWO 4$\frac{1}{4}$-OUNCE (127-ML) DRINKS AT 27% ALCOHOL BY VOLUME, 9.4 G/100 ML SUGAR, 0.14% ACID

- 2 ounces (60 ml) gin
- 2 ounces (60 ml) Campari
- 2 ounces (60 ml) sweet vermouth
- 2 orange twists

EQUIPMENT

4 metal shaking tins

4 piles of ice

2 hawthorn or julep strainers

4 coupe glasses

Freezer

PROCEDURE

Mix the rye, vermouth, and bitters together, then divide the mix into two equal volumes. Do the same with the gin, Campari and vermouth. Stir one of the Manhattans and one of the Negronis for about 15 seconds with similar quantities of ice, strain them into covered containers, and carefully place them and the coupe glasses into the freezer for 1 hour, during which time their temperature should drop by 5° or 10° C. Now stir the second Manhattan and the second Negroni, strain them into coupes, pull the first two coupes out of the freezer, and give all four drinks their garnishes. Taste them. At this point, both Negronis should taste great: they are both in their prime serving zone. The Manhattans should be a different story. The fresh one should be round and inviting. The one from the freezer should, by comparison, taste flat and closed, with too much oak. Keep tasting over the next 20 minutes or so. You will notice that as they warm up, the freezer Manhattan becomes your favorite Manhattan, while the "fresh" one remains a decent-tasting drink. The freezer Negroni will remain a good drink, while the fresh one will become increasingly difficult to drink.

UPSHOT: Temperature is an important ingredient in your drinks, and colder isn't necessarily better. Almost all stirred drinks are in their prime between −5°C (23°F) and −1°C (30°F), while some drinks can stand being warmer and some colder.

After you stir a Negroni and a Manhattan, put them in the freezer for a couple hours. Then taste them next to the same drinks, but just stirred. Keep tasting all four as they warm up. This will give you an idea of how different drinks respond to temperature changes.

CLIFF OLD-FASHIONED

Built Drinks: The Old-Fashioned

You make a built drink by pouring it over ice in its serving glass and giving it a quick stir. These are the least-diluted cocktails. The old-fashioned is the prototypical built drink, a simple concoction of whiskey, bitters, and sugar garnished with a twist. Like many simple things, the old-fashioned is a study in nuance. With so few ingredients, each one makes a big impact. Order up an old-fashioned and you'll have a good indication of your barkeep's sensibilities. (For a history of the old-fashioned, and some lengthy arguments about its proper contents, see Further Reading, page 379.)

The best way to make an old-fashioned is not necessarily the way that's most original or "authentic" but rather the way you like it best. I use simple syrup instead of granulated sugar—anathema to many. The only old-fashioned rule you must adhere to: produce a minimally diluted sipping drink that isn't too sweet and that tastes heavily of its base spirit. While I'm all for doing what makes you happy, don't put smashed fruit into the bottom of your drink and call it an old-fashioned. Smash your fruit and name the drink something else! Let us proceed with a variant I particularly enjoy.

Building the Cliff Old-Fashioned

The Cliff Old-Fashioned is named for my friend Clifford Guilibert. Cliff and I were making coriander soda for an event. Coriander soda is similar to ginger ale, refreshing and spicy, giving a hint of warmth to the back of the throat. After making a batch of syrup for the soda, Cliff suggested we use it in an old-fashioned. Brilliant idea!

BUILDING AN OLD-FASHIONED: 1) *If your ice doesn't fit your glass just spin it in.* **2)** *A rock in an un-chilled glass.* **3)** *Dash in the bitters.* **4)** *Add the base spirit.* **5)** *Add the syrup.* **6)** *Give a brief stir.* **7)** *Express the orange peel over the drink.* **8)** *Hold the peel stationary against the inside rim of the glass, then suavely rotate the glass against the peel and release.*

INGREDIENTS

MAKES ONE 3-OUNCE (90-ML) DRINK AT 32% ALCOHOL BY VOLUME, 7.7 G/100 ML SUGAR, 0% ACID

One 2-inch-by-2-inch clear ice cube

2 dashes Angostura bitters

2 ounces (60 ml) Elijah Craig 12-year bourbon (47% alcohol by volume)
You can use the bourbon of your choice, but choose one with lots of backbone that isn't too expensive. If you are made of cash, use anything you want. The best Cliff Old-Fashioned I ever made was for a charity event using Hibiki 12-year-old Japanese whiskey. We made two hundred of them that night. At $70 for a 750 ml bottle, those were some spendy drinks, but I wasn't paying.

⅜ ounce (11 ml) Coriander Syrup (recipe follows)
The syrup is the nontraditional aspect of this old-fashioned. You could use regular 1:1 simple syrup to make a more typical version. Many purists use granulated or cubed sugar and crush the sugar with the bitters—they like the graininess, and the fact that the sugar level evolves over time. I don't.

Orange twist

EQUIPMENT

One double old-fashioned glass, at room temperature

Straw or short mixing rod (optional)

PROCEDURE FOR A SIMPLE DRINK IN EXCRUCIATING DETAIL

I don't build old-fashioneds in chilled glasses. An unchilled glass represents a relatively large thermal mass at room temperature. When you make a drink in a glass at room temperature, you have to melt a good bit of ice to chill the glass down to the temperature of the drink. This extra bit of melting adds to the initial dilution of the drink, which I like. You can overcome this by stirring the drink more after you build it, but then the initial drink will be colder, which I don't like. Also, while chilled glasses look great when they are fresh, they attract condensation and don't look good on a drink like the old-fashioned, which is meant to be sipped.

Fine points like this—whether to chill a glass or not—are all a matter of personal preference. Chill your glasses or don't, but understand the consequences.

CONTINUES

Some bartenders build their drink directly into the glass before they add the rock, allowing them to stir the ingredients with a spoon to mix without diluting. If you work this way, you should use the time-honored practice of adding ingredients from cheapest to most expensive. Adding the cheap stuff first means you don't throw away the good stuff if you make a mistake while you are measuring. Almost all professionals work this way, even though they rarely make mistakes while jiggering.

The disadvantage of adding the ice last: you have to get a large, ungainly rock into the glass smoothly without splashing, and you have to know in advance that your rock is a good partner for your glass. I prefer to add the rock to the glass before the liquid so I can be certain the rock looks good in the glass. Your large, hand-cut ice cube should be given due respect. Put the ice cube into your double old-fashioned glass. Make sure the ice fits in the glass and reaches to the bottom. A rock that doesn't touch the bottom of the glass is an abomination. If the ice doesn't fit properly, spin it with a spoon, and the corners should melt, allowing the ice to reach the bottom. If the cube is really too big, knock the corners off with a knife first.

If you work this way—ice first—you have two drink-building options: build the drink in a mixing tin and stir to incorporate the ingredients before pouring them over the rock, or build the drink directly on the rock. Although the mixing glass is undoubtedly superior from a technical point of view, I actually prefer to build over the rock—it's just an aesthetically enjoyable process. Be sure not to splash the ingredients on the rock; that looks foolish. Work in this order: 2 dashes of bitters, then whiskey, then Coriander Syrup; you put the least dense ingredient in first so the denser ones will auto-mix when they pass through them. Stir the drink for about 5 seconds, then express (that is, quickly wring) the orange rind over the top of the drink and wipe the rim of the glass with it. Drop the twist into the drink if you want to continue adding aroma while you're drinking. It is shocking how much a difference adding versus discarding the peel makes; experiment side-by-side to see for yourself. Do not underestimate the effect of the twist. Try to make all your peels identical and try to express them the same way, so your drinks are consistent over time.

HERE IS THE CORIANDER SYRUP RECIPE:
INGREDIENTS

125 grams coriander seeds, preferably with a fresh, citrusy aroma (for soda syrup, reduce to 100 grams)

550 grams filtered water

500 grams granulated sugar

5 grams salt

10 grams crushed red pepper

PROCEDURE

Blend the coriander seeds and water in a blender for several seconds until the seeds are well broken. Transfer the mixture to a saucepan, add the sugar and salt, and heat over medium heat until it is simmering. Stir in the crushed red pepper. Turn off the heat and keep tasting until the spiciness from the pepper becomes apparent at the back of your throat when you taste the syrup. (It is impossible to quantify this part of the recipe because batches of crushed red pepper vary so widely.) Quickly strain the mixture through a coarse strainer to prevent further infusion, then pass it through a muslin cloth or fine chinois.

To use this syrup for soda, make the version with slightly less coriander (unless you want a version with as much punch as a ginger beer) and use 1 part syrup to 4 parts water before carbonating (or use 4 parts soda water). Garnish with lime or, preferably, clarified lime juice (see page 372).

Properly made, this syrup should have a Brix of 50, meaning it should be 50 percent sugar by weight, the same as regular simple syrup. The extra 50 grams of water in the recipe are absorbed by the coriander seed. At the bar we use a refractometer to correct the sugar level.

MAKING CORIANDER SYRUP: 1) *Blend the coriander and water.* **2)** *Add the blended mix to a pot with sugar and* **3)** *heat to a simmer.* **4)** *Add crushed red pepper and keep tasting for adequate spiciness.* **5)** *Quickly strain the syrup. I use two strainers—coarse inside of fine.* **6)** *The finished product.*

The last decision you need to make is whether to put a stir rod in the drink. The stir rod lets your guest add extra dilution quickly if he or she desires. Bars often use straws for stir rods because they are cheap, not because they want you to drink through them. If you go for the rod, avoid straws! They constantly move around and threaten eyes. Use the glass or metal variety, and always give your drinker a napkin on which to place it when it's no longer useful.

Made according to my instructions, the Cliff Old-Fashioned should have a volume of 3 ounces (90 ml). Of that 3 ounces, ⅝ ounce (19 ml, or 20%) will be water from melted ice dilution. The alcohol by volume as served will be roughly 31%—in other words, stiff.

Take a sip. The drink should be cool but not frigid. A frigid old-fashioned loses its character. Let the drink sit and observe it without touching. After a while you should see a layer of watery liquid forming around the large ice cube. Notice there is very little mixing of the meltwater with the drink, and all changes happen slowly. Give a slight swirl to the glass to mix the meltwater with the cocktail and sip again to taste the difference. The Cliff should be cooler now and not as intense because of the added dilution, but it will still be balanced. If you constantly swirl the glass, the ice will melt much faster, as fresh meltwater is transported away from the cube and a fresh layer of alcohol is presented to the cube. Don't overswirl: you want slow change, which is why you need a big ice cube and a lot of time. Let the drink rest a couple of minutes between sips so you can see how it evolves over time. That's the point of this experiment (one you will likely want to repeat): to see how a drink can stay balanced for a long time over a range of temperatures and dilutions. A good old-fashioned made with one big ice cube will be pleasurable for at least 20 minutes as it slowly waters down.

Remember, the essence of any built drink is slow change over time. Built drinks should start really boozy—they are sipping drinks, not pounders. After they sit awhile and become more watered down, they become more refreshing and whet the palate for another round. Keeping in mind that built drinks must stay balanced over a large range of dilutions, we can lay out some simple rules that will help you evaluate recipes and devise your own:

- Choose liquors that taste good over a wide range of dilutions. Some spirits are too rough to function well when they aren't diluted a lot; these are not good candidates for built drinks. Other liquors taste great out of the bottle but fall apart when diluted past a certain point; avoid them.
- Avoid acids. The flavors and tartness of acidic ingredients such as lemon and lime do not maintain their balance when they are diluted. A properly balanced mix of ethanol and sugar can stay in harmony as water is added or taken away, but acid doesn't work that way. Acids used for tartness can be properly used only in drinks with a fixed dilution.
- Brighten a built drink with essential oils instead. Lemon, orange, and grapefruit twists work well.
- Take advantage of bitters. Bitters are designed to taste good over an absurd range of dilutions and to join the other flavors of the cocktail together without overpowering them.
- Take the time to make big ice cubes. If you have a lot of time and energy, make or buy fancy clear presentation cubes.

Blended Drinks and Shaved Drinks: The Margarita

I know, I know. A proper margarita is a shaken cocktail served up in a coupe with or without a rim of salt . . . but many people enjoy a blender margarita, including me. A properly made blended drink is a wondrous thing. And yes, there is a trick to them.

A blender full of ice is an extremely efficient chilling machine. It will get your drink colder and more diluted more quickly than almost any other technology. Blenders break ice into very fine particles, increasing the surface area tremendously, and they spin very fast, providing large movement of cocktail across the ice. I don't use blenders at the bar because they are loud. I use something better: an ice shaver. Like the blender, an ice shaver will produce very fine ice crystals. Unlike a blender, the ice shaver doesn't agitate your drink. With a little stirring, shaved-ice drinks are almost identical to blended drinks in terms of dilution and chilling. While blenders, especially inexpensive ones, tend to leave unblended chunks of ice in your drink, the ice shaver makes beautifully uniform pieces. Manual cast-iron ice shavers like the one I use are blessedly silent and pleasing to look at. That said, blenders work fine at home, so don't feel compelled to give up counter space to a shaver unless you have more space than you know what to do with. Shaved-ice and blender recipes can be swapped back and forth.

Shaken drink recipes cannot be used directly in a blender. Blend a standard shaken recipe and the result will be unbalanced—overtart, undersweet, and overdiluted. You can balance a drink recipe for a blender by adding more sugar and less acid. Fixing the dilution is more difficult. Most shaken drink recipes just have too much liquid in them to blend well. What to do? Reduce the liquid!

The Shaken Margarita

**MAKES ONE 5⁹/₁₀-OUNCE (178-ML) DRINK AT 18.5% ALCOHOL
BY VOLUME, 6.0 G/100 ML SUGAR, 0.76% ACID**

INGREDIENTS

2 ounces (60 ml) tequila (40% alcohol by volume)

¾ ounce (22.5 ml) Cointreau

¾ ounce (22.5 ml) freshly strained lime juice

¼ ounce (7.5 ml) simple syrup

5 drops saline solution or a generous
pinch of salt

PROCEDURE

Shake the ingredients together, strain into a glass with or without a salt rim according to your preference, and drink.

The amount of liquid in this recipe is 3¾ ounces (112.5 ml). When you shake that margarita, it will probably pick up a total of around 2 ounces (60 ml) of water to make a drink a little under 6 ounces (178 ml) with an alcohol by volume of roughly 19%. If you blended that drink and were able to remove the ice crystals, you would see that it picked up closer to 3 ounces (90 ml) of water—too much. Make one in a blender by combining the above ingredients with about 5 ounces (150 g) of ice and you'll see what I mean. Here is a simple blender variant.

BLENDERS: BEHIND THE POWER CURVE

Even without liquor, spinning blender blades melt ice via friction. (I actually use my blender as a heater when I am clarifying juices.) How much ice they melt depends on how powerful your blender is and the volume of cocktail you are blending. Below a certain volume, blenders can't effectively add energy to the cocktail. Adding more volume at this stage increases the amount of ice you melt per milliliter of cocktail per second. At a certain point, the blender is adding the maximum amount of frictional energy possible per milliliter of cocktail. Increasing the volume further decreases the amount of ice the blades are melting per milliliter of cocktail per second. I use a Vita-Prep, a very powerful blender. On full power, with a 180-gram (6 ounce) load of ice and cocktail, it will melt 0.0007 grams of ice per second per gram load, or 1.9 milliliters of extra dilution in the cocktail after 15 seconds of blending, which is minimal. By contrast, the Vita-Prep on full power with a 500-gram load melts 0.029 g/g/sec—four times more. That adds up to almost 8 milliliters extra dilution per cocktail for a 15-second blend—not minimal. By the time the blended load is a full liter, the Vita-Prep is melting 0.0016 g/g/sec, or 4.3 milliliters per cocktail per 15-second blend. The upshot? Don't blend too long. Just long enough to get a good texture.

USING THE ICE SHAVER

Ice shavers come in many varieties. Anything that will produce good-textured shaved ice will work, including my kids' Snoopy Sno-Cone maker. I favor machines that shave ice off one large block; they produce a very consistent product and are easy to use. When we make shaved-ice drinks at the bar, we shave the ice to order into 5-ounce coupe glasses. To get the dilution right consistently, do the following:

- Divide your drink recipe into two roughly equal volumes. Add half to the coupe. You need to melt some of the ice right away as it falls into the booze for the presentation to work properly.

- Shave ice into the coupe until a beautiful mountain of ice domes over the top of the glass. You will add roughly 70 grams.

- Put the coupe in front of the customer and pour the remaining half of the drink over the ice. The ice should melt almost instantly, and the level of the drink should be almost at the rim of the coupe.

- Give the drink a gentle stir with a bar spoon to finish the chilling and thoroughly mix the ingredients. Some ice will be present, but most will be melted.

Shaving ice into a coupe glass. Note that there is already some liquid in the coupe, which melts the ice immediately. A dry coupe couldn't hold enough shaved ice to make the drink.

POURING A SHAVED ICE DRINK: THE MARG

POURING A SHAVED ICE DRINK: *The Marg.*

Blender Marg

MAKES ONE 5⅓-OUNCE (158-ML) DRINK AT 17.2% ALCOHOL BY VOLUME, 7.9 G/100 ML SUGAR, 0.57% ACID

INGREDIENTS

1 ounce (30 ml) Cointreau (Yep, you read right: more Cointreau than mezcal.)

¾ ounce (22.5 ml) La Puritita mezcal. (A robust blanco mezcal is required for this drink because so little is used. Tequila would be lost. I used a blanco because I didn't want any oak in this drink.)

½ ounce (15 ml) Yellow Chartreuse (Completely untraditional, really good.)

½ ounce (15 ml) freshly strained lime juice

10 drops Hellfire bitters or spicy nonacidic stuff of your choice

5 drops saline solution or a generous pinch of salt

About 4 ounces (120 grams) ice

PROCEDURE

Throw everything into a blender, blend just till the ice is fully crushed, and drink. There should be some ice left over. If there isn't, you've blended too long and the friction from the blades has led you to over-dilute your drink.

This blender margarita has an initial recipe volume of 2¾ ounces (82.5 ml) instead of the 3¾ ounces (112.5 ml) of the shaken recipe. It will pick up around 2½ ounces (75 ml) of water from the blending, making a drink with a final alcohol by volume of around 17.2%. This recipe works because both Cointreau and Yellow Chartreuse have high alcohol *and* high sugar contents, so the amount of liquid in the recipe is lower, even though the alcohol level is the same. The blender recipe therefore has 2¼ ounces (67.5 ml) of 40% alcohol-by-volume booze in it and about 12.75 grams of sugar—the equivalent of a little under ¾ ounce (0.7 ounce, 21 ml) of simple syrup—all with a total liquid volume of only 2¾ ounces (82.5 ml)!

The beauty of this recipe is that it can be generalized. Just keep the alcohol:sugar:acid volume ratio fairly constant. The specs for a generic blender sour are as follows.

BLENDER MARG

Generic Blender Sour

GENERAL INGREDIENTS

2¼ ounces (67.5 ml) of liquid that contains around 0.9 ounce (27 ml) of pure ethanol and 12.75 grams of sugar

½ ounce (30 ml) freshly strained lemon, lime, or other sour juice

4 ounces (120 ml) ice

2–5 drops saline solution or a generous pinch of salt

SPECIFIC INGREDIENTS

The trick to making blender recipes is to find or make a mixture of liqueurs and spirits and flavors that add up to give the proper ethanol:sugar:volume ratio. Eighty-proof spirits contain the requisite amount of alcohol per unit volume but contain no sugar. Adding simple syrup throws off the liquid balance. To get the sugar you can use high-alcohol liqueurs, like we did in the margarita, or very high proof spirits, like Lemon Hart 151-proof rum, or you can sugar your booze.

SUGARED BOOZE INGREDIENTS

MAKES 1140 ML (38 OUNCES) LIQUOR AT EITHER 44% OR 35% ALCOHOL BY VOLUME

212 grams superfine sugar (regular granulated sugar is okay but will take longer to dissolve)

1 liter either 80- or 100-proof spirits (40% or 50% alcohol by volume)

PROCEDURE

Add the sugar to the spirits in a covered container and agitate until the sugar is completely dissolved. This process will take a while. You can heat the spirits, as long as the liquor is covered so that you don't evaporate any alcohol. Don't boil, or your container will get pressurized and possibly blow up the bottle. Wait for the liquor to cool. You will end up with roughly 1120 ml (37 ounces) of sugared booze. If you are using liquor that comes in 750 ml bottles, use 159 grams superfine sugar.

- If you started with 100-proof spirits, use 2 ounces (60 ml) per drink. You might need a bar spoon of simple syrup, as the sugar in 2 ounces will be a little less than what is in the marg. Now add ¼ ounce (7.5 ml) liquid of your choice—orange juice, pomegranate juice, water, whatever—just make sure it isn't too sugary or boozy.

- If you started with 80-proof spirits, add 2¼ ounces of the sugared booze to the acid and salt. The alcohol level will be slightly low but will work fine. With 80-proof there's no room for extra ingredients.

EXAMPLE 1:

Rittenhouse Blender Sour

MAKES ONE 5¼-OUNCE (157-ML) DRINK AT 16.7% ALCOHOL BY VOLUME, 7.8 G/100 ML SUGAR, 0.61% ACID

INGREDIENTS

2 ounces (60 ml) sugared Rittenhouse rye (44% alcohol by volume; see above)

½ ounce (15 ml) freshly strained lemon juice

¼ ounce (7.5 ml) freshly strained orange juice

4 drops saline solution or a generous pinch of salt

4 ounces (120 grams) ice

PROCEDURE

Blend and drink, baby. Blend and drink.

EXAMPLE 2:

Blender Daiquiri

MAKES ONE 5¼-OUNCE (157-ML) DRINK AT 15% ALCOHOL BY VOLUME, 8.1 G/100 ML SUGAR, 0.57% ACID

INGREDIENTS

2¼ ounces (67.5 ml) sugared Flor de Caña white rum (35% alcohol by volume, see above)

½ ounce (15 ml) freshly strained lime juice

4 drops saline solution or a generous pinch of salt

4 ounces (120 grams) ice

PROCEDURE

Blend, drink, repeat.

RITTENHOUSE
BLENDER SOUR

Cocktail Calculus:
The Inner Workings of Recipes

I recently constructed a database of cocktail recipes, including both classics and my own, so I could analyze them for ethanol content, sugar, acidity, and dilution. Each drink category—built, stirred, shaken, blended, and carbonated (which we will discuss later)—has clear, well-defined relationships between the characteristics, regardless of the flavors in a particular recipe. This might seem obvious, but the implications are not. I discovered that given a set of ingredients and a style of drink, I can write a decent recipe without tasting along the way at all. I have tested this process dozens of times, and I am shocked at how close I can get to the desired result strictly through applying the math. Bitterness is a bit of a wild card—very hard to quantify. Thank God something is.

I'm not talking about swapping rum for gin or lemon for lime. I'm talking about this: given apple juice, bourbon, Cointreau, and lemons, can I make a recipe with the same basic profile as a daiquiri by plugging in a few numbers? Yes, I can. It won't taste like a daiquiri, but it will have the same feel. I developed several recipes in this book mathematically, but I won't tell you which ones for fear you'll be prejudiced against them.

I don't really know how I feel about this ability. It's a little disconcerting. I tell myself that I still need to understand how flavors work together, I still need a brain and a palate—and that's true. All the math in the world won't help you if you can't put good flavors together. And the math isn't always right, either. Some drinks are better with more than average sugar or acid, some with less. The math will only give you the backbone of the drink— its structure. The soul of the drink will be the aromatics and flavors you choose. But the math has been incredibly useful to me for judging existing cocktail recipes and for developing my own.

It is easy to replicate the basic profile of a recipe you like in a new drink so long as you know the alcohol, sugar, and acid contents of your ingredients and target alcohol, sugar, acid, and dilution numbers for the recipe profile you want to emulate. To that end, my recipes specify alcohol content, sugar content, acidity, and final beverage size. To calculate new recipes of your own based on my numbers, you'll need to be armed with a list of the alcohol, sugar, and acid content of basic ingredients that I've provided on pages 136–37.

- Ethanol is measured in percentage alcohol by volume.
- Sugar is measured in grams per 100 milliliters, abbreviated g/100 ml, which is roughly equivalent to "percentage." This might seem like a bizarre measurement, but weight-in-volume measurements like g/100 ml are the only way to deal with dissolved solids such as sugar that must be measured volumetrically.
- Acid is quoted as a simple percentage. Although the same solid-in-liquid problem exists for acid as for sugar, the difference between actual percentage and grams per 100 ml is very small at the low concentrations of acid present in drinks (usually roughly an order of magnitude lower than the concentration of sugar) and percentage numbers are simpler to think about.
- Volumes are measured in ounces (remember, 30 milliliters to the ounce in this book) and milliliters.
- Dilution is measured in percentage. If I quote a dilution of 50 percent, that means that every 100 ml of original cocktail recipe will be diluted with 50 ml of water from melted ice for a finished drink size of 150 ml. If I quote a 25 percent dilution, that means that every 100 ml original cocktail recipe produces 25 ml of dilution from melted ice for a finished drink size of 125 ml.

HOW TEMPERATURE, DILUTION, AND INGREDIENTS WORK TOGETHER

Ethanol: Cocktail styles ordered by alcohol from most boozy to least gives you this list: built, stirred, shaken, shaken egg white, and blender and carbonated tied for last. This same order, with the exception of carbonated and egg-white drinks, sorts cocktails from warmest to coldest. You might expect the opposite—drinks with lots of ethanol seem as if they would be served colder instead of warmer, since chilling is often used as a way to mitigate high alcohol content. When you drink straight shots of vodka, they are chilled to extremely low temperatures—well below any cocktail

CALCULATING DILUTION

After lots of testing, I came up with an equation to estimate dilution from stirring and shaking that takes only initial alcohol content into consideration. It works well for the range of alcohol content in cocktails. I discovered that I could safely ignore sugar content. In these equations, alcohol by volume must be input as a decimal (22 percent would be 0.22) and dilution is returned as a decimal percent. I derived these equations by measuring a series of cocktails and using Excel to fit a curve to my data.

Dilution of a stirred drink stirred quickly with 120 grams of ¼-inch cubes for 15 seconds:

$$\text{Dilution ratio} = -1.21 \times ABV^2 + 1.246 \times ABV + 0.145$$

Dilution of a shaken drink shaken with 120 grams of ¼-inch cubes for 10 seconds:

$$\text{Dilution ratio} = 1.567 \times ABV^2 + 1.742 \times ABV + 0.203$$

temperature—to kill the sting of the alcohol. But the opposite is true for cocktail recipes. Why? Because of the Fundamental Law of Traditional Cocktails (see page 84). Making a high ethanol drink means having less dilution. Since dilution and chilling are linked by the Fundamental Law, this low dilution corresponds to warmer temperatures. The nature of different drink styles is built into the physics of chilling with ice. We feel that high-alcohol built and stirred drinks taste good warmer, and the flavors of fine liquors in those drinks would be ruined by lower temperatures, but it is hard to decipher which came first, our preferences or the physics.

Drinks with higher initial alcohol content will dilute more per unit volume than drinks with lower initial alcohol content (and they will get colder as well, by the Fundamental Law). The limiting case is trying to chill a juice with ice (which doesn't get very diluted) and trying to chill pure ethanol with ice (which gets very, very diluted). The diluted pure ethanol will always have a higher alcohol percentage than the diluted water, even though it dilutes much more.

Sugar: Remembering that our perception of sweetness is radically dulled by cold, so you would expect sugar levels to be higher in colder shaken drinks than in warmer stirred drinks to achieve the same sweetness on our palate—and they are. Built drinks, which are warmest, have the least added sugar, but because they have so little dilution, they often end up with more sugar per unit volume than stirred drinks.

Acid: Your perception of acid isn't affected as much by temperature as your perception of sugar is, and acid flavor doesn't attenuate as quickly as sugar during dilution. Highly diluted cold drinks, like blended drinks, will have less acid per unit of sugar than warmer, less diluted shaken drinks. Stirred drinks typically contain less acid than shaken, blender, or carbonated drinks, not because of their temperature or sugar content, but because they are not usually supposed to be tart. Built drinks contain very little or no acid.

Flavor Concentration and Sugar-to-Acid Ratio: Flavor concentration is a measure of how much sugar and acid are present in a drink relative to its dilution. It is hard to quantify, because it deals with two different ingredients, sugar and acid. Usually more diluted drinks, such as carbonated and blended drinks, have an overall lower flavor concentration than higher-alcohol drinks. The ratio of sugar to acid that will achieve the particular balance of sweet and tart you desire in a recipe will change, as stated above, depending on a drink's service temperature and its dilution.

STRUCTURE OF DIFFERENT TYPES OF DRINKS

These guidelines for specific drink styles are based on my analysis of forty-five classic cocktails—built, stirred, shaken, and shaken with egg white—and ten carbonated and blender recipes of my own. In the chart on pages 128–29 and accompanying recipe sheet on pages 130–35 you can peruse all the values for yourself. All the numbers represent typical ranges, not hard-and-fast rules, and I ignore outliers in the typical ranges that I present.

BUILT DRINKS: Built drinks are typically almost entirely liquor, so their alcohol by volume is very dependent on the strength of their base liquor. Because they are sipping drinks served on a rock, built drinks must taste good over a range of dilutions. This range makes it impossible to come up with a good sugar-to-acid ratio—the proper ratio would constantly change as dilution changed. As a result, built drinks typically contain little or no acid.

> **Recipe volume:** 2⅓– 2½ ounces (70–75 ml)
> **Initial alcohol by volume:** 34–40%
> **Initial sugar and acid content:** roughly 9.5 g/100 ml sugar, no acid
> **Dilution:** roughly 24%
> **Finished drink volume:** 2⁹⁄₁₀–3¹⁄₁₀ ounces (88–93 ml)
> **Finished alcohol by volume:** 27–32%
> **Finished sugar and acid content:** roughly 7.6 g/100 ml sugar, no acid

STIRRED DRINKS: Stirred drinks usually have some acidity, but are not tart. They have a wider range of alcohol levels than other drinks do. The sixteen drinks I analyzed range between 21% and 29%, with one outlier, the widow's kiss, at 32%. The Negroni is the lowest-alcohol stirred drink. Perhaps that's why it is so versatile, tolerant of many levels of dilution. The numbers below assume you use a lively 15-second stir with 120 grams of ice that is 1¼ inches on a side.

> **Recipe volume:** 3–3¼ ounces (90–97 ml)
> **Initial alcohol by volume:** 29–43%

> **Initial sugar and acid content:** 5.3–8.0 g/100 ml sugar, 0.15–0.20% acid
> **Dilution:** 41–49%
> **Finished drink volume:** 4⅓–4¾ ounces (130–142 ml)
> **Finished alcohol by volume:** 21–29%
> **Finished sugar and acid content:** 3.7–5.6 g/100 ml sugar, 0.10–0.14% acid

SHAKEN DRINKS: Shaken drinks are often sour drinks, containing roughly equal parts simple syrup (or its equivalent) and lime or lemon juice (or its equivalent). Simple syrup contains ten times as much sugar per ounce as lime or lemon juice contains acid per ounce, so most shaken drinks contain about 10 times as much sugar as acid. Finished alcohol levels for shaken drinks float mostly between 15 and 20%. The recipe volumes quoted here are too large to fit in a coupe glass in some cases. Remember that these numbers give you the actual size of the drink you create, not what is poured into the glass. Once you factor in holdback and loss on pouring, the drink in your glass might be a quarter-ounce lower—sometimes more. My assumptions are based on a 10-second shake with 120 grams of 1¼-inch ice cubes.

> **Recipe volume:** 3¼–3¾ ounces (98–112 ml)
> **Initial alcohol by volume:** 23.0–31.5%
> **Initial sugar and acid content:** 8.0–13.5 g/100 ml sugar, 1.20–1.40% acid
> **Dilution:** 51–60%
> **Finished drink volume:** 5⅕–5⁹⁄₁₀ ounces (156–178 ml)

Finished alcohol by volume: 15.0–19.7%
Finished sugar and acid content: 5.0–8.9 g/100 ml sugar, 0.76–0.94% acid

SHAKEN DRINKS WITH EGG WHITE: A typical large egg white is around 1 ounce (30 ml), so drinks shaken with egg white start out an ounce (30 ml) more diluted than their shaken siblings. Because they are more diluted, egg-white drinks often have a higher sugar-to-acid ratio than other shaken drinks, although because the drinks are more diluted, the overall levels of acid and sugar tend to be lower. My assumptions here are a 10-second dry-shake to mix and froth the egg white into the cocktail followed by a 10-second shake with 120 grams of 1¼-inch ice cubes.

Recipe volume: 4⅓–4¾ ounces (130–143 ml)
Initial alcohol by volume: 18–23%
Initial sugar and acid content: 10.0–13.2 g/100 ml sugar, 0.73–1.00% acid
Dilution: 46–49%. Notice that these dilution rates are much lower than those of normal shaken drinks because of the lower starting alcohol content.
Finished drink volume: 6⅔–7 ounces (198–209 ml)
Finished alcohol by volume: 12.1–15.2%
Finished sugar and acid content: 6.7–9.0 g/100 ml sugar, 0.49–0.68% acid

BLENDED DRINKS: I analyzed the blended drinks from the previous section of this book. Remember, I cheated the laws of dilution slightly by dissolving sugar directly into booze to keep the alcohol level fairly high despite massive dilution. Because blended drinks are very diluted, they have slightly less acid per unit of sugar than you would find in a shaken drink. The volumes given here are for the liquid part of the drink only, not the unmelted ice crystals. They will contain an extra 1½ ounces (45 ml) of unmelted ice crystals when they are poured.

Recipe volume: ¾ ounces (82.5 ml)
Initial alcohol by volume: 28.6–32.8%

Initial sugar and acid content: 15.0–15.4 g/100 ml sugar, 1.08–1.09% acid
Dilution: 90%!
Finished drink volume: 5¼ ounces (157.5 ml) plus an additional 1½ ounces (45 ml) ice crystals
Finished alcohol by volume: 15.0–17.2%
Finished sugar and acid content: 7.9–8.1 g/100 ml sugar, 0.57% acid

CARBONATED DRINKS: We haven't discussed carbonation yet, so I'll leave the details for later. I have analyzed seven of my carbonated recipes to give you a cross-section of carbonated drink types. I developed all the higher-alcohol recipes (above 16%) years ago. My newer, more evolved recipes tend to float between 14 and 15%. Carbonated cocktails tend to have slightly lower sugar-to-acid ratios than shaken drinks, much like blender and egg-white drinks. Like other highly diluted drinks, they lower overall sugar and acid content as well. Carbonated drinks are diluted before they are chilled, so you don't have before and after numbers to remember.

Recipe volume: 5 ounces (150 ml)
Alcohol by volume: 14–16%
Sugar and acid content: 5.0–7.5 g/100 ml sugar, 0.38–0.51% acid

NEGRONI

COCKTAIL BALANCE AT A GLANCE

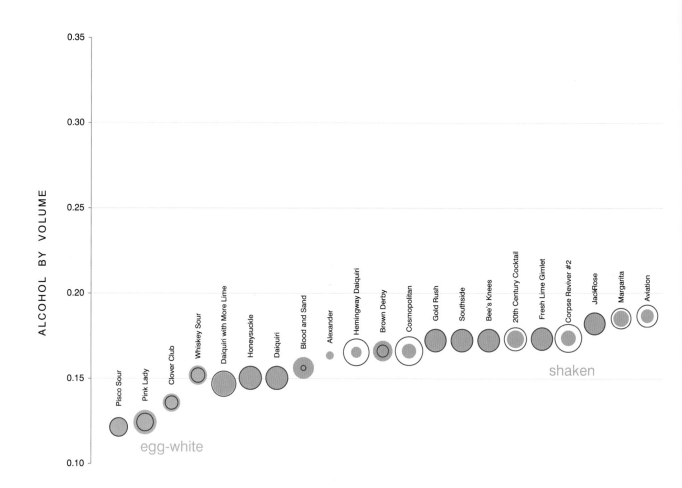

Cocktails within each drink type presented in order of alcohol content, spaced equally along x-axis for clarity. Shaded circles represent sugar content in grams per 100 ml. Rings represent acid content in percent.

○ Size of this ring represents 1% acid

● Size of this shaded circle represents 10 g/100 ml sugar

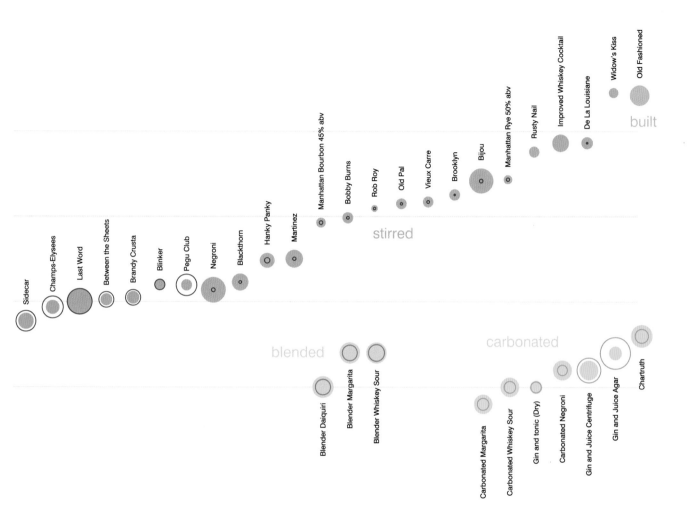

Sidecar

Champs-Elysees

Last Word

Between the Sheets

Brandy Crusta

Blinker

Pegu Club

Negroni

Blackthorn

Hanky Panky

Martinez

Manhattan Bourbon 45% abv

Bobby Burns

Rob Roy

Old Pal

Vieux Carre

Brooklyn

Bijou

Manhattan Rye 50% abv

Rusty Nail

Improved Whiskey Cocktail

De La Louisiane

Widow's Kiss

Old Fashioned

built

stirred

blended

Blender Daiquiri

Blender Margarita

Blender Whiskey Sour

carbonated

Carbonated Margarita

Carbonated Whiskey Sour

Gin and tonic (Dry)

Carbonated Negroni

Gin and Juice Centrifuge

Gin and Juice Agar

Chartruth

BUILT

OLD-FASHIONED

Mix volume: 72.6 ml
Finished volume: 90 ml
Start: 39.8% abv, 9.4 g/100 ml
   sugar, 0% acid
Finish: 32.1% abv, 7.6 g/100 ml
   sugar, 0% acid
2 oz (60 ml) bourbon (47% abv)
⅜ oz (11 ml) simple syrup
2 dashes Angostura bitters
Build over a large rock in an old-
   fashioned glass with an orange
   twist.

STIRRED

WIDOW'S KISS

Mix volume: 76.6 ml
Finished volume: 113.8 ml
Start: 47.9% abv, 5.5 g/100 ml
   sugar, 0% acid
Finish: 32.3% abv, 3.7 g/100 ml
   sugar, 0% acid
2 oz (60 ml) apple brandy (50%
   abv)
¼ oz (7.5 ml) Benedictine
¼ oz (7.5 ml) Yellow Chartreuse
2 dashes Angostura bitters
Stir and serve in a coupe glass.

DE LA LOUISIANE

Mix volume: 97.4 ml
Finished volume: 143.6 ml
Start: 43.2% abv, 6.6 g/100 ml
   sugar, 0.09% acid
Finish: 29.3% abv, 4.5 g/100 ml
   sugar, 0.06% acid
2 oz (60 ml) rye (50% abv)
½ oz (15 ml) Benedictine
½ oz (15 ml) sweet vermouth
3 dashes Peychaud's bitters
3 dashes Angostura bitters
3 dashes absinthe
Stir and serve in a coupe glass
   with a cherry.

IMPROVED WHISKEY COCKTAIL

Mix volume: 76.6 ml
Finished volume: 113 ml
Start: 43.2% abv, 9.5 g/100 ml
   sugar, 0% acid
Finish: 29.3% abv, 6.5 g/100 ml
   sugar, 0% acid
2 oz (60 ml) rye (50% abv)
¼ oz (7.5 ml) Luxardo Maraschino
¼ oz (7.5 ml) simple syrup
2 dashes Angostura bitters
Stir and serve over a large rock
   in an absinthe-rinsed old-
   fashioned glass with a lemon
   twist.

RUSTY NAIL

Mix volume: 75 ml
Finished volume: 110.4 ml
Start: 42.4% abv, 6 g/100 ml
   sugar, 0% acid
Finish: 28.8% abv, 4.1 g/100 ml
   sugar, 0% acid
2 oz (60 ml) Scotch (43% abv)
½ oz (15 ml) Drambuie
Stir and serve over a large rock in
   an old-fashioned glass with a
   lemon peel.

MANHATTAN WITH RYE

Mix volume: 88.3 ml
Finished volume: 129.2 ml
Start: 39.8% abv, 4.9 g/100 ml
   sugar, 0.18% acid
Finish: 27.2% abv, 3.4 g/100 ml
   sugar, 0.12% acid
2 oz (60 ml) rye (50% abv)
0.875 oz (26.66 ml) sweet
   vermouth
2 dashes Angostura bitters
Stir and serve in a coupe glass
   with a cherry or orange twist.

BIJOU

Mix volume: 90.8 ml
Finished volume: 132.9 ml
Start: 39.6% abv, 13.6 g/100 ml
   sugar, 0.2% acid
Finish: 27.1% abv, 9.3 g/100 ml
   sugar, 0.14% acid
1 oz (30 ml) gin (47.3% abv)
1 oz (30 ml) sweet vermouth
1 oz (30 ml) Green Chartreuse
1 dash orange bitters
Stir and serve in a coupe glass
   with a cherry and a twist of
   lemon.

BROOKLYN

Mix volume: 97.6 ml
Finished volume: 142.3 ml
Start: 38.3% abv, 6.1 g/100 ml
   sugar, 0.09% acid
Finish: 26.3% abv, 4.2 g/100 ml
   sugar, 0.06% acid
2 oz (60 ml) rye (50% abv)
½ oz (15.75 ml) Amer Picon
½ oz (14.25 ml) dry vermouth
¼ oz (6.75 ml) Luxardo
   Maraschino
1 dash Angostura bitters
Stir and serve in a coupe glass
   with a cherry.

VIEUX CARRÉ

Mix volume: 91.6 ml
Finished volume: 133.4 ml
Start: 37.6% abv, 5.9 g/100 ml
   sugar, 0.15% acid
Finish: 25.9% abv, 4.1 g/100 ml
   sugar, 0.1% acid
1 oz (30 ml) rye (50% abv)
1 oz (30 ml) Cognac (41% abv)
¾ oz (23.25 ml) sweet vermouth
¼ oz (6.75 ml) Benedictine
1 dash Angostura bitters
1 dash Peychaud's bitters
Stir and serve over a large rock in
   an old-fashioned glass.

OLD PAL

Mix volume: 105 ml
Finished volume: 152.8 ml
Start: 37.5% abv, 5.8 g/100 ml
    sugar, 0.13% acid
Finish: 25.7% abv, 4 g/100 ml
    sugar, 0.09% acid
2 oz (60 ml) rye (50% abv)
¾ oz (22.5 ml) Campari
¾ oz (22.5 ml) dry vermouth
Stir and serve in a coupe glass.

ROB ROY

Mix volume: 99.1 ml
Finished volume: 144.1 ml
Start: 37% abv, 3.7 g/100 ml
    sugar, 0.14% acid
Finish: 25.5% abv, 2.5 g/100 ml
    sugar, 0.09% acid
2.5 oz (75 ml) Scotch (43% abv)
¾ oz (22.5 ml) sweet vermouth
2 dashes Angostura bitters
Stir and serve in a coupe glass
    with a lemon twist.

BOBBY BURNS

Mix volume: 90 ml
Finished volume: 130.4 ml
Start: 36.1% abv, 6 g/100 ml
    sugar, 0.15% acid
Finish: 24.9% abv, 4.2 g/100 ml
    sugar, 0.1% acid
2 oz (60 ml) Scotch (43% abv)
¾ oz (22.5 ml) sweet vermouth
¼ oz (7.5 ml) Benedictine
Stir and serve in a coupe glass
    with a lemon twist.

MANHATTAN WITH BOURBON

Mix volume: 91.6 ml
Finished volume: 132.6 ml
Start: 35.7% abv, 5.3 g/100 ml
    sugar, 0.2% acid
Finish: 24.6% abv, 3.7 g/100 ml
    sugar, 0.14% acid
2 oz (60 ml) bourbon (45% abv)
1 oz (30 ml) sweet vermouth
2 dashes Angostura bitters

Stir and serve in a coupe glass
    with a cherry or orange twist.

MARTINEZ

Mix volume: 98.4 ml
Finished volume: 140.8 ml
Start: 32.2% abv, 9.5 g/100 ml
    sugar, 0.18% acid
Finish: 22.5% abv, 6.6 g/100 ml
    sugar, 0.13% acid
2 oz (60 ml) Old Tom gin (40%
    abv)
1 oz (30 ml) sweet vermouth
¼ oz (6.75 ml) Luxardo
    Maraschino
1 dash Angostura bitters
1 dash orange bitters
Stir and serve in a coupe glass
    with a lemon twist.

HANKY PANKY

Mix volume: 94 ml
Finished volume: 134.4 ml
Start: 32.1% abv, 8 g/100 ml
    sugar, 0.29% acid
Finish: 22.4% abv, 5.6 g/100 ml
    sugar, 0.2% acid
1½ oz (45 ml) sweet vermouth
1½ oz (45 ml) gin (47% abv)
1 bar spoon Fernet Branca
Stir and serve in a coupe glass
    with an orange twist.

BLACKTHORN

Mix volume: 91.6 ml
Finished volume: 130 ml
Start: 30% abv, 8.9 g/100 ml
    sugar, 0.15% acid
Finish: 21.1% abv, 6.3 g/100 ml
    sugar, 0.1% acid
1½ oz (45 ml) Plymouth gin
¾ oz (22.5 ml) sweet vermouth
¾ oz (22.5 ml) sloe gin
2 dashes orange bitters
Stir and serve in a coupe glass
    with an orange twist.

NEGRONI

Mix volume: 90 ml
Finished volume: 127.3 ml
Start: 29.3% abv, 13.3 g/100 ml
    sugar, 0.2% acid
Finish: 20.7% abv, 9.4 g/100 ml
    sugar, 0.14% acid
1 oz (30 ml) sweet vermouth
1 oz (30 ml) gin (47.3% abv)
1 oz (30 ml) Campari
Stir and serve in a coupe glass
    or over a large rock in an old-
    fashioned glass with a twist of
    orange or grapefruit.

SHAKEN
PEGU CLUB

Mix volume: 106.6 ml
Finished volume: 172 ml
Start: 33.8% abv, 6.7 g/100 ml
    sugar, 1.27% acid
Finish: 21% abv, 4.2 g/100 ml
    sugar, 0.78% acid
2 oz (60 ml) gin (47.3% abv)
¾ oz (22.5 ml) lime juice
¾ oz (22.5 ml) Curaçao
1 dash orange bitters
1 dash Angostura bitters
Shake and serve in a coupe glass
    with a lime wheel.

BLINKER

Mix volume: 86.5 ml
Finished volume: 140 ml
Start: 34.7% abv, 6.7 g/100 ml
    sugar, 0.62% acid
Finish: 21.4% abv, 4.1 g/100 ml
    sugar, 0.39% acid
2 oz (60 ml) rye (50% abv)
¾ oz (22.5 ml) grapefruit juice
1 bar spoon raspberry syrup
Shake and serve in a coupe glass.

BRANDY CRUSTA

Mix volume: 97.5ml
Finished volume: 156.3 ml
Start: 32.5% abv, 7.6 g/100 ml
    sugar, 0.92% acid

Finish: 20.2% abv, 4.7 g/100 ml
  sugar, 1/28% acid
2 oz (60 ml) Cognac (41% abv)
½ oz (15 ml) Curaçao
½ oz (15 ml) lemon juice
¼ oz (7.5 ml) Luxardo Maraschino
Shake and serve in a sugar-
  rimmed coupe glass with a big
  lemon spiral.

BETWEEN THE SHEETS

Mix volume: 97.5 ml
Finished volume: 156.2 ml
Start: 32.2% abv, 7.2 g/100 ml
  sugar, 0.92% acid
Finish: 20.1% abv, 4.5 g/100 ml
  sugar, 1/28% acid
1½ oz (45 ml) Cognac (41% abv)
¾ oz (22.5 ml) Curaçao
½ oz (15 ml) white rum (40% abv)
½ oz (15 ml) lemon juice
Shake and serve in a coupe glass
  with a discarded lemon twist.

LAST WORD

Mix volume: 90.1 ml
Finished volume: 144.2 ml
Start: 32% abv, 15.4 g/100 ml
  sugar, 1 1/2% acid
Finish: 20% abv, 9.6 g/100 ml
  sugar, 0.94% acid
¾ oz (22.5 ml) lime juice
¾ oz (22.5 ml) Green Chartreuse
¾ oz (22.5 ml) Luxardo
  Maraschino
¾ oz (22.5 ml) Plymouth gin
2 drops saline solution
Shake and serve in a coupe
  glass.

CHAMPS-ÉLYSÉES

Mix volume: 105.8 ml
Finished volume: 168.8 ml
Start: 31.4% abv, 8.3 g/100 ml
  sugar, 1.28% acid
Finish: 19.7% abv, 5.2 g/100 ml
  sugar, 0.8% acid
2 oz (60 ml) Cognac (41% abv)

¾ oz (22.5 ml) lemon juice
½ oz (15 ml) Green Chartreuse
¼ oz (7.5 ml) simple syrup
1 dash Angostura bitters
Shake and serve in a coupe glass
  with a discarded lemon twist.

SIDECAR

Mix volume: 112.5 ml
Finished volume: 178.2 ml
Start: 29.9% abv, 9.4 g/100 ml
  sugar, 1.2% acid
Finish: 18.9% abv, 6 g/100 ml
  sugar, 0.76% acid
2 oz (60 ml) Cognac (41% abv)
¾ oz (22.5 ml) Cointreau
¾ oz (22.5 ml) lemon juice
¼ oz (7.5 ml) simple syrup
Shake and serve in a coupe glass
  with a discarded orange twist.

AVIATION

Mix volume: 105 ml
Finished volume: 166 ml
Start: 29.5% abv, 8 g/100 ml
  sugar, 1.29% acid
Finish: 18.7% abv, 5.1 g/100 ml
  sugar, 0.81% acid
2 oz (60 ml) Plymouth gin
¾ oz (22.5 ml) lemon juice
½ oz (15 ml) Luxardo Maraschino
¼ oz (7.5 ml) crème de violette
Shake and serve in a coupe
  glass.

MARGARITA

Mix volume: 112.8 ml
Finished volume: 178ml
Start: 29.3% abv, 9.4 g/100 ml
  sugar, 1.2% acid
Finish: 18.5% abv, 6 g/100 ml
  sugar, 0.76% acid
2 oz (60 ml) blanco tequila (40%
  abv)
¾ oz (22.5 ml) lime juice
¾ oz (22.5 ml) Cointreau
¼ oz (7.5 ml) simple syrup
5 drops saline solution

Shake and serve in a coupe glass
  (salted rim optional).

JACK ROSE

Mix volume: 105.8 ml
Finished volume: 166.6 ml
Start: 28.7% abv, 13.5 g/100 ml
  sugar, 1.28% acid
Finish: 18.2% abv, 8.5 g/100 ml
  sugar, 0.81% acid
2 oz (60 ml) apple brandy (50%
  abv)
¾ oz (22.5 ml) Grenadine
¾ oz (22.5 ml) lemon juice
1 dash Angostura bitters
Shake and serve in a coupe glass.

CORPSE REVIVER #2

Mix volume: 92.5 ml
Finished volume: 144.3 ml
Start: 27.1% abv, 8.9 g/100 ml
  sugar, 1.61% acid
Finish: 17.4% abv, 5.7 g/100 ml
  sugar, 1.03% acid
¾ oz (22.5 ml) lemon juice
¾ oz (22.5 ml) gin (47% abv)
¾ oz (22.5 ml) Cointreau
¾ oz (22.5 ml) Lillet Blanc
3 dashes absinthe or Pernod
Shake and serve in a coupe glass
  with a discarded orange twist.

FRESH LIME GIMLET

Mix volume: 105 ml
Finished volume: 163.7 ml
Start: 27% abv, 13.5 g/100 ml
  sugar, 1.29% acid
Finish: 17.3% abv, 8.7 g/100 ml
  sugar, 0.82% acid
2 oz (60 ml) gin (47.3% abv)
¾ oz (22.5 ml) lime juice
¾ oz (22.5 ml) simple syrup
Shake and serve in a coupe glass
  with a lime wheel.

20TH-CENTURY COCKTAIL

Mix volume: 112.5 ml
Finished volume: 175.4 ml

Start: 27% abv, 10.1 g/100 ml
  sugar, 1.32% acid
Finish: 17.3% abv, 6.5 g/100 ml
  sugar, 0.85% acid
1½ oz (45 ml) gin (47% abv)
¾ oz (22.5 ml) lemon juice
¾ oz (22.5 ml) white crème de
  cacao
¾ oz (22.5 ml) Lillet Blanc
Shake and serve in a coupe glass.

BEE'S KNEES

Mix volume: 105 ml
Finished volume: 163.6 ml
Start: 26.9% abv, 13.5 g/100 ml
  sugar, 1.29% acid
Finish: 17.2% abv, 8.7 g/100 ml
  sugar, 0.83% acid
2 oz (60 ml) gin (47% abv)
¾ oz (22.5 ml) honey syrup
¾ oz (22.5 ml) lemon juice
Shake and serve in a coupe glass
  with a lemon wheel.

SOUTHSIDE

Mix volume: 105 ml
Finished volume: 163.6 ml
Start: 26.9% abv, 13.5 g/100 ml
  sugar, 1.29% acid
Finish: 17.2% abv, 8.7 g/100 ml
  sugar, 0.83% acid
2 oz (60 ml) gin (47% abv)
¾ oz (22.5 ml) lemon juice
¾ oz (22.5 ml) simple syrup
Shake with a handful of mint
  leaves and serve in a coupe
  glass with mint.

GOLD RUSH

Mix volume: 105 ml
Finished volume: 163.6 ml
Start: 26.9% abv, 13.5 g/100 ml
  sugar, 1.29% acid
Finish: 17.2% abv, 8.7 g/100 ml
  sugar, 0.83% acid
2 oz (60 ml) bourbon (47% abv)
¾ oz (22.5 ml) lemon juice
¾ oz (22.5 ml) honey syrup

Shake and serve over a large rock
  in an old-fashioned glass.

COSMOPOLITAN

Mix volume: 105 ml
Finished volume: 162.5 ml
Start: 25.7% abv, 8.4 g/100 ml
  sugar, 1.63% acid
Finish: 16.6% abv, 5.5 g/100 ml
  sugar, 1.05% acid
1½ oz (45 ml) Absolut Citron
  vodka
¾ oz (22.5 ml) Cointreau
¾ oz (22.5 ml) cranberry juice
½ oz (15 ml) lime juice
Shake and serve in a coupe glass
  with a discarded (optionally
  flamed) orange twist.
Note: The fellow who came up with
  the Cosmo, Toby Cecchini, told
  me that the recipe above is a
  bastardized farce. He says the
  actual recipe is this even more
  acidic version:

COSMOPOLITAN TC
(NOT CHARTED)

Mix volume: 139 ml
Finished volume: 215 ml
Start: 25.9% abv, 7.2 g/100 ml
  sugar, 1.85% acid
Finish: 16.7% abv, 4.7 g/100 ml
  sugar, 1.19% acid
2 oz (60 ml) Absolut Citron vodka
1 oz (30 ml) Cointreau
¾ oz (22.5 ml) lime juice
½ oz (15 ml) cranberry juice
Shake and serve in a coupe glass
  with an orange twist.

BROWN DERBY

Mix volume: 105 ml
Finished volume: 162.5 ml
Start: 25.7% abv, 11.8 g/100 ml
  sugar, 0.69% acid
Finish: 16.6% abv, 7.6 g/100 ml
  sugar, 0.44% acid
2 oz (60 ml) bourbon (45% abv)

1 oz (30 ml) grapefruit juice
½ oz (15 ml) honey syrup
Shake and serve in a coupe glass
  with a discarded grapefruit
  twist.

HEMINGWAY DAIQUIRI

Mix volume: 112.6 ml
Finished volume: 174.1 ml
Start: 25.6% abv, 6.4 g/100 ml
  sugar, 1 1/22% acid
Finish: 16.5% abv, 4.1 g/100 ml
  sugar, 0.98% acid
2 oz (60 ml) white rum (40% abv)
¾ oz (22.5 ml) lime juice
½ oz (15 ml) grapefruit juice
½ oz (15 ml) Luxardo Maraschino
2 drops saline solution
Shake and serve in a coupe glass
  with a lime wheel.

ALEXANDER

Mix volume: 97.5 ml
Finished volume: 150.4 ml
Start: 25.2% abv, 4.7 g/100 ml
  sugar, 0% acid
Finish: 16.4% abv, 3.1 g/100 ml
  sugar, 0% acid
2 oz (60 ml) Cognac (41% abv)
1 oz (30 ml) heavy cream
¼ oz (7.5 ml) Demerara syrup
Shake and serve in a coupe glass
  with grated nutmeg.

BLOOD AND SAND

Mix volume: 90 ml
Finished volume: 137.7 ml
Start: 23.9% abv, 12.3 g/100 ml
  sugar, 0.28% acid
Finish: 15.6% abv, 8 g/100 ml
  sugar, 0.19% acid
1 oz (30 ml) Scotch (43% abv)
¾ oz (22.5 ml) Cherry Heering
¾ oz (22.5 ml) sweet vermouth
½ oz (15 ml) orange juice
Shake and serve in a coupe glass
  with an (optionally flamed)
  orange twist.

CLASSIC DAIQUIRI

Mix volume: 105 ml
Finished volume: 159.5 ml
Start: 22.9% abv, 13.5 g/100 ml
    sugar, 1.29% acid
Finish: 15% abv, 8.9 g/100 ml
    sugar, 0.85% acid
2 oz (60 ml) white rum (40% abv)
¾ oz (22.5 ml) lime juice
¾ oz (22.5 ml) simple syrup
Shake and serve in a coupe glass.

HONEYSUCKLE

Mix volume: 105 ml
Finished volume: 159.5 ml
Start: 22.9% abv, 13.5 g/100 ml
    sugar, 1.29% acid
Finish: 15% abv, 8.9 g/100 ml
    sugar, 0.85% acid
2 oz (60 ml) white rum (40% abv)
¾ oz (22.5 ml) lime juice
¾ oz (22.5 ml) honey syrup
Shake and serve in a coupe glass
    with a lime wheel.

DAIQUIRI (MORE LIME)

Mix volume: 108 ml
Finished volume: 163.4 ml
Start: 22.2% abv, 13.2 g/100 ml
    sugar, 1.42% acid
Finish: 14.7% abv, 8.7 g/100 ml
    sugar, 0.94% acid
2 oz (60 ml) white rum (40% abv)
0.875 oz (25.5 ml) lime juice
¾ oz (22.5 ml) simple syrup
Shake and serve in a coupe glass.

EGG WHITE

WHISKEY SOUR

Mix volume: 130.1 ml
Finished volume: 197.9 ml
Start: 23.1% abv, 10.9 g/100 ml
    sugar, 0.81% acid
Finish: 15.2% abv, 7.1 g/100 ml
    sugar, 1/23% acid
2 oz (60 ml) rye (50% abv)
¾ oz (22.5 ml) simple syrup
0.625 oz (17.5 ml) lemon juice

2 drops saline solution
1 oz (30 ml) egg white
Shake without ice to mix and foam
    egg white, then shake with ice
    and serve in a coupe glass.

CLOVER CLUB

Mix volume: 135 ml
Finished volume: 201.4 ml
Start: 20.3% abv, 10 g/100 ml
    sugar, 0.73% acid
Finish: 13.6% abv, 6.7 g/100 ml
    sugar, 0.49% acid
2 oz (60 ml) Plymouth gin
1 oz (30 ml) egg white
½ oz (15 ml) Dolin dry vermouth
½ oz (15 ml) raspberry syrup
½ oz (15 ml) lemon juice
1 oz (30 ml) egg white
Shake without ice to mix and
    foam egg white, then shake
    with ice and serve in a
    coupe glass. Garnish with a
    raspberry.

PINK LADY

Mix volume: 142.5 ml
Finished volume: 209.4 ml
Start: 18.3% abv, 13.2 g/100 ml
    sugar, 0.95% acid
Finish: 12.4% abv, 9 g/100 ml
    sugar, 0.64% acid
1½ oz (45 ml) Plymouth gin
1 oz (30 ml) egg white
¾ oz (22.5 ml) lemon juice
½ oz (15 ml) Grenadine
½ oz (15 ml) simple syrup
½ oz (15 ml) Lairds Applejack
    Bottled in Bond
Shake without ice to mix and foam
    egg white, then shake with ice
    and serve in a coupe glass.

PISCO SOUR

Mix volume: 135 ml
Finished volume: 197.5 ml
Start: 17.8% abv, 11/2 g/100 ml
    sugar, 1% acid

Finish: 12.1% abv, 7.2 g/100 ml
    sugar, 0.68% acid
2 oz (60 ml) pisco (40% abv)
1 oz (30 ml) egg white
¾ oz (22.5 ml) lime juice
¾ oz (22.5 ml) simple syrup
Shake without ice to mix and
    foam egg white, then shake
    with ice and serve in a coupe
    glass topped with 3 drops
    Angostura bitters or Amargo
    Chuncho.

BLENDED

BLENDER WHISKEY SOUR

Mix volume: 157.7 ml
Finished volume: 157.7 ml
Start: 16.7% abv, 7.8 g/100 ml
    sugar, 0.61% acid
Finish: 16.7% abv, 7.8 g/100 ml
    sugar, 0.61% acid
2½ oz (75 ml) water
2 oz (60 ml) sugared 100-proof
    rye (44% abv)
½ oz (15 ml) lemon juice
¼ oz (7.5 ml) orange juice
4 drops saline solution
Blend with 120 grams ice, strain
    out large chunks, and serve in
    a coupe glass.

BLENDER MARGARITA

Mix volume: 158 ml
Finished volume: 158 ml
Start: 17.2% abv, 7.9 g/100 ml
    sugar, 1.27% acid
Finish: 17.2% abv, 7.9 g/100 ml
    sugar, 1.27% acid
2½ oz (75 ml) water
1 oz (30 ml) Cointreau
¾ oz (22.5 ml) white mezcal (40%
    abv)
½ oz (15 ml) Yellow Chartreuse
½ oz (15 ml) lime juice
10 drops Hellfire bitters
Blend with 120 grams ice, strain
    out large chunks, and serve in
    a coupe glass.

BLENDER DAIQUIRI

Mix volume: 157.7 ml
Finished volume: 157.7 ml
Start: 15% abv, 8.1 g/100 ml
    sugar, 1.27% acid
Finish: 15% abv, 8.1 g/100 ml
    sugar, 1.27% acid
2½ oz (75 ml) water
2¼ oz (67.5 ml) sugared 80-proof
    rum (35% abv)
½ oz (15 ml) lime juice
4 drops saline solution
Blend with 120 grams ice, strain
    out large chunks, and serve in
    a coupe glass.

CARBONATED
CHARTRUTH

Mix volume: 165 ml
Finished volume: 165 ml
Start: 18% abv, 8.3 g/100 ml
    sugar, 1.21% acid
Finish: 18% abv, 8.3 g/100 ml
    sugar, 1.21% acid
3¼ oz (97 ml) water
1¾ oz (54 ml) Green Chartreuse
½ oz (14 ml) clarified lime juice
Chill and carbonate.

GIN AND JUICE, AGAR
CLARIFIED

Mix volume: 165.1 ml
Finished volume: 165.1 ml
Start: 16.9% abv, 5 g/100 ml
    sugar, 1.16% acid
Finish: 16.9% abv, 5 g/100 ml
    sugar, 1.16% acid
2⅝ oz (80 ml) agar-clarified
    grapefruit juice
2 oz (59 ml) gin (47.3% abv)
⅞ oz (26 ml) water
2 drops saline solution
Chill and carbonate.

GIN AND JUICE, CENTRIFUGE
CLARIFIED

Mix volume: 165 ml
Finished volume: 165 ml
Start: 15.8% abv, 7.2 g/100 ml
    sugar, 0.91% acid
Finish: 15.8% abv, 7.2 g/100 ml
    sugar, 0.91% acid
1⅞ oz (55 ml) gin (47.3% abv)
1⅞ oz (55 ml) centrifuge-clarified
    grapefruit juice
1⅜ oz (42 ml) water
⅜ oz (10 ml) simple syrup
4 dashes champagne acid
Chill and carbonate.

CARBONATED NEGRONI

Mix volume: 165.1 ml
Finished volume: 165.1 ml
Start: 16% abv, 7.3 g/100 ml
    sugar, 0.38% acid
Finish: 16% abv, 7.3 g/100 ml
    sugar, 0.38% acid
2¼ oz (67.5 ml) water
1 oz (30 ml) sweet vermouth
1 oz (30 ml) gin (47.3% abv)
1 oz (30 ml) Campari
¼ oz (7.5 ml) clarified lime juice or
    champagne acid
2 drops saline solution
Chill and carbonate. Serve with a
    discarded grapefruit twist.

GIN AND TONIC (DRY)

Mix volume: 164.6 ml
Finished volume: 164.6 ml
Start: 15.4% abv, 4.9 g/100 ml
    sugar, 0.41% acid
Finish: 15.4% abv, 4.9 g/100 ml
    sugar, 0.41% acid
⅞ oz (87 ml) water
1¾ oz (53.5 ml) gin (47.3% abv)
⅜ oz (12.8 ml) Quinine Simple
    Syrup
⅜ oz (11 1/4 ml) clarified lime
    juice
2 drops saline solution
Chill and carbonate.

CARBONATED WHISKEY SOUR

Mix volume: 162 ml
Finished volume: 162 ml
Start: 15.2% abv, 7.2 g/100 ml
    sugar, 0.44% acid
Finish: 15.2% abv, 7.2 g/100 ml
    sugar, 0.44% acid
⅝ oz (78.75 ml) water
1¾ oz (52.5 ml) bourbon (47%
    abv)
⅝ oz (18.75 ml) simple syrup
⅜ oz (12 ml) clarified lemon juice
2 drops saline solution
Chill and carbonate.

CARBONATED MARGARITA

Mix volume: 165.2 ml
Finished volume: 165.2 ml
Start: 14.2% abv, 7.1 g/100 ml
    sugar, 0.44% acid
Finish: 14.2% abv, 7.1 g/100 ml
    sugar, 0.44% acid
2½ oz (76 ml) water
2 oz (58.5 ml) blanco tequila (40%
    abv)
⅝ oz (18.75 ml) simple syrup
⅜ oz (12 ml) clarified lime juice
4 drops saline solution
Chill and carbonate.

COCKTAIL INGREDIENT PERCENTAGES

NOTE: The alcohol levels that I have listed for commercial spirits are accurate. Because it is difficult to measure the sugar levels in any liquor that contains both sugar and alcohol (like Chartreuse), I have relied on published sources and educated guesses to provide sugar levels. The figures for acid in wine-based liqueurs are likewise educated approximations. My fruit-juice sugar and acid levels are averages supplied by the U.S. government and by commercial growers for standard single-strength (as opposed to concentrated) juices; for the Wickson apple I used my own refractometer measurements. Of course, the sugar and acidity of fruits will vary wildly. The alcohol levels of modified spirits are as close as I could estimate.

No straight spirits are included in this list. Their alcohol levels are listed on the bottles, and they typically contain no sugar and minimal titratable acid, even when aged in oak.

TYPE	INGREDIENT	ETHANOL	SWEETNESS	TITRATABLE ACID
Vermouths	Carpano Antica Formula	16.5%	16.0%	0.60%
	Dolin Blanc	16.0%	13.0%	0.60%
	Dolin Dry	17.5%	3.0%	0.60%
	Dolin Rouge	16.0%	13.0%	0.60%
	Generic dry vermouth	17.5%	3.0%	0.60%
	Generic sweet vermouth	16.5%	16.0%	0.60%
	Lillet Blanc	17.0%	9.5%	0.60%
	Martinelli	16.0%	16.0%	0.60%
Liqueurs	Amaro CioCiaro	30.0%	16.0%	0.00%
	Amer Picon	15.0%	20.0%	0.00%
	Aperol	11.0%	24.0%	0.00%
	Benedictine	40.0%	24.5%	0.00%
	Campari	24.0%	24.0%	0.00%
	Chartreuse, Green	55.0%	25.0%	0.00%
	Chartreuse, Yellow	40.0%	31.2%	0.00%
	Cointreau	40.0%	25.0%	0.00%
	Crème de cacao, white	24.0%	39.5%	0.00%
	Crème de violette	20.0%	37.5%	0.00%
	Drambuie	40.0%	30.0%	0.00%
	Fernet Branca	39.0%	8.0%	0.00%
	Luxardo Maraschino	32.0%	35.0%	0.00%
Bitters	Angostura	44.7%	4.2%	0.00%
	Peychauds	35.0%	5.0%	0.00%
Juices	Ashmead's Kernel apple	0.0%	14.7%	1.25%
	Concord grape	0.0%	18.0%	0.50%
	Cranberry	0.0%	13.3%	3.60%
	Granny Smith apple	0.0%	13.0	0.93%

TYPE	INGREDIENT	ETHANOL	SWEETNESS	TITRATABLE ACID
	Grapefruit	0.0%	10.4%	2.40%
	Honeycrisp apple	0.0%	13.8%	0.66%
	Orange	0.0%	12.4%	0.80%
	Strawberry	0.0%	8.%	1.50%
	Wickson apple	0.0%	14.7%	1.25%
Acids	Champagne acid	0.0%	0.0%	6.00%
	Lemon juice	0.0%	1.6%	6.00%
	Lime acid orange	0.0%	0.0%	6.00%
	Lime juice	0.0%	1.6%	6.00%
	Orange juice, lime strength	0.0%	12.4%	6.00%
Sweeteners	70 Brix caramel syrup (sweetness is low because of sugar breakdown during caramelization—a guess)	0.0%	61.5%	0.00%
	Butter syrup	0.0%	42.1%	0.00%
	Coriander syrup	0.0%	61.5%	0.00%
	Demerara syrup	0.0%	61.5%	0.00%
	Djer syrup	0.0%	61.5%	0.00%
	Honey syrup	0.0%	61.5%	0.00%
	Maple syrup	0.0%	87.5%	0.00%
	Any nut orgeat	0.0%	61.5%	0.00%
	Commercial orgeat	0.0%	85.5%	0.00%
	Quinine simple syrup	0.0%	61.5%	0.00%
	Simple syrup	0.0%	61.5%	0.00%
Others	Cabernet sauvignon	14.5%	0.2%	0.55%
	Coconut water	0.0%	6.0%	0.00%
	Espresso	0.0%	0.0%	1.50%
	Sour orange juice	0.0%	12.3%	4.50%
Modified spirits	Café Zacapa	31.0%	0.0%	0.75%
	Chocolate Vodka	40.0%	0.0%	0.00%
	Jalapeño Tequila	40.0%	0.0%	0.00%
	Lemongrass Vodka	40.0%	0.0%	0.00%
	Milk-Washed Rum	34.0%	0.0%	0.00%
	Peanut Butter and Jelly Vodka	32.5%	16.5%	0.25%
	Sugared 100 proof	44.0%	18.5%	0.00%
	Sugared 80 proof	35.0%	18.5%	0.00%
	Tea Vodka	34.0%	0.0%	0.00%
	Turmeric Gin	41.2%	0.0%	0.00%

PART 3

New TECHNIQUES *and* IDEAS

Alternative Chilling

Ready to move beyond frozen water in your cocktail chilling pursuits? Let's start with things you can do in your freezer, and then we will move to liquid nitrogen and dry ice.

USING YOUR HOME FREEZER LIKE A COCKTAIL NINJA

My freezer at home hovers around −10.5°F (−23.5°C), a fantastic temperature for most of my freezer tricks. I have another freezer that stays between −4°F (−20°C) and 0°F (−18°C) at its coldest setting, barely adequate for the recipes you'll read below. A couple degrees make a big difference when freezing mixtures of alcohol, sugar, and water. You should determine how cold your freezer is. To find out, simply put a bottle of straight booze greater than 40 percent alcohol by volume in your freezer and let it sit there overnight. It won't freeze. Measure the temperature of the liquid with a digital thermometer. Voilà: your average freezer temperature. If your freezer is too warm, 0°F (−18°C), you probably have it set too high. Turn it down. Colder is better. Your ice cream and frozen foods will thank you. I have read a number of sources recommending that you keep your freezer temperature on the warm side to conserve energy. Apparently the authors haven't done their research on the pernicious effects of temperatures above 0°F (−18°C) on the longevity and quality of frozen foodstuffs. Keeping your freezer warm is penny wise, pound foolish.

SLUSHY SLUSH

In the traditional cocktail section I describe how to make blended *drinks* with ice—they are liquids with some ice crystals in them. Here we will make slushies. If you want to replicate a 7-Eleven–style frozen treat, freeze your entire cocktail batch ahead of time in a reusable container (I use plastic quart containers or Ziploc bags, depending on the recipe), minus hyper-perishable ingredients such as lime juice but including dilution water, and blend it to serve. Blenders don't usually do a good job with single drinks, so you'll want to make at least two at a time. The trick here is to get the dilution right. You want to get the finished alcohol by volume below 15.5 percent. Go for 14 percent, lower than most shaken drinks. You need to get the alcohol by volume down so that your freezer can do its job and freeze the mixture. Sugar reduces the freezing point of drinks as well, so don't let it get above 9 grams per 100 milliliters. For a drink recipe based on a 2-ounce (60-ml) pour of 40 percent alcohol-by-volume liquor, you'll need to add about 3½ ounces (105 ml) of water-based ingredients with no more than the equivalent of ¾ ounce (22.5 ml) of simple syrup (about 14 grams of sugar).

It takes a long time to freeze these drinks properly, so make your batch the night before you want to serve it. When you wake up in the morning, look at the cocktail. If it doesn't look like it is freezing nicely, your freezer is probably −4°F (−20°C) or higher—not cold enough. Don't despair. You will just need to adjust the technique a bit. Three to six hours before you plan to blend the drinks, add the lime juice. Once the mix is diluted a bit more, it should freeze properly.

The slushy procedure can be applied to many, many drinks. Here's a simple daiquiri to get you started:

LEGIT SLURPEE

Frozen Daiquiri

MAKES TWO 5⅗-OUNCE (169-ML) DRINKS AT 14.2% ALCOHOL BY VOLUME, 8.4 G/100 ML SUGAR, 0.93% ACID

INGREDIENTS

4 ounces (120 ml) white rum (40% alcohol by volume), preferably clean-tasting and cheap, such as Flor de Caña

4 ounces (120 ml) filtered water

1½ ounces (45 ml) simple syrup

4 drops saline or 2 pinches of salt

1¾ ounces (52.5 ml) freshly strained lime juice

PROCEDURE

The day before you plan to make the drinks, combine the rum, water, simple syrup, and saline or salt. Pour the mixture into a wide-mouth plastic container or Ziploc bag and freeze. At drink time, put the daiquiri mix directly from the freezer into a blender, add the lime juice, blend to a slush, and serve. College students: you're welcome. And pace yourselves.

LEFT: *This is the correct texture for a blended drink when it comes out of your freezer.* ***RIGHT:*** *Frozen daiquiri.*

EBONY AND IVORY

Two drinks—one dark, one light—served side by side, living together in perfect harmony. Either drink is also good enough to stand on its own merits as a solo act.

Both drinks are fundamentally frozen vermouth. Ebony is a mixture of Carpano sweet vermouth (if you can't find Carpano, substitute another fine and sweet vermouth) with a bit of vodka to tone down the sweetness, plus a hint of lemon juice (you won't need much lemon; vermouths are wine-based and contain acidity already). I like this drink a lot. Even though it's very cold, the flavor of the vermouth shines through purely without being cloying.

Ivory is based on Dolin Blanc, a sweet white vermouth. It's also lightened with a touch of vodka, but it's acidified with lime. It is much brighter and more refreshing than you expect a vermouth drink to be.

Vermouth oxidizes quickly and badly once diluted, so these drinks must be frozen in Ziploc bags with all the air removed and not in a large container where they can be exposed to a lot of air. Pour the mixes into bags and seal the bags 90 percent of the way across the top. Lay the bags down and push all the air out of them, then pinch the final portion of the bag lock shut.

EBONY AND IVORY

Ebony

MAKES TWO 4⅘-OUNCE (145.5-ML) DRINKS AT 14.4% ALCOHOL
BY VOLUME, 8.4 G/100 ML SUGAR, 0.74% ACID

INGREDIENTS

5 ounces (150 ml) Carpano vermouth (16% alcohol by volume, roughly 16 g sugar/100 ml, roughly 0.6% acidity)

1½ ounces (45 ml) vodka (40% alcohol by volume)

2½ ounces (75 ml) filtered water

4 drops saline solution or 2 pinches of salt

Flat ¾ ounce (21 ml) freshly strained lemon juice

PROCEDURE

The day before you plan to make the drinks, combine the vermouth, vodka, water, and saline or salt. Pour the mixture into a Ziploc bag, exclude all air from the bag, and freeze. At drink time, put the mix directly from the freezer into a blender, add the lemon juice, blend to a slush, and serve by itself or with Ivory.

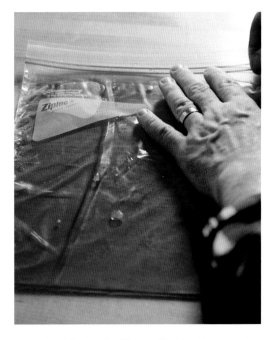

Removing air from the Ziploc. Air is bad because it will oxidize your diluted vermouth.

Ivory

MAKES TWO 4⅗-OUNCE (138-ML) DRINKS AT 13.9% ALCOHOL
BY VOLUME, 7.9 G/100 ML SUGAR, 0.81% ACID

INGREDIENTS

5½ ounces (165 ml) Dolin Blanc vermouth (16% alcohol by volume, roughly 13 g/100 ml sugar, roughly 0.6% acid)

1 ounce (30 ml) vodka (40% alcohol by volume)

2 ounces (60 ml) filtered water

4 drops saline solution or 2 pinches of salt

Flat ¾ ounce (21 ml) freshly strained lime juice

PROCEDURE

The day before you plan to make the drinks combine the vermouth, vodka, water, and saline or salt. Pour the mixture into a Ziploc bag, exclude all air from the bag, and freeze. At drink time, put the mix directly from the freezer into a blender, add the lime juice, blend to a slush, and serve by itself or with Ebony.

THE JUICE SHAKE

Here is another low-tech trick you can perform with your home freezer: the juice shake. When someone asks me to develop a recipe that people can make at home without fancy equipment, this is one of my go-to techniques.

It is difficult to make balanced cocktails with most juices—they just aren't concentrated enough. After you've added enough juice to get the flavor right, you can't afford to dilute the drink further with ice. Apple juice, grapefruit juice, strawberry juice, and watermelon juice contain too much water to be useful in a cocktail chilled with ice. The simple chilling solution: make juice ice cubes and shake away.

At first blush, the juice shake couldn't be much easier. Get a bunch of ice-cube trays, jigger your juice into the trays, and freeze. When it comes time to make the drink, jigger your alcohol and any other mixers into a cocktail shaker, add juice cubes, and shake. There is one wrinkle.

When you shake with regular ice, it doesn't really matter how much ice you add. Shaking will melt roughly the same amount of water no matter how much ice you use (refer back to the Traditional Cocktails section, page 63, if you forget why this is true). Juice shakes don't work this way. Juice ice is a combination of sugars, acids, flavors, and water. As you shake, the first part of the juice cubes to melt is richer in sugar, acid, and flavor than the stuff that melts later, so adding too many juice cubes throws off the balance of the drink.

With the juice shake, you need to add exactly the right number of juice cubes to achieve the flavor you want. Next you'll practice a technique I call "shaking to completion": shake the drink until the juice cubes have completely disintegrated. You will hear the cubes break up in the shaker, and you'll know you are done when the drink sounds slushy and the tins are preposterously cold. While this process is more taxing than a typical shake, it isn't as difficult as you might think, because frozen juice is much softer than frozen water.

I use the juice shake to make shaken-sour-style drinks (think daiquiris, whiskey

The juice shake gets your tins extremely cold.

sours, margaritas). Most often these drinks taste best with a final alcohol level of 15.5 to 20 percent, a sugar content of 6.5 to 9 percent and a total acidity of 0.84 to 0.88 percent. Depending on the recipe you are making and the kind of juice you have, you may not want all your dilution to come from juice. The drink might taste too juicy. If so, just add some regular ice as well. As long as you don't add too much regular ice, the exact amount you use isn't important, because the juice ice will melt while you are shaking long before the regular ice will. An ounce give or take is no problem.

You can approximate the juice shake with a blender, using frozen fruit instead of frozen fruit juice. The blender technique is supersimple, and I have to admit that these drinks often taste good, but the pectin and other solids in the fruits lend a smoothielike texture that I find unsatisfying. Don't let me stop you from blending whole frozen fruit, though it does pain me a little bit. You'll need to add a bit more frozen fruit than you would juice, because of the solids—and you'll also need to add a bit of ice to loosen the texture and make the drinks more . . . drinkable.

The juice shake makes drinks colder than shaking with plain ice does. Everyone likes them, and anyone can make them. Here are some suggestions.

*Approximating the juice shake
with frozen fruit in a blender*

Strawberry Bandito

Tequila and strawberries is unimaginative, I hear you say? Well, it is delicious. Strawberries can vary greatly in sweetness and acidity. You will have to adjust the ratios below if your berries (or the juice, if you are buying it premade) are different from mine. If you desire a slushier drink, keep your tequila in the freezer prior to shaking. I make mine with jalapeño-infused tequila, but regular blanco tequila tastes good as well.

MAKES ONE 4⅗-OUNCE (140-ML) DRINK AT 17.1% ALCOHOL BY VOLUME, 9.0 G/100 ML SUGAR, 0.96% ACID

ABOVE: *Strawberry juice cubes*

INGREDIENTS

2 ounces (60 ml) strawberry juice (8 g/100 ml sugar, 1.5% acid) (or you can use 2½ ounces [75 grams] frozen strawberries and 15 grams ice, but you didn't hear it from me)

2 ounces (60 ml) Jalapeño Tequila, page 207 (40% alcohol by volume)

¼ ounce (7.5 ml) freshly strained lime juice

Short ½ ounce (12.5 ml) simple syrup

2 drops saline solution or a pinch of salt

PROCEDURE

Several hours before you want to make the drinks, measure two 1-ounce (30-ml) pours of strawberry juice into individual ice-cube trays for every drink you'll make, and freeze. This drink has a lot of unfrozen ingredients, so for a colder drink, refrigerate your tequila, lime juice, and simple syrup before making. At drink time, combine the tequila, lime juice, simple syrup, and saline solution or salt in a mixing tin, add 2 strawberry ice cubes, and shake until the ice cubes are fully melted and there are no large particles left. Strain using a hawthorn strainer into a chilled coupe glass. Enjoy.

STRAWBERRY BANDITO

SHAKEN DRAKE

Shaken Drake

This drink is a mix of unclarified grapefruit juice and kümmel, the German version of caraway liqueur. Kümmel is sweeter than its better-known Scandinavian cousin, aquavit, but not too sweet. Grapefruit and caraway were born to be together. A bar spoon of maple syrup rounds out the bitterness of the grapefruit. Even though all three ingredients contain sugar, the result does not taste overly sweet. The bitterness of grapefruit requires more than the average amount of salt in this cocktail. If you like salt-rimmed glasses, they work here. If you cannot find kümmel, substitute aquavit and up the maple syrup slightly.

MAKES ONE 4⅗-OUNCE (139-ML) DRINK AT 15.6% ALCOHOL BY VOLUME, 10.2 G/100 ML SUGAR, 1.03% ACID

INGREDIENTS

2 ounces (60 ml) freshly squeezed and strained grapefruit juice (10.4 g/100 ml sugar, 2.4% acid)

1½ ounces (45 ml) Helbing Kümmel liqueur (35% alcohol by volume)

½ ounce (15 ml) vodka (40% alcohol by volume)

1 bar spoon (4 ml) grade B maple syrup (87.5 g/100 ml sugar)

5 drops saline solution or a generous pinch of salt

PROCEDURE

Several hours before you want to make the drinks, measure two 1-ounce (30-ml) pours of strained grapefruit juice into individual ice-cube trays for every drink you'll make, and freeze. At drink time, combine the kümmel, vodka, maple syrup, and saline solution or salt in a mixing tin, add 2 grapefruit-juice ice cubes, and shake like the devil until the ice cubes turn to slush and there are no large particles left (30 seconds at least). Your tins will get very cold. Strain using a hawthorn strainer into a chilled coupe glass.

OPPOSITE: Shaken Drake with grapefruit-juice cubes in the foreground.

Scotch and Coconut

My friend chef Nils Noren loves the combination of coconut water and Scotch. I stole his combination for this unusual shaken drink. I use Ardbeg 10 in this recipe, because I want the smokiness of a peaty Islay Scotch to go with the coconut water. Coconut water on its own is a bit musky and needs some fruit flavors to round it out, so I add Cointreau, which brings both sweetness and the flavor of orange without adding acidity. I rarely like fruit acids with Scotch. Even so, I found this drink needed just a skosh of acid, so I added a small amount of lemon juice. After the drink is made, it receives a twist of orange peel—the oils bring brightness without extra acidity—and a star anise pod. While I normally hate inedible things floating in my drinks, the aroma of star anise marries so well with coconut water that I make an exception here.

Your choice of coconut water is important, because most of the stuff on the market is pretty bad. Try to get one that hasn't been pasteurized at high heat. Ideally, make your own. Coconut water for drinking does not come from the hard brown coconuts normally found in the supermarket. Go to an Asian greengrocer and seek out immature coconuts sold specifically for their water. They are usually beige or white and pithy on the outside, having had a portion of their thick husk cut away before sale. Knock two holes in the top of the coconut—one for pouring and one to let in air—and pour out the coconut water. Strain it before use to get out the bits of husk.

MAKES ONE 4^7/$_{10}$-OUNCE (142-ML) DRINK AT 18.6% ALCOHOL BY VOLUME, 5.9 G/100 ML SUGAR, 0.32% ACID

INGREDIENTS

2½ ounces (75 ml) fresh coconut water (6.0 g/100 ml sugar)

1½ ounces (45 ml) Ardbeg 10-year Scotch (46% alcohol by volume)

½ ounce (15 ml) Cointreau (40% alcohol by volume 25 g/100 ml sugar)

¼ ounce (7.5 ml) freshly strained lemon juice

2 drops saline solution or a pinch of salt

1 star anise pod

1 orange twist

PROCEDURE

Several hours before you want to make the drinks, measure two 1¼-ounce (37.5-ml) pours of strained fresh coconut water into individual ice-cube trays for every drink you'll make, and freeze. At drink time, combine the Scotch, Cointreau, lemon juice, and saline solution or salt in a mixing tin, add 2 coconut water ice cubes, and shake until the ice cubes turn to slush and there are no large particles left. Your tins will get very cold. Strain with a hawthorn strainer into a chilled coupe glass. Express the orange twist over the top of the drink, then wipe the rim of the glass with the orange side of the peel before discarding it. Float the star anise pod on top.

OPPOSITE: MAKING THE SCOTCH AND COCONUT:
1) Punch two holes in a juice coconut, one for draining and one for air. 2) Drain and strain the coconut water.
3) Pour the coconut water into ice-cube trays and freeze. 4) The finished drink ready for its garnish.

STIRRED DRINKS EN MASSE

Many drinks that you stir rather than shake you can make in large quantities ahead of time with no reduction in quality. In fact, you can probably make them better. Shaken drinks can't be made ahead of time because you actually have to shake them to get their characteristic texture. You'll recall from the traditional cocktail section that stirring a cocktail does only two things: stirring chills and stirring dilutes. Using your freezer, you can easily separate these two actions. You can choose the exact dilution you want and chill it to the exact temperature you want. Making drinks ahead of time and chilling them in the freezer also lets you crank out a lot of really good drinks really quickly. At the bar I make fully diluted cocktails and chill them to precise serving temperatures with a fancy freezer. Unfortunately, regular freezers are too cold to store diluted stirred cocktails. First, the freezer will cause your drinks to crystallize, and second, stirred drinks don't taste good too cold anyway. The solution: put only the booze in the freezer and dilute with ice water at the last minute.

Let's revisit the Manhattan. I want you to know how to make lots of Manhattans with aplomb. One evening years ago my wife and I stopped in at the Howard Johnson's in Times Square. Whatever for, you ask? Not because we knew Jacques Pépin had devised its crispy clam strip recipe after finishing a stint as Charles de Gaulle's personal chef. No, we were drawn in by a vintage hand-lettered sign in the window that asked, "May we recommend a pitcher of Manhattans?" Why, yes, you may! Alas, the waitress stared at us blankly; the sign was a relic from some earlier Ho Jo golden age, and cocktail pitchers were no longer to be had in 1995. If only they'd had this recipe.

MAKING MANHATTANS BY THE PITCHER: 1) *Put the mix into a plastic bottle and squeeze out the air before capping to prevent oxidation. Put the bottle in the freezer.* **2)** *The freezer-cold Manhattan mix.* **3)** *Measure out the ice water.* **4)** *Pour the freezer-cold mix and ice water into a chilled pitcher.* **5)** *Stir.* **6)** *Serve.*

Manhattans by the Pitcher

MAKES SEVEN 4²/₅-OUNCE (132-ML) DRINKS (OR ANY MULTIPLE YOU CHOOSE) AT 26% ALCOHOL BY VOLUME, 3.2 G/100 ML SUGAR, 0.12% ACID

INGREDIENTS

14 ounces (420 ml) Rittenhouse rye (50% alcohol by volume)

6¼ ounces (187.5 ml) Carpano Antica Formula vermouth (16.5% alcohol by volume, roughly 16% sugar, 0.6% acid)

¼ ounce (7.5 ml) Angostura bitters

10½ ounces (315 ml) ice water (water that has been chilled with ice; don't add the ice)

Garnish of your choice

EQUIPMENT

1 liter plastic soda bottle

Chilled serving pitcher

Chilled coupe glasses

PROCEDURE

Combine the rye, vermouth, and Angostura bitters in the soda bottle. Squeeze the excess air out of the bottle, cap it, and put it in the freezer for a minimum of 2 hours. You are using the soda bottle to exclude air so the vermouth won't alter as you store the drink. Also, even though I don't love storing liquor in plastic, plastic won't explode in the freezer like glass will if you make a mistake and overfill a container.

At drink time, combine the cocktail base from the freezer with the ice water in the chilled serving pitcher and stir briefly. If your freezer was roughly −4°F (−20°C), the finished drink should be around 26°F (−3.3°C), just a bit cooler than you would get from stirring. Fill and garnish the coupes and pour Manhattans to your heart's content.

BOTTLED COCKTAILS READY TO DRINK, PROFESSIONAL STYLE

If you have an accurate enough freezer, you can make ready-to-drink stirred cocktails in advance, bottle them, and serve them whenever you want. At the bar we always have stirred-type drinks bottled and ready to go, and the Manhattan is one we are never without. We make batches of thirty drinks using a recipe very similar to the Manhattans by the Pitcher recipe, but even simpler:

MAKES 30 4¼-OUNCE (136-ML) DRINKS AT 26% ALCOHOL BY VOLUME, 3.2 G/100 ML SUGAR, 0.12% ACID

INGREDIENTS

Three 750 ml bottles Rittenhouse rye

One 1 liter bottle Carpano Antica Formula vermouth

1 ounce (30 ml) Angostura bitters

1700 ml filtered water

Combine the ingredients and divide among thirty 6.35-ounce (187-ml) champagne-style bottles. Now comes the hard part. You cannot simply cap the Manhattans, chill them, and hope for the best. The oxygen in the headspace of the bottle is sufficient to change the flavor of the vermouth within a few hours, as vermouth in a diluted cocktail is horribly unstable. To get around this problem we add a small amount of liquid nitrogen to the bottle after we fill it with cocktail. If you can't obtain liquid nitrogen it's okay, your cocktails will just oxidize a little.

The bottles we use have crown caps like beer bottles, so while the nitrogen is bubbling off and dis-

placing the air from the bubble, we place a crown cap lightly on the lip of the bottle. After the LN stops smoking, we know it has completely evaporated, and we seal the bottle with a cap sealer (which you can get at any home-brew shop for almost nothing). Once the bottles are thus purged, their contents will last indefinitely.

And here's the second hard part: chilling accurately. At the bar we use a very accurate Randall FX freezer set to 22°F (−5.5°C). Typical refrigerator temperatures (40°F/4.4°C) make a drink that is too warm, and typical freezer temperatures (−4°F/−20°C) will freeze these drinks outright. Even if the drinks don't freeze, any temperature below 20°F (−6.7°C) will severely degrade the flavor and the aroma of a Manhattan until it warms back up a bit. If I'm taking Manhattans to a party, I use ice and salt to chill them. Careful: salt-and-ice mixtures can get way too cold. Start by adding 10 percent by weight of salt to ice and mixing thoroughly. Measure the temperature of the slurry. If it is too cold, add ice or water. Too warm? Add salt. Bring more salt with you so you can continue to add more as the ice melts. Remember to wipe off the salt before you serve! (See photographs on next page.)

The advantages of the bottled cocktail are many. The drinks can be cranked out very quickly; you can make a drink colder than stirring can without over-diluting; the drinks are consistent 100 percent of the time; no cocktail is lost to the ice in the form of hold-back from stirring; and, most importantly, people get to pour cocktails themselves into chilled coupes at their own pace, rather than trying to get an overfilled glass to their lips without spilling.

CONTINUES

OPPOSITE: BOTTLING MANHATTANS THE PROFES-SIONAL WAY: 1) Pre-dilute the Manhattans and put in bottles with a little liquid nitrogen to get rid of oxygen. 2) Wait for the nitrogen to completely stop smoking, your sign that it has all boiled off. 3) Cap.

SERVING BOTTLED COCKTAILS: 1) Add copious salt to ice with a bit of water. Nestle the bottles into the salted ice and allow the cocktails to chill. 2) Chill and garnish coupes. 3) Pour. 4) Enjoy.

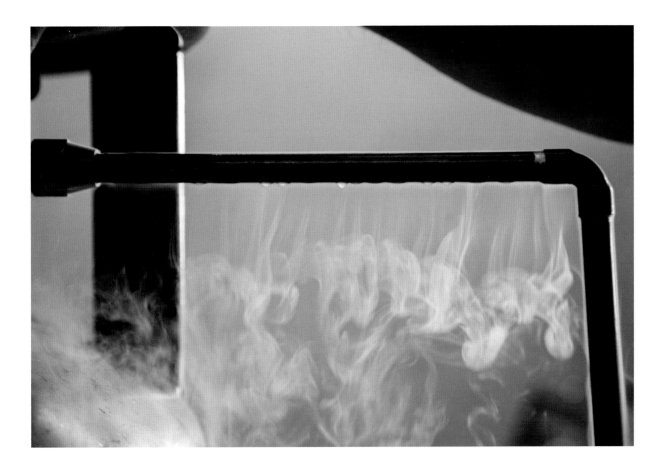

CHILLING WITH CRYOGENS:
LIQUID NITROGEN AND DRY ICE

Now for some stuff you can't do with a freezer. We are moving into tech land.

For some basic information on liquid nitrogen and dry ice, including some vital safety procedures, see the Equipment section, page 21. Don't use these materials unless you know what you are doing, which means you have been trained by someone with practical experience and you feel very comfortable with the materials. Reading this book does not constitute adequate training. Never, ever serve a cryogenic agent such as LN or dry ice to a person. When you serve a drink prepared with a cryogen, the cryogen itself must no longer be present—just the chilling effect. Safety mistakes can end your life or the life of someone around you, and the mistakes aren't always obvious.

Rocking and rolling drinks with liquid nitrogen.

LIQUID NITROGEN

Liquid nitrogen is what it says it is: liquefied nitrogen. It is –196°C (–321°F) at atmospheric pressure. Liquid nitrogen constantly boils off into nitrogen gas, which joins the atmosphere—of which nitrogen is already the major component. Now, you can chill a cocktail with liquid nitrogen. It is fairly easy to tell when all the nitrogen has boiled off and you aren't in danger of serving someone liquid nitrogen by mistake. But chilling cocktails with LN is problematic.

Even though liquid nitrogen is absurdly cold, every gram has only the same chilling power as 1.15 grams of ice: it takes more LN than you expect to get any real chilling done. On the flip side, if you add too much, you will make things dangerously cold very quickly. Liquid nitrogen floats, so you can't just pour it on top of a drink and expect it to chill. The LN will float to the top and create a frozen crust on the liquid, leaving the bottom unaffected: not what you want. Stirring is ineffective at getting the LN to really mix in well unless you stir vigorously, which is difficult, because as you mix LN with relatively warm liquids, it boils violently, splashing cocktails everywhere, getting LN all over your arms, and making so much fog that you can't see what the hell is going on. For all these reasons, I don't recommend chilling individual cocktails with LN. LN is useful for chilling large batches of cocktails, with a technique I call rock and roll.

Rock and Roll: Get two large containers, larger than you think you need. They should each hold more than four times the total amount of cocktail you are going to chill—preferably six times. If the containers are made of stainless steel, never touch them with your bare hands while you are chilling or you risk frostbite. If you use plastic containers, be careful not to leave the liquid nitrogen stationary in them too long, or they'll crack from the cold. Don't ever use glass—it might shatter. I usually use plastic and take the chance on cracking.

If you don't agitate liquid nitrogen it will just float on top of your cocktail, creating a frozen crust without chilling the bulk of the drink.

Pour the cocktail into one of the large containers, and pour about two-thirds of the cocktail's volume's worth of liquid nitrogen on top. Quickly (remember, we don't want plastic containers to get brittle and crack) pick up the container and pour the contents through the air into the second container. Nitrogen fog will be everywhere. Quickly pour the mix back into the first container. Keep going back and forth, rocking and rolling the drink between the two containers until the fog stops, which is your indication that the LN has evaporated. If you have added too little LN, the drink will be too warm; add more LN and repeat. If the drink is too cold (as in, it has solidified), run tap water over the outside of the container while stirring the slush. It will melt quickly.

If your rocking and rolling sprayed cocktail everywhere, making a king-hell mess, you either have terrible aim or your containers weren't big enough. Remember, I told you to get bigger containers than you thought you needed! Trick: if you sense that the cocktail is about to boil over and make a mess—and your senses will get attuned to such things—simply stop midpour and begin pouring only half the contents back and forth at a time until the boiling calms down. Pouring only half reduces the amount of mixing and therefore the violence of the boiling as well.

When I chill this way, I'm usually making a lot of a particular drink that can't be made with typical shaking techniques. Batches of cocktail

VISUAL CUES TO CHILLED DRINKS (LEFT TO RIGHT): 1) This cocktail is so cold it will likely be painful to consume. 2) This cocktail won't hurt you (unless you drink it quickly) but it is too cold to taste balanced. 3) This cocktail is still too cold for a cocktail with oaked spirit, but is good if you know you won't serve the drink for several minutes. 4) Ready to drink.

for carbonated drinks, for instance, are often rock-and-rolled. Ditto drinks that you would use the juice shake for at home. Be careful not to overchill your drinks. They'll taste bad and you'll burn someone's tongue. If the drink is a solid, it is way too cold. If the drink is merely slushy, you won't hurt anyone, but the drink will be colder than optimum.

Liquid Nitrogen for Glasses: There is no better way to chill glasses than with liquid nitrogen. It feels great, looks great, and chills only the part of the glass where the drink will be, avoiding the base and stem and thereby the sweaty ring on your table. Keep an open vacuum-insulated thermos full of LN behind the bar. It will last several hours. Pour a little LN out of the thermos into a glass and give the glass a swirl, as if you are observing a fine wine. In several seconds the glass will be cold. Pour the extra LN back into the thermos, onto the floor, or into the next glass you need to chill. In a few seconds the glass will frost over most appealingly.

LEFT: *When you chill a glass with liquid nitrogen, make sure to choose one that won't spray liquid as you swirl.* **RIGHT:** *A perfectly chilled glass.*

But remember, this is LN, and there are some less-than-obvious safety rules to observe. Never chill a glass in front of a person's face. If you inadvertently spill LN or the glass cracks, he or she could be injured. Never chill glasses over open ingredients or over an ice bin, lest a glass break and ruin your products.

Choose appropriate glassware. Only use glasses that curve in toward the lip, such as champagne flutes, wineglasses, and coupes. Don't try martini glasses: when you swirl, LN will spray out of the glass and could get in your eyes. Some glasses—even those with inward-curving lips—crack when exposed to liquid nitrogen because of stresses developed during rapid chilling. Glasses that are uniformly thick, have flat bottoms or corners, or have thin walls and a thick base tend to crack. Pint glasses and rocks glasses are poor choices. Many stemmed glasses and most champagne flutes don't crack with LN, but test them first. Test two or three glasses of a particular pattern. If none of your test glasses crack, buy more of those glasses and you should be fine. My experience has been that any particular glass pattern is going to break either consistently or almost never. Quality is not an indicator of chillability.

At a pro bar, chilling glasses with LN means I don't need a dedicated glass-chilling freezer. It may be easier at home to chill your glasses in the freezer, but at the bar, LN helps us maximize our back-of-house space.

DRY ICE

Dry ice is solidified carbon dioxide. It's called dry because carbon dioxide doesn't exist as a liquid at normal atmospheric pressure; it turns directly from a solid to a gas through a process called sublimation. Dry ice appears to be the friendlier cryogen—it's easier to get, less likely to give you cryogenic burns, and, as a solid, easier to handle than liquid nitrogen. Furthermore, even though dry ice is much warmer than liquid nitrogen—a balmy −109.3°F (−78.5°C)—don't let that warmer temperature fool you. Pound for pound, dry ice has almost twice the cooling power of liquid nitrogen, because it takes a lot of energy to turn CO_2 directly from a solid into a gas (136.5 calories per gram), whereas liquid nitrogen needs only a measly 47.5 calories per gram to vaporize.

As dry ice chills, it lightly carbonates whatever liquid it is in, which is why I primarily use dry ice at events to prechill drinks that I'm about to carbonate.

The problem with dry ice: it is hard to use its cooling power effectively. Unlike liquid nitrogen, which can surround and coat solids or mix with other liquids to create a large surface area for effective rapid chilling, dry ice is a solid, so it is hard to get it to chill drinks rapidly. Drop a chunk of dry ice into a glass of liquid. Initially you get bubbling and foaming and a nice carbon dioxide fog. Pretty soon, however, the liquid calms down and your chilling rate radically slows. Look at the liquid: there's still dry ice in there, but a layer of liquid has frozen over the dry ice, insulating it. Beat on the nugget with something to break off the frozen layer and the chilling speeds up again.

Safety rule: never use dry ice to carbonate drinks in a sealed container unless you are an engineer qualified to design pressure vessels with over-pressure safety valves. Search the Internet for pictures of unsuspecting boneheads who dropped dry ice into a soda bottle and sealed it, just to have it blow up in their hands. On second thought, don't.

An obscure but enjoyable use for dry ice is to keep large batches of drinks chilled at events. You need an immersion circulator (a device with a heater that keeps liquids at very accurate temperatures) that can be set below freezing, a large plastic tub, a bunch of cheap vodka, and some dry ice. You fill the tub with the vodka, which has a very low freezing point, set up the immersion circulator in it, and then add dry ice. A pump in the circulator will keep the vodka moving around; it stirs the bath for you, ensuring even chilling. If the temperature drops below −16°C, the heater in the circulator will kick on and prevent overchilling.

Let's say you are serving a high-proof drink in shot form and you want to serve it at −16°C (0°F). Put the bottles of liquor in the tub, add the cheap vodka, and put in the immersion circulator. Turn on the circulator and set it to −16°C. Now throw in some chunks of dry ice. All you need to do is look at the bath every once in a while to see if you need to throw in some more chunks of dry ice. I have served thousands of well-chilled shots and drinks at events this way. This technique can also be used to keep carbonated cocktails cold during an event.

Safety again: always make sure there is no way for someone to ingest the dry ice accidentally.

EUTECTIC FREEZING

If you have more time than money, and you really want to maintain a large batch of cocktails at an exact temperature for an event like a wedding, consider eutectic freezing. You already know that adding salt to water lowers its freezing point. What you might not know is that at a certain concentration, a mixture of salt and water freezes and melts at a single temperature, called the eutectic point. These eutectic solutions can be used to maintain a constant temperature the same way regular water ice can maintain 0°C. (At other salt concentrations, the solution's temperature will keep rising as it melts, without any plateau.)

Different salt solutions have different eutectic points. The eutectic point for table salt, NaCl, is a solution of 23.3 percent salt by weight at −6.16°F (−21.2°C), a temperature too low for most cocktail work. The magic salt for this trick is KCl, potassium chloride. A mixture containing 20 percent KCl and 80 percent water by weight has a eutectic point of 13°F (−11°C). The best part is that KCl is really, really cheap. Some people use it as road salt, so you can often find large bags of it at home improvement stores.

To hold batched cocktails at a temperature just a bit shy of ideal serving temperature, make that mixture of 20 percent KCl, 80 percent water. Make sure the KCl dissolves completely and then freeze it in your freezer. When the time comes to keep your drinks cold, throw the KCl ice in a cooler with your drinks (in bottles, of course; the KCl would kill the taste), and your drinks will stay at temperature for hours. Alternatively, freeze the KCl in plastic soda bottles and drop them directly in your drink in the cooler. These bottles will take a while to cool, and you'll have to move them around a bit every now and again to maintain a constant temperature.

THAI BASIL DAIQUIRI

Nitro-Muddling and Blender-Muddling

Muddling is the process of crushing ingredients with a stick prior to making a cocktail. You muddle to release fresh flavors right before a drink is made. The problem is, muddling breaks up herbs, activating polyphenol oxidases (PPOs), vicious enzymes that cause fruits and herbs to turn brown and taste oxidized. That's why muddled herbs never taste as fresh as you want them to. When you crush a piece of mint and it goes black and starts to taste like a swamp, PPOs are the main culprits.

Nitro-muddling and blender-muddling are techniques I developed to combat PPOs. Let's say I want to muddle fresh basil into gin. To nitro-muddle, I freeze the basil solid with liquid nitrogen, crush it with a muddler, and then thaw it out by pouring the gin on top. The finely crushed gin-soaked basil won't turn brown; it stays beautifully, shockingly green. It also tastes incredibly fresh and powerful. If I don't have liquid nitrogen on hand, I can blender-muddle instead by simply blending the basil with gin and straining it out. My gin will be green and fresh, but not quite as fresh as the nitro-muddled version.

Herb on right has been crushed with a muddler. The polyphenol oxidase enzymes have started doing their dirty work.

HOW NITRO-MUDDLING AND BLENDER-MUDDLING FIGHT PPOS

The enemy of your enemy is your friend, and PPOs have two main enemies. The first enemy is booze. Forty percent alcohol by volume liquor will totally and permanently disable PPOs. The other enemy is the ascorbic acid (vitamin C) in lemon and lime juice—an antioxidant that retards their action.

Your two allies, booze and vitamin C, aren't enough to save you with regular muddling. You can't muddle effectively when your herbs are floating around in liquor. Even if you could, regular muddling doesn't get the booze into the leaves before the PPOs can do their dirty work. You need a secret weapon: liquid nitrogen or a blender.

Liquid nitrogen freezes herbs quickly. PPOs cannot work their devilry when they are frozen. The herbs also become so brittle in liquid nitrogen that they can be finely pulverized with a regular muddler, almost to a dust. I call this nitro-muddling (take note: regular freezers don't freeze fast enough or cold enough for this trick). After you nitro-muddle, you thaw the herbs *with liquor*. The liquor inactivates the PPOs as the herbs thaw out, and the herbs stay green. If you nitro-muddle an herb such as basil or mint and let it thaw without adding liquor, it will turn black almost instantly. After you thaw the herb in liquor and add antioxidant citrus juice, your victory against PPOs is complete.

Blenders, by contrast, puree herbs into alcohol so fast that the alcohol does its job of wiping out the PPOs before the enzymes have a chance to work.

Whether you nitro-muddle or blender-muddle, you finish your drink by adding the rest of the ingredients, shaking as usual, and straining the drink through a tea strainer into a chilled serving glass. Straining is important. Nobody wants pieces of pureed or pulverized herbs sticking to the sides of the glass or their teeth. If you nitro-muddle, you should strain the drink *after* you shake. Rattling the herbs around in the shaking tin with ice extracts more flavor. If you use a blender, you can strain before you shake, because the blender had already done all the flavor-extraction work you are going to get.

Nitro- and blender-muddled drinks have stunning colors and absurdly punchy fresh herb tastes. Because these techniques are such thorough flavor extractors, I sometimes use them even when PPOs aren't much of a problem. Roses don't oxidize quickly, but I've nitro-muddled a fresh rose into gin and carbonated it. The drink came out pink and fresh and, well, smelling like a rose.

TOP: *Liquid nitrogen makes this herb so brittle it shatters.* **BOTTOM:** *On the right is a pile of nitro-muddled Thai basil, with the texture you are aiming for.*

NITRO-MUDDLING VERSUS BLENDER-MUDDLING: WHICH IS BETTER?

Nitro-muddling is without question the superior technique, which is unfortunate, because most people don't have access to liquid nitrogen. I think the air that is whipped into the herb-liquor mix during blending makes blending a little less effective at preventing oxidation than nitro-muddling is. If I had to put a number on it, I'd say blender-muddling is about 90 percent as effective as nitro-muddling. The good news is that blender-muddling, which is within most people's reach, is still a really good technique. If I had never tasted a nitro-muddled drink, I wouldn't be grousing about blender-muddling. Because most readers will be using a blender, I'll give blender tips first.

BLENDER-MUDDLING TIPS

You cannot blender-muddle a single drink. The liquor has to cover the blender blades completely or the herbs will be exposed to oxygen and turn brown and swampy. Depending on the blender, I usually endorse a two-drink minimum. Just make sure the blades are covered. If your liquid level is a little low, you can also add your acid before you blend, but add the other ingredients *after* the blending (keep the proof high). At the beginning, blend slowly for a couple of seconds to get the herbs broken up, and then blend on high for only a few seconds. Overblending whips in a lot of air and is counterproductive.

You don't have to shake your drinks right after you blender-muddle. You can blend, add the rest of the ingredients, and shake later if you wish. How long you can wait after you blend but before you shake depends on the herb you are using. Mint doesn't last more than 15 minutes before I start noticing changes. Basils last about 45 minutes before I start noticing big changes. Nonbrowning herbs like lovage last longer.

When you blender-muddle, you can also strain your drink through a fine chinois *before* you shake it. Straining ahead of time makes the drinks easier to serve (you can just use a regular hawthorn strainer after shaking).

Almost all recipes that can be nitro-muddled can be blender-muddled. The only exceptions are very high-water-content, low-flavor-intensity ingredients such as lettuce. Lettuce nitro-muddles well, but you have to use so

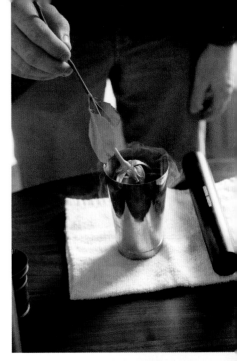

TOP: *Freezing a rose.*
BOTTOM: *Nitro-muddled rose cocktail.*

When you are nitro-muddling, be sure not to grip your mixing tin, only touch the top, and use a shatterproof muddler.

much lettuce to get a strong lettuce-flavored drink that blender-muddled lettuce becomes an unstrainable soup.

NITRO-MUDDLING TIPS

Very important: use liquid nitrogen only if you have been trained and are aware of all the hazards involved (see the Equipment section, page 21, and the section on Alternative Chilling, page 140).

I always nitro-muddle in stainless steel shaking tins. Stainless steel can get cold enough to hurt you when you add liquid nitrogen to it. Only handle the tin by the rim near the top where it has not frozen. You can visually confirm where the tin is too cold to hold. Don't try to be a hero.

Add herbs to the tin before you add the liquid nitrogen. Pour in a small amount of liquid nitrogen. Carefully swirl the tin and wait a couple seconds for the herbs to get crunchy (remember, don't touch the cold bottom of the tin). The liquid nitrogen will be boiling furiously. If you have not added enough liquid nitrogen, it will evaporate before the herbs are totally frozen. Add a bit more.

If you have added too much, the nitrogen will stop boiling violently—it will calm down—while there is still a big pool of LN in the bottom of the tin. You don't want to muddle this way. First, you will have splashing problems, and you don't want to splash liquid nitrogen. Second, if there is a lot of liquid in the bottom of the tin, it is very hard to get a good grinding action with the muddler. Third, your ingredients and tin will get so cold from the liquid nitrogen that the dilution of the cocktail will be thrown way off when you shake—not enough ice will melt. If you do add too much LN, don't worry; just pour off the extra before you begin muddling. What you want is for the LN to calm down when there are just a few millimeters (3/16 inch) left at the bottom of the tin. Then you can grind the herbs to a frozen dust.

Make sure you use a good muddler. The muddler must fit into the bottom of the tin but needs to be big enough to create a good grinding action. Don't use some puny ¾-inch muddling stick; 1½ inches (38 mm) in diameter is ideal. Material is also important. Don't use glass—it will get brittle and crack. Ditto with many plastics and rubber. Wood works well—I've nitro-muddled plenty of drinks with old-school "French"-style rolling pins—but my favorite is the Bad Ass Muddler (BAM) from Cocktail

Blender-muddling is almost as good as nitro-muddling. With blender-muddling you should strain out particles (as shown above right) before you shake. The pre-shaken mix is shown here in a glass for clarity.

Kingdom. Although it *is* made of plastic, the BAM has never chipped or cracked, and I've made thousands of nitro-muddled drinks with it.

When you muddle, muddle hard, with plenty of wrist action. Don't mess around. Poorly muddled herbs lead to weak-colored drinks. It takes some practice, but eventually your drinks will have the same colors you see in these pictures. After the herbs are muddled, there should still be a wisp of liquid nitrogen left in the tin. Pour your liquor in right away. You should get a small *pfwoof!* as the last of the liquid nitrogen boils off. I'll repeat: the LN needs to be completely gone before you proceed. Stir with a bar spoon for a few seconds and then add your acid (if the recipe contains acid), then the other ingredients. The mix shouldn't be too cold. If it is, run some cold tap water over the outside of the tin to warm it up a bit, or your dilution after shaking will be too low. Add ice and shake like a demon. Strain through a tea strainer to catch all the herb crud. You are done.

Note that the correct nitro-muddling procedure guarantees that you will never serve liquid nitrogen to a guest. Before you shake, you have to verify that the LN is gone, and even if you somehow mess that up (which you can't), if you try to shake with tins containing LN, the LN will boil and

blow the tins apart. (Do not try it. I'm serious. You do not want LN flying around.)

GENERAL TIPS

Blender-muddling and nitro-muddling share a drawback: they make a mess of your barware. Little bits of herb stick to everything. At home this is not much of a problem. In a professional bar, I recommend keeping one set of tins, a tea strainer, and a hawthorn separate just for your nitro-muddling work. I hate, hate, hate when a piece of herb contaminates a drink where it doesn't belong, and even with vigilant rinsing, you are bound to slip up once in a while.

RECIPES

Note: in the following recipes I give proportions for one drink. These recipes will work for nitro-muddling. If you are blender-muddling, you must make at least two. Three or more is better. Keep those blender blades covered at all times.

NITRO-MUDDLING, STEP-BY-STEP: 1) Pick your herbs and drop them in your tin. 2) Add some liquid nitrogen to freeze the herbs. 3) This is too much liquid nitrogen. With this much liquid you can't muddle well, and your drink will end up underdiluted. 4) This looks good. 5) Muddle. 6) It should look like this. 7) Add liquor, let thaw, then add acids and syrups and shake with ice. 8) Strain drink through a tea strainer into a chilled coupe glass.

TBD: Thai Basil Daiquiri

This is the drink I developed the foregoing techniques for. The TBD is a typical daiquiri with the addition of Thai basil. Thai basil is not like Italian globe basil in taste; it has a fantastic anise note that I love. Even people who hate anise tend to like it. This drink is as green and fresh-tasting as I can muster.

MAKES ONE 5⅓-OUNCE (160-ML) DRINK AT 15% ALCOHOL BY VOLUME, 8.9 G/100 ML SUGAR, AND 0.85 % ACID

INGREDIENTS

5 grams (7 large) Thai basil leaves

2 ounces (60 ml) Flor de Caña white rum (40% alcohol by volume) or other clean white rum

¾ ounce (22.5 ml) freshly strained lime juice

Short ¾ ounce (20 ml) simple syrup

2 drops saline solution or a pinch of salt

PROCEDURE

Nitro-muddle the Thai basil in a shaking tin. Then add the rum and stir. Add the lime juice, simple syrup, and saline solution or salt. Check to make sure the mix isn't freezing cold. Shake with ice and strain through a tea strainer into a chilled coupe glass.

Alternatively, make a double recipe and blender-muddle the Thai basil with the rum and lime juice, stir in the simple syrup and saline or salt, then strain through a fine strainer, shake with ice, and strain into two chilled cocktail coupes.

THAI BASIL DAIQUIRI

Spanish Chris

Mezcal and tarragon go very well together but aren't satisfying on their own. What they need is a bit of Maraschino liqueur. The three together make a formidable troika. Spanish Chris, by the way, is a guy who lurks in the basement of the Booker and Dax development lab.

MAKES ONE 5-OUNCE (149-ML) DRINK AT 15.3% ALCOHOL BY VOLUME, 10 G/100 ML SUGAR, AND 0.91% ACID

INGREDIENTS

3.5 grams (a small handful) fresh tarragon leaves

1½ ounces (45 ml) La Puritita mezcal or any fairly clean blanco mezcal (40% alcohol by volume)

½ ounce (15 ml) Luxardo Maraschino (32% alcohol by volume)

¾ ounce (22.5 ml) freshly strained lime juice

½ ounce (15 ml) simple syrup

3 drops saline solution or a generous pinch of salt

PROCEDURE

Nitro-muddle the tarragon in a shaking tin. Add the mezcal and Maraschino and stir. Add the lime juice, simple syrup, and saline solution or salt. Check to make sure the mix isn't freezing cold. Shake with ice and strain through a tea strainer into a chilled coupe glass.

Alternatively, make a double recipe and blender-muddle the tarragon with the mezcal, Maraschino, and lime juice. Stir in the simple syrup and saline or salt, then strain through a fine strainer, shake with ice, and strain into two chilled cocktail coupes.

The Flat Leaf

You might think that parsley wouldn't make a delicious drink, but you'd be wrong. This drink is intensely green and fresh without being savory. It tastes like spring, but you'll like it in winter too.

The citrus in this drink is bitter orange, a variety of *Citrus aurantium*. Bitter orange is also known as sour orange—and sour it is. This drink doesn't call for the fancy version of sour oranges known as Sevilles—those of beautiful peel and relatively non-bitter juice. It calls for the bitter oranges with ugly peels and true bitterness that are found quite often in Latin American groceries, labeled "arancia." I fell in love with these oranges over twenty years ago when I was dating my future wife. I started making sour/bitter orangeade from the trees that grew outside her parents' house in Phoenix, Arizona. The streets there were lined with bitter orange trees and nobody used the fruit. Pity. If you can't find sour oranges, substitute lime juice.

Make a wonderful variant of this drink with lovage. I love lovage. The flavor is as if parsley and celery had a bitter baby. The celery note has a bit of celery seed to it, which is, I think, why I like it so much.

MAKES ONE 5½-OUNCE (164-ML) DRINK AT 17.7% ALCOHOL BY VOLUME, 7.9 G/100 ML SUGAR, AND 0.82% ACID

INGREDIENTS

4 grams fresh parsley leaves or 4 grams (a small handful) fresh lovage leaves

2 ounces (60 ml) gin (47.3% alcohol by volume)

1 ounce (30 ml) freshly squeezed and strained bitter/sour orange juice or ¾ ounce (27.5 ml) freshly strained lime juice

½ ounce (15 ml) simple syrup

3 drops saline solution or a generous pinch of salt

PROCEDURE

Nitro-muddle the parsley or lovage in a shaking tin. Then add gin and stir. Add the fruit juice, simple syrup and saline or salt. Check to make sure the mix isn't freezing cold. Shake with ice and strain through a tea strainer into a chilled coupe glass.

Alternatively, make a double recipe and blender-muddle the parsley or lovage with the gin and sour fruit juice. Stir in the simple syrup and saline or salt, then strain through a fine strainer, shake with ice, and strain into two chilled cocktail coupes.

The Carvone

This drink combines aquavit, the Swedish superdrink whose main flavor is caraway seed, and mint. It is a study in chirality. In chemistry, a chiral molecule is one that can be composed in two ways that share the same structure but are mirror images of each other—they are not identical in the same way that your right and left hands are not identical. The chemical compound carvone is the main flavor compound in both caraway and mint, but carvone is chiral. R(–)carvone is the predominant smell of spearmint. Its mirror image, S(+)carvone, is the predominant smell of caraway (and also of dill, another Swedish obsession). I am usually resistant to pairing flavors because they have chemical similarities. Why should ingredients that are chemically similar necessarily taste good together? This case is different. I was sitting in on an organic chemistry lecture when I learned the story of chiral carvone. I immediately thought, Good story, better flavor combo.

This drink is different from the others in the nitro-muddled group because it contains no acid, so the mint won't last as long.

MAKES ONE 3⁹⁄₁₀-OUNCE (117-ML) DRINK AT 20.4 % ALCOHOL BY VOLUME, 6.8 G/100 ML SUGAR, AND 0% ACID

INGREDIENTS

6 grams (a good handful) fresh mint leaves

2 ounces (60 ml) Linie aquavit (40% alcohol by volume)

Short ½ ounce (13 ml) simple syrup

3 drops saline solution or a generous pinch of salt

1 lemon twist

PROCEDURE

Nitro-muddle the mint in a shaking tin. Then add the aquavit and stir. Add the simple syrup and saline or salt. Check to make sure the mix isn't freezing cold. Shake with ice and strain through a tea strainer into a chilled coupe glass, and finish with the lemon twist.

Alternatively, make a double recipe and blender-muddle the mint with the aquavit. Stir in the simple syrup and saline or salt, then strain through a fine strainer, shake with ice, and strain into two chilled cocktail coupes. Finish with 2 lemon twists.

Red-Hot Pokers

If on a cold night in colonial America you walked into a tavern and ordered a hot drink, the proprietor would make you a flip: a concoction of beer or cider mixed with liquor and sugar, heated by the plunge of a red-hot poker fresh out of the fireplace. Sometime around the Civil War, this drink began to change. Eggs were added to the mix, and flips began to be served both hot and cold. The hot flips were made by adding boiling water instead of a poker, and something crucial was lost. A few years ago I decided it was time to bring the red-hot poker into the twenty-first century so we could taste what we had been missing.

You see, the red-hot poker is not merely a heating device, like a pot. A red-hot poker shoved directly into a drink is a high-temperature flavor maker. The poker makes awesome toasty, burned, and caramel flavors that are hard to achieve in liquids. Compare two hot drinks, one made by heating water in a pan and one made with a red-hot poker, and you will never want the pan version again. The smell of a drink made with a red-hot poker is enchanting, filling the room like no other aroma at a bar. When somebody is making a poker drink, everybody perks up a bit—especially when it's nippy outside.

RED-HOT JOURNEY: TRADITIONAL POKERS

For my first poker experiment, I purchased a piece of equipment called a soldering copper, a heavy, pointed, octagonal slug of copper on the end of an iron stick. Coppers aren't expensive, and anyone can get their hands on one. You heat the slug end with a torch, or by leaving it for ten minutes on a gas burner set on high. When you heat it to a dull red and plunge it into a drink, the results can be spectacular.

The copper tends to add a flavor to the drinks it heats. Sometimes the copper flavor is not noticeable, and sometimes it's even beneficial. For years I made glögg, a hot Swedish Christmas wine, with my chef friend Nils Noren, using soldering coppers, and I think the glögg was the better for it. Other

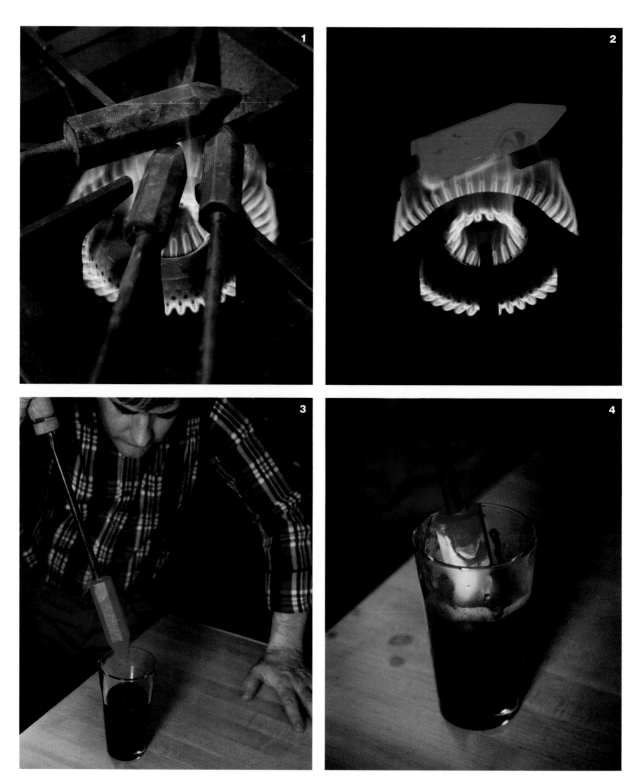

Using soldering coppers to poke a drink. **1)** *Heat the coppers over high heat till . . .* **2)** *They are glowing red.* **3)** *Plunge the copper into the drink.* **4)** *For a moment it will seem like nothing is happening. The Leidenfrost effect will prevent rapid bubbling—then the drink will go crazy.*

times the copper makes drinks taste like you are sucking on a penny. I tried switching to cast iron, which was likely the original official red-hot-poker material. It made drinks taste awful, like an iron supplement, and wasn't nearly as efficient at heating as the copper is. (I wonder what the colonials did to make their irons not taste so bad. Or maybe they became accustomed to the bad taste?) I tried using slugs of stainless steel, but they didn't work very well: stainless is a relatively poor conductor, and relatively poor at storing heat.

HOT ROCKS

Next I tried using hot stones. A friend of mine cut up some dolsots, the famous Korean stone bowls used to make gobdol bibimbop. (Everyone should own some of these fantastic bowls.) I would put a wok ring around a burner, fill it with dolsot chunks, and heat them until they had a faint, dull glow—roughly 800°F (430°C). I dropped a couple of hot rocks into a drink, and the drink would boil and bubble nicely. The procedure worked quite well, and I considered creating a category of drinks called "on the hot rocks," before I thought better of serving people drinks containing 800°F (430°C) chunks. Should you be foolish enough to attempt the hot-rock trick,

Dropping a hot rock in a drink: At first, the boiling is moderate—more Leidenfrost effect—then the drink goes nuts, and mellows out again.

beware: most rocks will explode when heated, sending tiny pieces of rock shrapnel whizzing through the room. You must use rocks that are heat-proof, like the dolsots and some soapstones.

SELF-IMMOLATION

After accepting the limitations of the hot-rocks approach, I began making self-heated pokers using a variety of immersion heaters. I built my first ones from heating elements (very much like an electric range element) that I bent into short helixes. Just like range elements, they glowed a bright red when cranked high. They were a good deal hotter than my previous pokers and rocks. One of them was particularly hot, and when I plunged it into cocktails, they instantly ignited.

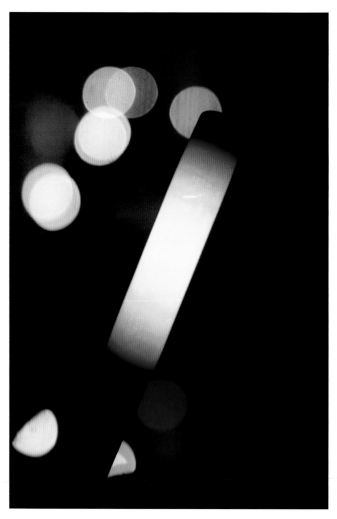

The first time a red-hot poker is plugged in it gets hotter than it ever will again, and since it has not yet oxidized, it gets incandescent and shimmery.

Lighting a drink on fire has two pronounced effects: it dials up the awesome meter, and it reduces the alcohol content of the drink faster than boiling does. I had a quandary now: should I make pokers that ignite cocktails or revert to cooler poking technology? I had to choose one or the other. I couldn't have a poker that only ignited drinks sometimes, because ignition changes the alcohol content. To produce a balanced drink, I had to know beforehand whether the drink was going to catch fire or not. If it wasn't, I had to add more water, or boil a lot longer, to achieve the same alcohol level. If it was going to ignite, I could heat for a shorter amount of time or add more booze—which would mean more of the flavor of the base spirit without increasing the finished alcohol by volume of the drink.

It surprises most people that you actually *want* to reduce the alcohol content of a hot drink. When you bring a hot drink with too much alcohol up to your face, it has a very unpleasant, stinging nose.

Never ignite or always ignite? You probably guessed that I chose always. Not just because I like fire, but because it allowed me to increase the amount of flavor in the finished drink.

To guarantee instant ignition on any red-hot-poker cocktail, you need to achieve pretty high temperatures—

Using the poker.

roughly 1600° to 1650°F (870° to 899°C). Temperatures that high are perilously close to the auto-meltdown temperatures of the immersion heaters you can buy. Add the problem that red-hot pokers go through vicious thermal cycling as they get plunged into drink after drink, which quickly weakens their metal alloy and insulation, and I had stumbled into quite a design problem.

I moved away from the bent-helix pokers, which looked a bit like a prop in a Frankenstein film, not clean and professional. I landed instead on a high-temperature cartridge heater. These devices are designed to heat holes in industrial metal dies and molds. The cartridge heaters I use are hollow sticks of a high-temperature corrosion-resistant nickel-steel alloy called Incoloy surrounding a long, thin, wire resistance heater wound into a small helix that is wound into a bigger helix (to increase the length of the heating element contained in a small volume). Magnesium-oxide insulation is packed into the cylinder around the resistance wire to prevent electrical shorts.

After many dozens of tests, I discovered that a ¾-inch-diameter cartridge heater 4 inches long with 500 watts of power hits the sweet spot. Any less power and the heater won't achieve the temperatures you need. Any more power and the heater will melt down—sometimes spectacularly, in a shower of sparks. Originally I used higher-power heaters—1500 watts—and used thermocouple thermometers to control the temperature so the heaters wouldn't destruct. With that much power, the pokers can make a drink every 30 seconds; the 500-watt pokers take 90 seconds to refresh themselves. The problem: 1500 watters always broke quickly and dramatically, in a blazing display that freaked out bystanders. The thermocouple thermometers always failed and led to heater meltdown. I started testing different sizes and wattages of heater to find one that didn't require thermal regulation but still did its job. Five-hundred watts at ¾-inch diameter and 4 inches long was the answer.

The end result of all the experimentation isn't super-rosy, however, because even the 500-watters always break eventually; they just last a lot longer. Occasionally the insulation inside the heater cracks and fails from the incessant thermal cycling, which leads to arcs and sparks. This is rare in a 500-watt poker, and not dangerous like an arcing scenario with a 1500-watt poker, but it is startling. The most common mode of failure is the fade-away. Over time the resistance wire inside the heater oxidizes. As it oxidizes the resistance goes up, the amount of power it produces goes down, and the drinks no longer ignite. The poker just fades away. A poker that lives in a busy bar and is turned on seven days a week from 6 p.m. to 1 or 2 a.m. has a lifetime of about a month before it fades. At home they would last a lot longer, but I'll never sell them. Imagine the insurance. I'm not going to tell you how to make them, either, because I can't recommend that you do.

POKING TECHNIQUE

I always use regular pint glasses to make red-hot-poker drinks. For years I never had one shatter—they would crack if you made drink after drink without letting them cool, but they would never shatter—until one did. I now recommend holding the glass in a metal wire holder so you won't get burned if it breaks.

How long you burn the drink is a matter of preference. It is hard to describe good poking technique. Every bartender has to figure it out for him- or herself. Almost all bartenders who use the poker like it—it gives them the sense that they are really controlling something primal and creating new flavors rather than mixing preexisting ones.

HOW TO SERVE HOT LIQUOR

Serve hot liquor in wide-mouth cups, such as teacups and coffee cups, not in vessels with tall walls like a coffee mug. Tall cups should be avoided like the plague: they concentrate alcohol vapors and throw off your perception of the drink. A teacup with a wide surface area tends to help the alcohol vapor dissipate over a wide area, making it less concentrated at any one point. Drinks that smell fantastic in a teacup can be unbearable even to get near when poured into tall-walled glasses.

FLAVOR PROFILES FOR RED-HOT-POKER DRINKS

No matter what kind of poking technique you use, there are some flavor points you must keep in mind. When you are poking, add a bit more sugar than you would for a standard hot drink. While in general hot drinks require less sugar (sucrose) than cold ones do, because your perception of sucrose increases as temperature increases, poking caramelizes some of the sugar in the drink, adding nice flavor but reducing sweetness. If you are using a sweetener with a lot of fructose, such as agave, the sweetness will actually decrease as the temperature goes up, even if you don't burn any of it. Use more.

Some bitter notes, such as the hops in beer, are intensified by poking. When I use beer, I tend to use a less bitter type with some fruitiness—often an abbey ale—and I rarely use lagers. Bitters like Angostura work very well in poker drinks.

Most poker drinks benefit from aged spirits. Rarely does a poker drink contain only white liquor. Whiskey, dark rum, Cognac, brandy, and applejack—these are the bread and butter of hot drinks. Many liqueurs and amaros are also excellent modifiers or even main players in hot drinks. Fernet Branca, a digestif liquor that almost every bartender has a perverse love for but that I dislike with a passion, tastes good hot. Jägermeister may advise that its product should be served very cold, but I've made more than a few nice flaming Jägers in my time.

When acids are called for, use them sparingly. Lots of acid doesn't taste good hot. The most common acid used in poker drinks is lemon juice. We are accustomed to drinking lemon in hot tea, so it just seems to fit many of the drinks. Occasionally I will use hot lime, but heat intensifies the bitterness of lime, which can be offputting. Sour orange works well hot.

TWO RECIPES

If using an igniting poker, burn these recipes for 7 to 10 seconds. If you are using a nonigniting poker, boil for 15 to 20 seconds. Pan recipes follow poker recipes.

YOUR OTHER OPTIONS FOR HOT DRINKS

Assuming that (a) you are not going to make your own electric red-hot poker and (b) you are not going to go buy a soldering copper, what are your options? The best faux-poker technique I have come up with works only on recipes that contain simple syrup, and involves swapping the simple syrup for sugar you burn in a pan before you flambé the liquor (see the recipes). If you don't mind flames in the kitchen, this technique never fails to impress.

Burning sugar in a pan. Let it get dark but not too dark. The third photo is good; the fourth, burnt.

5) *Add and ignite the base spirit—there will be flames.* 6) *Quickly add lower-proof liquids.* 7) *Remove from heat, and stir to dissolve the caramelized sugar that will be stuck to the bottom of the pan.* 8) *Serve.*

RED-HOT ALE, POKER STYLE

Red-Hot Ale

This is an old-school style poker drink. Like your great-great-great-great-grandparents used to make.

POKER STYLE

MAKES ONE 4⅗-OUNCE (138-ML) DRINK AT 15.3% ALCOHOL BY VOLUME, 3.5 G/100 ML SUGAR, 0.33% ACID. FINISHED VOLUME AND ALCOHOL BY VOLUME VARY WITH POKING TIME.

INGREDIENTS

1 ounce (30 ml) Cognac

3 ounces (90 ml) malty but not hoppy abbey ale (I like Ommegang Abbey Ale for this recipe)

¼ ounce (7.5 ml) simple syrup

¼ ounce (7.5 ml) freshly strained lemon juice

3 dashes Rapid Orange Bitters (page 211) or commercial orange bitters

2 drops saline solution or a pinch of salt

1 orange twist

PROCEDURE

Mix everything but the orange peel and poke. Serve in a teacup and express the orange peel over the top.

PAN STYLE

MAKES ONE 4⅗-OUNCE (138-ML) DRINK AT 15.3% ALCOHOL BY VOLUME, SUGAR INCALCULABLE, 0.33% ACID. FINISHED VOLUME AND ALCOHOL BY VOLUME VARY WITH BURN TIME.

INGREDIENTS

2½ teaspoons (12 g) granulated sugar

1 ounce (30 ml) Cognac

3 dashes Rapid Orange Bitters (page 211) or commercial orange bitters

3 ounces (90 ml) abbey ale

¼ ounce (7.5 ml) freshly strained lemon juice

2 drops saline solution or a pinch of salt

1 orange twist

PROCEDURE

First add the sugar to a sauté pan. Heat over a high flame until the sugar begins to caramelize. You are looking for a particular shade of dark brown—dark but not yet burned. Immediately add the Cognac and let it flame by tipping the pan toward the fire, or by igniting it with a long-handled butane lighter if you are heating electrically. Be careful! Large flames will erupt. Allow the Cognac to burn for a few seconds and add the bitters while the pan is still flaming. Then add the ale and let the mixture extinguish. Finally, add the lemon juice and saline or salt and remove the pan from the heat. Stir with a spoon to dissolve the caramel (don't get burned by the steam). Serve in a teacup and express the orange peel over the top.

Red-Hot Cider

Another old-style cocktail, this time with cider.

POKER STYLE

MAKES ONE 4⅗-OUNCE (138-ML) DRINK AT 15.3 ALCOHOL BY VOLUME, 6.5 G/100 ML SUGAR, 0.31% ACID. FINISHED VOLUME AND ALCOHOL BY VOLUME VARY WITH POKING TIME.

INGREDIENTS

- 1 ounce (30 ml) apple brandy (50% alcohol by volume. I use Laird's Bottled in Bond)
- 3 ounces (90 ml) hard apple cider (use decent stuff; I usually use a Norman-style cider)
- ½ ounce (15 ml) simple syrup
- ¼ ounce (7.5 ml) freshly strained lemon juice
- 2 dashes Rapid Orange Bitters (page 211) or commercial orange bitters
- 2 drops saline solution or a pinch of salt
- 1 cinnamon stick

PROCEDURE

Mix the liquid ingredients and poke with the cinnamon stick in the poking glass. Serve in a teacup with the cinnamon stick.

PAN STYLE

MAKES ONE 4⅗-OUNCE (138-ML) DRINK AT 15.3 ALCOHOL BY VOLUME, SUGAR INCALCULABLE, 0.31% ACID. FINISHED VOLUME AND ALCOHOL BY VOLUME VARY WITH BURN TIME.

INGREDIENTS

- 3 teaspoons (12.5 grams) granulated sugar
- 1 ounce (30 ml) apple brandy (50% alcohol by volume)
- 1 cinnamon stick
- 2 dashes Rapid Orange Bitters (page 211) or commercial orange bitters
- 3 ounces (90 ml) hard apple cider
- ¼ ounce (7.5 ml) freshly strained lemon juice
- 2 drops saline solution or a pinch of salt
- 1 orange peel

PROCEDURE

Put the sugar in a sauté pan over high heat. Heat until the sugar turns brown, but don't let it burn. Add the brandy and the cinnamon stick and ignite. Allow to burn for a few seconds (caution!). Add the bitters while it is burning, then add the cider. Finally add the lemon juice and saline or salt, then remove from the heat. Stir with a spoon to dissolve the caramel (don't get burned by the steam). Serve in a teacup and express the orange peel over the top.

Rapid Infusions, Shifting Pressure

Infusion refers to two intertwined processes. It can mean extracting the flavor of a solid into a liquid or impregnating a solid with the flavor of a liquid—or both. When you make coffee, the process is one-way: the coffee liquid is good, the spent grounds are not. Making coffee is about extracting the flavor of the coffee grounds into the liquid. When you brandy cherries, you get two delicious products, brandy-flavored cherries and cherry-

ISI WHIPPERS

flavored brandy. When you infuse, you're basically making your own new liquor with a distinct flavor. I like to use infusions in simple, uncomplicated drinks that highlight that flavor.

The infusion process can take seconds or minutes, or it can take days or weeks, depending on how it is done. Most traditional infusions in the cocktail world take days or weeks to complete. Scads of books have been written about traditional infusion techniques, and I commend some of them to you in the bibliography. Here we will focus on rapid infusion, which takes seconds or minutes to complete.

Rapid infusion is a pair of modern techniques that work by manipulating pressure and usually rely on one of two pieces of kitchen equipment: the iSi cream whipper and the chamber vacuum machine. These devices work in opposite ways; they are like mirror images. The iSi whipper increases pressure, then returns it to atmospheric pressure; while the vacuum machine reduces pressure and then returns it to atmospheric pressure. Either tool can make either kind of infusion, but the iSi is the champ for making liquid infusions, and the vacuum machine shines at creating infused solids.

I'll discuss the iSi first, because in the bar, liquid infusions are much more important than infused solids. Liquors and bitters made by infusion can be the stars of a cocktail. Infused solids are typically relegated to garnish status.

Rapid Nitrous Infusion with the iSi Whipper

RAPID NITROUS INFUSION

Nitrous infusion is a technique I developed using the cream whipper and nitrous oxide, a water/ethanol/fat-soluble, sweet-tasting anesthetic gas. You might know it as laughing gas or N_2O. In rapid infusion, nitrous dissolves into liquid and forces that liquid into a solid under pressure, where it can extract the flavors from the solid. After a short time you release the pressure and the nitrous boils, pushing the newly flavored liquid out of the solid.

RAPID NITROUS INFUSIONS
VERSUS TRADITIONAL INFUSIONS

Rapid infusion is neither better nor worse than traditional long-term infusion, just different. Rapid infusion tends to extract bitter, spicy, and tannic components less than long-term infusion does. If you have a product that is too bitter, too spicy, or too tannic for you, go rapid. Rapidly infused cocoa nib liquor will have less bitterness than one steeped for weeks, and will require less sugar. Rapidly infused jalapeño tequila will have more jalapeño flavor and less spiciness than slowly infused jalapeño tequila will. A consequence of this flavor shift is that rapid infusion tends to extract less total flavor from a given amount of solid ingredient, so rapid infusions call for larger quantities of solid ingredients than traditional ones do.

Before we begin, a word about the primary piece of equipment we'll be using in this section, the whipper.

THE ISI CREAM WHIPPER: ITS DESIGN AND USES

The Austrian company iSi pioneered the whipped cream whipper. It is a stainless-steel vessel with a screw-on top that you can pressurize with gas cartridges (chargers) through a one-way valve in the top, and you can release pressure—or dispense whipped cream—using a trigger valve and nozzle, also in the top. You can use either carbon dioxide or nitrous oxide chargers in the whipper, but for whipped cream and infusions, you use only nitrous oxide (carbon dioxide makes everything taste carbonated). To make whipped cream, add cream to the whipper, screw the top shut, and then pressurize with N_2O. The N_2O will dissolve into the cream under high pressure. When you want to dispense whipped cream, you hold the whipper upside down and squeeze the trigger valve to spray the nitrous-charged cream out of the whipper. When the cream leaves the pressurized vessel, the nitrous expands, making whipped cream. Whippers are sized according to the amount of cream they hold. The two most common sizes are the half-liter whipper (actual volume 772 ml) and the liter whipper (1262 ml).

The culinary significance of the cream whipper got a major bump in the late nineties when Catalan chef Ferran Adrià started using it at his restaurant, El Bulli, to make all sorts of foams—one of the earliest rumblings of the modern cuisine movement. Soon anyone worth his or her

Make sure the inside of the whipper head is clean.

modern-cooking salt had a whipper, one of the few pieces of kitchen tech equipment that almost everyone could afford. I had a couple, but I rarely used them. It turned out that I really, really dislike most foams. They are typically underflavored and misapplied. Done right, foams are great—but they are rarely done right.

My whippers sat around gathering dust until that life-changing day a number of years ago when, as I sat pondering the mechanics of infusion, I theorized that my cream whipper could make excellent rapid infusions. I was right, and now my whippers are always dust-free.

THE TECHNIQUE OF NITROUS INFUSION IN A NUTSHELL (WITH SOME TIPS TO AVOID EMBARRASSMENT)

Before you start, make sure you have all your ingredients, some empty containers, a strainer and a timer, and enough nitrous chargers. Now examine your whipper. Is the gasket in place? If not, the whipper won't seal and you'll lose your gas (I've been embarrassed more than once by not checking). Verify that the valve on the inside of the whipper head is clean. If your valve is dirty (as it often is when you make infusions one after the other), the whipper won't seal: more embarrassment. Finally, smell the whipper. If some joker screwed the whipper shut when it was wet, it will fester and stink to high heaven. Enjoy getting that stink out and keep reminding yourself never to store a whipper screwed shut.

Now we can get down to business. Let's say you want to extract the flavor of turmeric into gin. Turmeric is a good choice because it's porous, aromatic, colorful, and flavorful, the four characteristics you should look for in your solid infusion ingredient. Gin is a good choice because it's strong in alcohol, clean-flavored (so it will marry well with our turmeric), and clear (so it won't mess with turmeric's awesome hue). First put the turmeric in the whipper, then pour in the gin. Seal the whipper and charge it with nitrous. If your recipe calls for two chargers, as this one does, shake the whipper for a couple seconds and then quickly add the second charger. Right now the pressure inside the whipper will be as high as 360 psi (24.8 bars)! Shake the whipper. As you shake, the gas will dissolve into the liquid, and the pressure inside the whipper will drop down to between 72 and 145 psi (5 and 10 bars), depending on the recipe and the size of the whipper. The pressure will also force the gin and the nitrous oxide solution into the turmeric.

Now you wait, but not long—usually just 1 to 5 minutes; 2½ in this case. Set your timer. Rapid infusions happen so fast that even 15 seconds' difference in infusion times can make a taste impact. While you wait, the gin that was forced into the turmeric is rapidly extracting flavors from it. I tend to shake the whipper every once in a while when I'm waiting. I don't know if it helps, but I'm a fidgety guy.

When the infusion time is up, point the whipper straight up and vent it as fast as you can. Hold a container over the nozzle to catch anything that sprays out. As you vent, the nitrous expands and boils out of the solution, forcing turmeric-flavored gin out of the turmeric and back into the rest of the gin, completing the infusion. The whipper should vent completely. If the valve gets clogged with particles during venting, you'll know it. If that happens, pump the valve violently to unclog it. As a last resort, hold the whipper over a container and carefully unscrew it. iSi has machined a pressure release into the threads of the lid so that fluid should flow down the edge of the whipper rather than blowing the lid off the top—one reason to go name brand.

After the pressure has vented, unscrew the top of the whipper and listen. You will hear bubbling as the nitrous continues to come out of the solution and push flavors out of the turmeric. When the bubbling dies down, strain the gin, and you are done. I don't know why, but the flavor seems to change, often getting stronger, for a couple of minutes after you strain, possibly because of nitrous leaving the liquor. I usually rest an infusion for 10 minutes before I use it.

Recently iSi has released an attachment for its whippers to make the infusion process easier—specifically to address the problems of messy venting and valve clogging. The technique is the same whether you go old-school or use the new attachments.

If your whipper clogs and you can't unclog it by pulsing the vent valve, you can unscrew the lid as a last resort. This process is messy, and can be quite difficult if the whipper is under full pressure.

NITROUS INFUSION VARIABLES AND HOW TO CONTROL THEM

CHOOSING THE SOLID

Even when your goal is to extract the flavor of a solid into a liquid, infusion always starts the other way around: infusing a liquid into a solid. Choosing the right solid is important. Any candidate for rapid infusion must be porous—it must have air holes in it. Coffee, cocoa nibs, galangal, ginger, peppers—most plant products, in fact—have pores and produce good infusions. During the rapid-infusion process, the pressure of the N_2O injects those pores with liquid. The larger and more numerous the pores in your solid, the more liquid you can inject into it and the more flavor you can get out.

Pores are only useful if the liquid can get to them, so some ingredient preparation is usually necessary. Cutting ingredients into thin slices or grinding them creates more surface area and makes more pores available to the liquid. Exactly how thinly you slice or finely you grind your ingredients will have a major impact on the flavor you achieve with rapid infusion, much more so than with traditional long-term infusions. Don't overlook the importance of consistent cutting and grinding when you work rapidly.

SOLID-TO-LIQUID RATIOS

Rapid infusions usually require more solids than traditional infusions, because rapid infusion extracts fewer flavor components from an ingredient than traditional infusion does. That sounds bad, but it isn't. When I'm doing rapid infusion, I'm doing it because I want to decrease the extraction ratio of a particular flavor I don't like—say, the detergent note of lemongrass. If I waited for the rapid infusion to extract as much flavor as a traditional infusion would, I'd also have that detergent note. Instead, I do a shorter infusion with a lot of lemongrass. Look at the ratios in the recipes below and you will get an idea of how much of a particular ingredient to use.

TEMPERATURE

Temperature drastically affects infusion rates. You'll perform most rapid infusions at room temperature. Room temperature doesn't alter the flavors of ingredients by cooking them and is fairly consistent. Cold temperatures are bad. If you infuse cold, the rate of infusion is lowered *and* there isn't as much bubbling during venting, so you draw out much less flavor. Elevated temperatures produce a double whammy in the opposite direction: lots of fast flavor extraction and lots of bubbling, usually too much. However, when I want to extract a lot of flavor, including bitter notes, I'll heat the infusions up (see the section on Rapid Bitters, page 210). Because the iSi whipper is sealed, volatile aromas are not lost when you heat it, unlike heating an infusion in a pot. Just make sure to cool the whipper before you vent.

PRESSURE

The pressure inside the whipper is a very important variable. I almost always want as much pressure as I can get. Higher pressure equals faster, more balanced infusions. You'd think that you could lower the infusion pressure and just infuse longer, but the results are almost never the same. The speed of infusion is vital to the balance of flavor. The higher the pressure and therefore the faster the flavor transfer rate, the more you preferentially select for aromas, top notes, fruitiness, and the more you downplay bitterness and muddiness. I don't know why this is true, but it is.

If you want consistent results with a rapid infusion recipe, you must generate the same pressure in the whipper time after time. You'll need to follow recipes exactly, including the amount of ingredients you add, the size of the whipper you use, and the number of chargers you use. All those factors affect pressure, as we'll see here.

ISI PRESSURE CHART

ALL TESTS AT 20°C. CHARGERS STANDARD 10ML, 7.5 GRAM N$_2$O. PERFORMED BY ISI AUSTRIA 7/18/2013				PRESSURE IN BOTTLE FROM 1ST CHARGER				PRESSURE IN BOTTLE FROM 2ND CHARGER			
				INITIAL		AFTER SHAKING 10X		INITIAL		AFTER SHAKING 10X	
SIZE	LIQUID VOLUME	HEAD SPACE VOLUME	LIQUID	PSI	BAR	PSI	BAR	PSI	BAR	PSI	BAR
half-liter	500	272	water	197	13.6	119	8.2	ISI DOES NOT RECOMMEND			
liter	1000	262	water	200	13.8	77	5.3	260	17.9	14.8	10.2
liter	1000	262	water	197	13.6	without shaking add 2nd charger		354	24.4	161	11.1
half-liter	500	272	oil	186	12.8	94	6.5	ISI DOES NOT RECOMMEND			
liter	1000	262	oil	186	12.8	74	5.1	247	17	106	7.3
half-liter	500	272	40% ethanol	181	12.5	100	6.9	ISI DOES NOT RECOMMEND			
liter	1000	262	40% ethanol	175	12.1	68	4.7	236	16.3	135	9.3
half-liter	500	272	90% ethanol	173	11.9	44	3.0	ISI DOES NOT RECOMMEND			
liter	1000	262	90% ethanol	171	11.8	25	1.7	196	13.5	49	3.4

iSi will not endorse the practice of loading two chargers into a half-liter whipper unless the whipper is fully vented before the second charger is applied, even though the two-charge practice is 100 percent safe. iSi's whippers can handle many times the actual pressure involved and have a safety mechanism that prevents them from charging to dangerous levels. Even if that safety mechanism fails, the initial failure of an iSi whipper under extreme pressure is a bulging-out at the bottom of the bottle—not catastrophic. Why does iSi take this stand on using two chargers? Apparently, because cheaper, no-name whippers do not have safety pressure releases like the iSi whippers do, and they can pose a threat of explosion if overcharged. The French government has specific use guidelines regarding cream whippers that apply to all whippers, regardless of brand, and iSi, as an EU-based company, adheres to those guidelines worldwide.

I don't, and you don't have to either. Just don't go over pressure with no-name units. Aside from the possible dangers, I've had nothing but bad luck. They tend to leak a lot.

For two-charger recipes, iSi recommends that you charge once, shake, fully vent the whipper, then add the second charger. This technique produces a lower ultimate pressure than my preferred two-charger approach, but it's still better than just one.

The pressure in the whipper is generated by the gas chargers you add. Each charger contains 7.5 grams of nitrous oxide. You can add one charger or two, sometimes three. How much pressure those chargers generate depends on three factors. The size of the whipper is the first determinant. A liter whipper at capacity will have a much lower pressure per charger than a half-liter whipper, because the gas has more volume to fill. Second, the quantity of product in the whipper directly influences the final pressure, because it takes up volume. Full whippers produce higher pressures than empty whippers, by a good margin. A half-liter whipper with 500 milliliters of 40 percent alcohol by volume vodka will have a one-charger pressure of 100 psi (6.9 bars). An empty half-liter whipper with one charger only has a pressure of 78 psi (5.4 bars). And solids in a whipper take up space just as liquids do, so increasing the solids also increases the pressure. Last, the type of liquid you use affects the pressure. Higher-ethanol concentrations produce lower pressures because N_2O is more soluble in ethanol than in water. For the pressures inside iSi whippers under various conditions, see the iSi Pressure Chart on page 195.

Last note on pressure: though iSi doesn't recommend putting two chargers into a half-liter whipper, or three chargers into a 1-liter whipper, both conditions are totally safe. For the reasons, see the sidebar on iSi Pressure Safety.

TIME

Rapid bitters and tinctures infuse from 5 minutes to 1 hour, and with these recipes your timing can be a bit sloppy. When you are making infused liquors, by contrast, infusion time is usually between just 1 and 2 minutes, so it's vital that you get your timing just right. When an entire infusion lasts only a minute or two, seconds matter. Use a timer. Precision is especially crucial on strong bitter ingredients such as coffee and cacao nibs, because the line between an infusion that is too weak and one that is too bitter is very thin.

VENTING

Venting is the process of squeezing the trigger and releasing the pressure in the iSi, allowing the liquid inside to return to atmospheric pressure. Venting is what generates flavor-extracting bubbles. Vent fast. The faster you vent, the more violent the bubbling and the more flavor you extract. Liquid often sprays out of the whipper when you are venting; that's okay. It's good, in fact.

Problems arise when small particles clog the valve and the whipper stops venting. You have to make sure you vent all the way, and in a timely fashion. As long as the whipper is under pressure it is still rapidly infusing. If you get hung up while venting, your infusion time increases. Worse, partial venting doesn't create as violent a bubbling effect as full-fast venting. You are bound to clog the whipper from time to time, but to help prevent clogs, don't agitate the infusion just prior to venting. Allow the particles to sink to the bottom. Most important, aim the whipper up while you vent so you aren't pouring particles through the valve. You'll know you have a clog on your hands when the rush of gas and foam stops suddenly instead of gradually petering out. Vigorously pump the handle to dislodge the clog. As a last resort, open the whipper slowly under pressure—but do this over a large bowl.

Don't strain the mixture right away after you vent unless the recipe tells you to. Bubbles are still extracting flavor. Remember to listen to them—they are the sound of infusion.

Now some recipes.

TOP: *Rapid venting is what you want.*
BOTTOM: *Listen to the bubbles.*

Rapid Liquors and Cocktails

All the recipes in this section are written for half-liter whippers. As I mentioned above, you cannot change the volumes of these recipes without changing the results. If you wish to use a liter-sized whipper, you can double the recipes and add an extra charger (make sure to shake between adding chargers). The results will not be exactly the same but will still be good. If you like the results of the scaled-up recipe, make it that way all the time. If the infusion is too strong, reduce time or omit the extra charger. If the recipe is too weak, increase the infusion time.

The Glo-Sour and Turmeric Gin

Many people are interested in turmeric these days because they've heard tell that it's a health-bringing superspice. I couldn't care less about that, but it tastes great and has a fantastic color. Most people are accustomed to working with powdered dry turmeric; this recipe requires the fresh stuff. Turmeric is a rhizome, and the rhizomes look like little juicy orange ginger roots with a papery brown skin. You can get fresh turmeric in any store that caters to an Indian clientele.

Traditional slow turmeric infusions are difficult to work with: they are funky and have weird aftertastes. Rapid iSi makes a much more balanced infusion. The drink is bright orange and very refreshing. Speaking of bright orange, turmeric stains surfaces indelibly. Do not peel turmeric; wash it under running water, wearing gloves. Wanna mess with someone? Have him peel some turmeric with his bare hands. They'll be orange until the skin wears off. Leave your gloves on while you cut the turmeric, and do not cut fresh turmeric on any surface you care about. I wrap my cutting board with plastic and put a piece of folded paper on top of that. Store this liquor in glass, as plastic containers will be ruined. If stored in the fridge, this infusion tastes best within about a week, after which the flavors begin to go flat and the color will dull.

The cocktail I've included for this infusion is a lime sour riff. For the bitters I call for, you can use a commercial variety or make the Rapid Orange Bitters. If you omit the bitters altogether, the resulting cocktail will still be good but will be missing that certain something. When light hits this drink it almost glows, hence Glo-Sour.

TURMERIC GIN

INGREDIENTS FOR TURMERIC GIN

500 ml Plymouth gin

100 grams fresh turmeric, thinly sliced into 1.6 mm (¹⁄₁₆-inch) disks

EQUIPMENT

Two 7.5-gram N₂O chargers

PROCEDURE

Rapid infuse with the chargers for 2 minutes 30 seconds (shake between adding chargers). Vent, and listen to the bubbles. Allow the bubbling to mellow out for a couple of minutes, then strain. Press on the turmeric to extract the majority of the gin. (Remember to wear gloves; this liquor will stain.) Rest 10 minutes.

YIELD: 94% (470 ML)

INGREDIENTS FOR GLO-SOUR

MAKES ONE 5¹⁄₃-OUNCE (160-ML) SHAKEN DRINK AT 15.9% ALCOHOL BY VOLUME, 8.0 G/100 ML SUGAR, 0.84% ACID

2 ounces (60 ml) Turmeric Gin

¾ ounce (22.5 ml) freshly strained lime juice

Flat ¾ ounce (20 ml) simple syrup

3 drops saline solution or a generous pinch of salt

1–2 dashes Rapid Orange Bitters (page 211)

PROCEDURE

Combine the ingredients in a cocktail shaker, add ice, shake, and strain into a chilled coupe glass.

GLO-SOUR

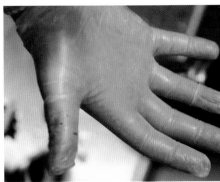
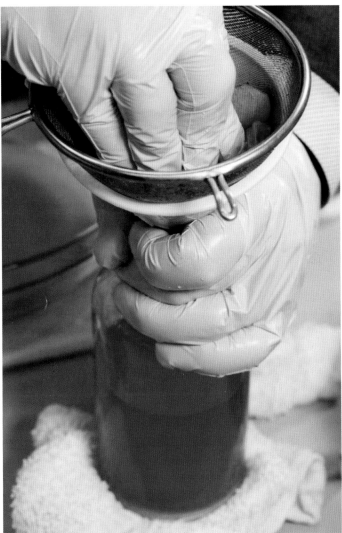

Turmeric is a brightly colored rhizome. Careful—it will stain everything indelibly. Wear gloves and wrap your cutting boards with plastic.

CHOPPED LEMONGRASS

The Lemon Pepper Fizz and Lemongrass Vodka

This drink is very refreshing. Lemongrass is sometimes difficult to work with because long infusions bring out a detergentlike flavor. My rapid iSi technique eliminates this problem. Note that this recipe asks for a lot of lemongrass: it's the only way to get the spirit to balance the way I like it to. The whipper will be almost full of lemongrass. The lemongrass infusion should be used fresh—within a day or two. Keep it in the freezer and it should last a week.

I carbonate this cocktail to highlight the freshness of the lemongrass. For more information on carbonation, see the section on bubbles (page 288). Although I use a carbonation rig to add bubbles, you could use a Sodastream, or in a pinch the same whipper you used for infusing. The cocktail calls for Black Pepper Tincture and clarified lemon juice. If that's too big a hassle for you, carbonate the base cocktail and then add regular lemon juice and a couple grinds of fresh black pepper. I won't tell.

INGREDIENTS FOR LEMONGRASS VODKA

300 ml vodka (40% alcohol by volume)

180 grams fresh lemongrass, sliced into disks

EQUIPMENT

Two 7.5-gram N$_2$O chargers

PROCEDURE

Combine the ingredients in a half-liter whipper. Charge with one charger, shake, then add the second charger and shake. Total infusion time is 2 minutes. Vent and allow the bubbling to stop, then strain. Press hard on the lemongrass to extract all the vodka. Allow to rest for 10 minutes before using.

YIELD: 90% (270 ML)

INGREDIENTS FOR LEMON PEPPER FIZZ

MAKES ONE 5½-OUNCE (166-ML) CARBONATED DRINK AT 14.3% ALCOHOL BY VOLUME, 7.1 G/100 ML SUGAR, 0.43% ACID

58.5 ml Lemongrass Vodka

12 ml clarified lemon juice (or regular lemon juice added at the end)

18.75 ml simple syrup

Dash Rapid Black Pepper Tincture (page 215)

2 drops saline solution or a pinch of salt

76 ml filtered water

PROCEDURE

Combine all the ingredients, chill to -10°C, and carbonate. Serve in a chilled champagne flute.

Café Touba and Coffee Zacapa

This drink is based on Café Touba, the famous hot coffee drink from Senegal. I discovered café Touba on a chef education trip I took to that country. The devotees of the Mourides, a Senegalese Muslim brotherhood, are famous for their small cups of sweetened frothy coffee spiked with djer, a peppery West African spice that comes from the *Xylopia aethiopica* tree. In English, djer is called grains of Selim, and it is difficult to get unless you have access to a West African grocery. Djer fulfills a similar role in Café Touba that cardamom fulfills in some versions of Middle Eastern coffee. If you can't get it, you can substitute cardamom and cubebs, but the flavor will be different. Like its inspiration, this cocktail is frothy, but it's also cold and alcoholic. The real Café Touba—remember, it's a Muslim favorite—contains no alcohol.

The cocktail's base is freshly ground coffee infused into dark rum. Rapid coffee infusions are fantastic but tricky. The size of the grind really alters the infusion results. My first experiments with rapid coffee infusions used an espresso grind, and the results were great—when they were great. My infusions were plagued with inconsistency. When I used my own grinder, I could get repeatable results, but when I did demos away from home and had other people grind the coffee for me, the results were erratic. The problems were especially bad if the grind was too fine. Not only would the liquor be way too bitter, but it would clog my filters, making straining difficult. Now I use a grind slightly finer than drip coffee, and my results are more consistent and filtration is much easier.

You must use freshly ground coffee for this recipe—preferably coffee that is just minutes old. I use Ron Zacapa 23 Solera as the rum base for this liquor because it has a well-balanced, unfunky nose and because you need an aged rum to stand up to and complement the coffee. Any full-bodied aged rum that isn't too sweet-smelling or funky will work.

Note that I overinfuse this liquor so that it tastes too much like coffee, and then I dilute it with unmodified rum. I overinfuse for two reasons: I can make a full 750 ml batch of Coffee Zacapa in a 500 ml whipper, and I can fine-tune the coffee flavor to taste. I use a technique called milk washing in this recipe, which removes some of the harshness of the coffee and adds whey protein (see the section on washing liquor, page 265). The whey protein from the milk washing gives the Café Touba cocktail its frothy head. You can omit the milk-washing step and instead add a ½ ounce (15 ml) cream to the drink right before you shake it.

*TOP: Café Touba. **MIDDLE:** Djer—the characteristic spice in Senegalese Café Touba—is known in English as grains of Selim. **BOTTOM:** This is how fine you should grind your coffee.*

INGREDIENTS FOR COFFEE ZACAPA

100 grams whole fresh coffee beans, roasted on the darker side

750 ml Ron Zacapa 23 Solera or other aged rum, divided into a 500 ml portion and a 250 ml portion

100 ml filtered water

185 ml whole milk (optional)

EQUIPMENT

Two 7.5-gram N$_2$O chargers

PROCEDURE

Grind the coffee in a spice grinder until it is slightly finer than drip grind. Combine 500 ml (16^2⁄$_3$ ounces) rum and the coffee in a half-liter whipper, charge with one charger, shake, and then add the second charger. Shake for another 30 seconds. Total infusion time should be 1 minute 15 seconds. Vent. Unlike with most infusions, don't wait for the bubbling to stop. If you do, the liquor will be overinfused. Instead, rest for just 1 minute, then pour through a fine cloth into a coffee filter. A coffee filter alone will clog too quickly. The mixture should filter within 2 minutes. If not, your grind was too fine. Stir the drained grounds in the filter, then add water evenly over the grounds and let the water drip through (this is called sparging). That water will replace some of the rum that was trapped in the grounds during the infusion. The liquid that comes out of the grounds from sparging should be about 50 percent water and 50 percent rum.

At this point you should have lost roughly 100 ml (3^1⁄$_3$ ounces) of liquid to the grounds. About half of that lost liquid is water and the other half rum, so your final product has a slightly lower alcohol by volume than you started with.

Taste the infusion. If it is strong (which is good), add the additional 250 ml (8^1⁄$_3$ ounces) of rum to the liquor. If the infusion can't be toned down without losing the flavor of the coffee, your grind size was too coarse; don't add any more rum and reduce the amount of milk you use for milk washing to 120 ml (4 ounces).

If you are milk washing (good for you!), read the next paragraph. If you're not, skip ahead and start making the drink.

While stirring, add the Coffee Zacapa to the milk and not the other way around, lest the milk instantly curdle. Stop stirring and allow the mixture to curdle, which it should do within about 30 seconds. If it doesn't curdle, add a little 15% citric acid solution or lemon juice bit by bit until it does, and don't stir when it's curdling. Once the milk curdles, gently use a spoon to move the curds around without breaking them up. This step will help capture more of the casein from the milk and produce a clearer product. Allow the mix to stand overnight in the fridge in a round container; the curds will settle to the bottom and you can pour the clear liquor off the top. Strain the curds through a coffee filter to get the last of the liquor yield. Alternatively, spin the liquor in a centrifuge at 4000 g's for 10 minutes right after it curdles. That's what I do.

Approximate final alcohol by volume: 31%

YIELD: APPROXIMATELY 94% (470 ML)

INGREDIENTS FOR DJER SYRUP

400 grams filtered water

400 grams granulated sugar

15 grams djer (grains of Selim), or 9 grams green cardamom pods and 5 grams black pepper

PROCEDURE

Combine all the ingredients in a blender and blend on high till the sugar is dissolved. Strain through a fine strainer to get rid of the big particles, then through a moist paper towel or strainer bag (don't use a coffee filter; it will take forever).

CONTINUES

INGREDIENTS FOR CAFÉ TOUBA

MAKES ONE 3⁹⁄₁₀-OUNCE (115-ML) SHAKEN DRINK AT 16.1% ALCOHOL BY VOLUME, 8.0 G/100 ML SUGAR, 0.39% ACID

2 ounces (60 ml) Coffee Zacapa

½ ounce (15 ml) Djer Syrup

3 drops saline solution

½ ounce (15 ml) cream (if you have not milk-washed the rum)

PROCEDURE

Shake all the ingredients with ice in a cocktail shaker and strain into a chilled coupe glass. The drink should be creamy and frothy.

Jalapeño Tequila

Most hot-pepper infusions are high on spice but low on pepper flavor. Rapid iSi brings out more of the pepper flavor, yielding a more complex spirit. You'll need more pepper than you would for a traditional long-term infusion. I remove the veins and seeds from the peppers in this recipe; capsaicin, the compound that gives peppers their punch, is concentrated around them. I want some heat, but I also want a lot of aroma and flavor. To do that, I nix the seeds and up the quantity of jalapeños.

INGREDIENTS

45 grams green jalapeño pepper seeded, deveined, and very thinly sliced

500 ml blanco tequila (40% alcohol by volume)

EQUIPMENT

Two 7.5-gram N$_2$O chargers

PROCEDURE

Combine the ingredients in a half-liter whipper. Add one charger and shake; add the second charger and shake. Total infusion time should be 1 minute 30 seconds. Vent. Allow to rest 1 minute. Strain and press on the pepper slices to extract the majority of the tequila. Allow to rest 10 minutes before use.

YIELD: OVER 90%

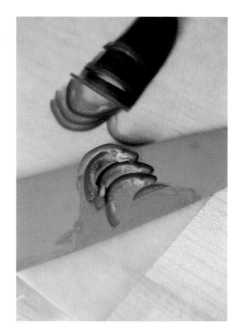

LEFT: Seed and devein jalapeños and slice them like this.
***BELOW**: Jalapeño tequila.*

*OPPOSITE: **1)** After you vent your whipper, you should strain the liquor through a cloth and then through a coffee filter. Then re-wet the grounds and filter again. **2)** Taste the liquor, **3)** it should require the addition of fresh rum. Pour the coffee liquor into milk while stirring. **4)** The milk will curdle. **5)** Gently stir the curds around till the liquid gets clearer and clearer then let the liquid settle. **6)** Strain.*

Schokozitrone and Chocolate Vodka

Cocoa nibs are porous and infuse quite well. The problem is that most cocoa nibs are crap. Companies often foist off the lowest-quality beans they have on unsuspecting nibs buyers. Never, ever use off-brand nibs for making infusions. Taste the nibs you have. Do they taste sour and acrid? Then your infusion will taste sour and acrid. I use only Valrhona cocoa nibs; the quality is always good.

As long as you can get good nibs, they are a much better bet for producing chocolate liquors than cocoa powder (messy and difficult to strain) or solid chocolate (nonporous) is. The chocolate vodka made with this recipe has a strong chocolate flavor but very little bitterness; you don't need to add a lot of sugar.

The cocktail I make with this liquor is a bit polarizing, because I add lemon. Some people just don't like citrus and chocolate. I do. I got the idea for this particular combination from my wife, Jennifer, who picked up the habit of enjoying chocolate ice cream with lemon sorbet while living in Frankfurt, Germany, as a teen. The drink has a couple dashes of chocolate bitters to round out the flavors. I give my recipe here, but you can substitute a commercial chocolate bitters or mole bitters, or omit the bitters altogether.

INGREDIENTS FOR CHOCOLATE VODKA

500 ml neutral vodka (40% alcohol by volume)

75 grams Valrhona cocoa nibs

EQUIPMENT

Two 7.5-gram N_2O chargers

PROCEDURE

Add the vodka and nibs to a half-liter cream whipper. Charge with one charger and swirl/shake for several seconds, then charge with an additional 7.5 grams of N_2O. Continue to agitate for a full minute. Let rest for 20 seconds, then vent and open the whipper. Allow the liquor to stay in the whipper for another minute or so, until the bubbling starts to subside. Strain out the nibs and pass the vodka through a coffee filter. Allow to rest several minutes before using. Chuck the leftover nibs—all that remains in them is the bitterness.

YIELD: OVER 85%, 14$\frac{1}{5}$ OUNCES (425 ML)

INGREDIENTS FOR SCHOKOZITRONE

**MAKES ONE 4⅓-OUNCE (128-ML) STIRRED DRINK AT
19.2% ALCOHOL BY VOLUME, 7.4 G/100 ML SUGAR, 0.70% ACID.**

2 ounces (60 ml) Chocolate Vodka

½ ounce (15 ml) freshly strained lemon juice

½ ounce (15 ml) simple syrup

2 dashes Rapid Chocolate Bitters (page 000)

2 drops saline solution or a pinch of salt

Candied ginger

PROCEDURE

Combine the vodka, lemon juice, simple syrup, bitters, and saline or salt in a mixing tin. Stir briefly with ice, strain into an old-fashioned glass, and garnish with the candied ginger. This recipe requires very little sugar, even though the cocoa nibs are unsweetened, because the infusion technique leaves the bitterness behind.

Rapid Bitters and Tinctures

What soy sauce is in China and ketchup is in America, bitters are in cocktails: ubiquitous. They are alcoholic infusions of aromatic herbs, spices, and peels with the addition of bittering agents—usually the bark, roots, or leaves of some fantastically bitter plant such as gentian, quassia, cinchona (from which quinine is derived), or wormwood. Bitters are a holdover from the days when people considered booze to be medicine. In their nineteenth-century heyday, bitters were patent medicines compounded by apothecaries from a variety of plant materials chosen for supposed medical or digestive benefits. Now people make bitters because they like the taste, and over the past decade there has been a bitters explosion. People recreate nineteenth-century recipes and compound new ones. Bartenders and home experimenters make their own bitters so their cocktails bear their unique stamp. Bitters are fun.

Traditional bitters can take weeks to produce, but with my rapid bitters technique you will have bitters in under an hour. In addition to being fast, rapid bitters have fantastic aromatic qualities. Unlike most nitrous infusion recipes, rapid bitters are sometimes heated and are infused for longer—20 minutes to an hour—because we *want* to extract bitterness and we want concentrated flavors. Bitters are powerful, so they are best tasted by the drop.

Unlike bitters, which are a mélange of flavors, tinctures are strong concentrated infusions of a single flavor. Like bitters, they are used by the drop and dash. I rarely heat tinctures, because I am often using ingredients whose flavors are adversely affected by heat.

Because traditional bitters and tinctures take so long to make, recipe development takes a long time—months. With rapid infusion you can cycle through ten iterations of a recipe in *one day*. I hope you will use the recipes below as a template to develop your own. These recipes are written for half-liter whippers.

INGREDIENTS FOR RAPID ORANGE BITTERS: *At center: cloves. Inner circle clockwise from bottom left: caraway seeds, cardamom seeds, gentian root and quassia bark. Outer arc from left to right: dried grapefruit peel, fresh orange peel, dried orange peel, and dried lemon peel.*

Rapid Orange Bitters

This is a very powerful and balanced aromatic orange bitters, which I use in the Glo-Sour recipe on page 200. Rapid infusion allows the aroma to be on a par with the bitterness. I use some fresh orange peel for brightness; a combination of dried Seville orange, lemon, and grapefruit peels brings a rounded pancitrus feeling. A tiny bit of cloves adds warmth. Cardamom and caraway, both good friends of citrus, lend their spiciness. Gentian root and quassia chips are classic bittering agents, putting the bitter in the bitters. I like this recipe a lot. Some of the ingredients may be hard to source locally. Here in New York we have at least two shops that specialize in dried spices, barks, roots, and plant materials. If you aren't so lucky, turn to the Internet.

As you make this recipe, don't be alarmed when you open the whipper and find almost no liquid inside. The peels will have inflated and absorbed almost all the booze. That's okay. You will press the bejeezus out of the peels to get your bitters. The yield is low, but the product goes a long way.

After these bitters are made, the pectin from the citrus rinds can sometimes form a weak gel. If you have Pectinex Ultra SP-L (an enzyme discussed in detail in the Clarification section, page 235), you can add a couple grams to break the gel down. If you don't have SP-L, just shake the bitters a bit and let them settle.

INGREDIENTS

0.2 grams whole cloves (3 cloves)

2.5 grams green cardamom seeds, removed from pod

2 grams caraway seeds

25 grams dried orange peel (preferably Seville)

25 grams dried lemon peel

25 grams dried grapefruit peel

5 grams dried gentian root

2.5 grams quassia bark

350 ml neutral vodka (40% alcohol by volume)

25 grams fresh orange peel (no pith, orange only)

EQUIPMENT

One 7.5-gram N_2O charger

PROCEDURE

If possible, crack the cloves, cardamom seeds, and caraway seeds before adding them and the rest of the dry ingredients to a blender and processing till everything is the size of peppercorns. Put the dry mix, the vodka, and the fresh orange peel in a half-liter iSi whipper and charge with the charger. Shake for 30 seconds. Leave the iSi under pressure and place it in a pan of simmering water for 20 minutes. Cool the whipper in ice water till it's at room temperature. Vent the whipper, open it, and look inside. The peels will have absorbed almost all the liquid. Do not despair. Put the solids in a nut milk bag, superbag (culinary straining bag), or cloth napkin and squeeze hard over a container. You should get 185 ml of bitters. You can increase yield by pressing harder, adding more liquid at the outset, or rewetting the peels. All of these will increase yield but reduce quality.

If necessary, pass through a coffee filter.

YIELD: 52% (185 ML)

MAKING RAPID ORANGE BITTERS: 1) *After you begin the infusion, put the sealed whipper in simmering water for 20 minutes, chill the whipper in ice water till it is at room temperature, then vent.* **2)** *Almost all the liquid will be absorbed.* **3)** *Squeeze the liquid out of the solids.* **4)** *The finished bitters are cloudy.* **5)** *If desired, add the enzyme Pectinex Ultra SP-L and let the bitters settle,* **6)** *then filter.*

Rapid Chocolate Bitters

Some drinks want a little drop of bitter chocolate flavor. These bitters are equally at home in tequila or mezcal recipes and in sour drinks with strong vegetal or herbal flavors. Drinks that you don't think want chocolate, like appletinis (yes, I make an appletini; see the section on apples, page 334), are helped by these bitters. I use them in the Schokozitrone cocktail, page 209.

You must use high-quality cocoa nibs for this recipe; I use Valrhona. I kept the recipe very simple—cocoa nibs and gentian for bitterness and mace for spice—which means it works in a wide variety of cocktails. These bitters aren't heated, since heating would make a mess of the nibs. This recipe takes longer than any of my other rapid recipes, clocking in at 1 hour.

INGREDIENTS

3.0 grams mace (3 whole)

350 ml neutral vodka (40% alcohol by volume)

100 grams Valrhona cocoa nibs

1.5 grams dried gentian

1.5 grams quassia bark

EQUIPMENT

Two 7.5-gram N_2O chargers

PROCEDURE

Break up the mace. Add everything to a half-liter iSi whipper and charge with one charger. Shake for several seconds and then add the second charger. Shake for 20 seconds more. Allow the mixture to infuse for 1 hour, shaking occasionally, then vent. It will bubble and foam a lot. Listen and wait for the bubbling to subside. Strain the liquid out and press on the solids to extract the most liquid. Pass the bitters through a coffee filter.

YIELD: 85% (298 ML)

INGREDIENTS FOR CHOCOLATE BITTERS (CLOCKWISE FROM TOP RIGHT): *Cocoa nibs, quassia bark, gentian, and mace.*

Rapid Hot Pepper Tincture

If you need to add heat to a drink, this is a good choice. It is a mix of red peppers, which bring fruitiness, and green peppers, which lend a vegetal note. To best extract the capsaicin—the main compound that provides the heat of peppers—I use food-grade 200-proof dehydrated alcohol (100% alcohol by volume). Capsaicin is much more soluble in alcohol than in water. I've tried the recipe using regular 40% alcohol by volume vodka and didn't like it as much: not enough punch. Unfortunately, pure food-grade ethanol is hard to find. You can substitute 195 proof (97.5% alcohol by volume), which you may be able to get at the liquor store, depending on your state's laws. After the infusion I add some water to reduce the proof and help rinse the infusion of the peppers.

If you store this tincture for more than several months, the heat level will stay the same but the aroma will change. The red fruity notes will fade and the green vegetal notes will predominate. Eventually the tincture will taste as if it had been made entirely with green peppers. It will still taste good.

ABOVE: *This is how peppers should be sliced for rapid hot pepper tinctures.*
LEFT: *I seed and devein peppers whenever I use them for cocktails.*

INGREDIENTS

8 grams red habanero peppers, seeded, deveined, and very finely sliced

52 grams red serrano peppers, seeded, deveined, and very finely sliced

140 grams green jalapeño peppers, seeded, deveined, and very finely sliced

250 ml pure ethanol (200 proof; 195 is fine)

100 ml filtered water

EQUIPMENT

Two 7.5-gram N_2O chargers

PROCEDURE

Combine the sliced peppers and alcohol in a half-liter whipper. Add one charger and shake; add the second charger and shake. Infuse for a total of 5 minutes, shaking occasionally. Vent. Add the water and allow to rest 1 minute (listen for bubbles). Strain the liquid and press on the peppers to extract liquid. Allow to rest 10 minutes before use.

YIELD: OVER 90% (315 ML)

Rapid Black Pepper Tincture

I like black pepper in drinks, but I don't like the way floating pepper looks or how hard pepper is to dose consistently. This tincture is a good way to go. I use a variety of peppers and near-peppers in this tincture, and it has quite a kick. The base is Malabar black pepper, grown in India and known for its pungency. On top of the Malabar I add Tellicherry pepper, another Indian peppercorn that is world-renowned. It has a more complex and interesting aroma than the Malabar but doesn't provide as much of a wallop. A small amount of dried green peppercorns brings some freshness. I also add a small amount of grains of paradise, which are not botanically related to black pepper but have been used like black pepper for centuries. Last I add cubebs, another pepper-similar spice, which was wildly popular in the Middle Ages. Unlike grains of paradise, cubebs are closely related to black pepper. They have an aromatic resiny note I like a lot. All these ingredients can be purchased at spice shops or online.

I use this tincture in the Lemon Pepper Fizz on page 203.

INGREDIENTS

200 ml neutral vodka (40% alcohol by volume)

15 grams Malabar black peppercorns

10 grams Tellicherry black peppercorns

5 grams green peppercorns

3 grams grains of paradise

2 grams cubebs

EQUIPMENT

Two 7.5-gram N_2O chargers

PROCEDURE

Pulse all the dry ingredients in a spice grinder. The pepper should remain somewhat coarse; the finer the grind, the fierier the infusion will be. Place the peppers and the alcohol in a half-liter whipper and rapid infuse with one charger. Shake for several seconds and then add the second charger. Shake for several seconds more and allow to infuse for 5 minutes. Vent and allow the bubbling to stop. Strain through a coffee filter and press out any extra liquid.

YIELD: 80% (160 ML)

LEFT: Black pepper tincture ingredients clockwise from top: grains of paradise, Malabar black peppercorns, Tellicherry black peppercorns, green peppercorns, and cubebs. **RIGHT:** Grind the pepper mixture to this texture. The finer the grind, the hotter the tincture will be.

Rapid Hops Tincture

Hops are the flowers that give beer its bitterness. You can get a bewildering array of hops at any home-brew shop. If you are a beer aficionado, you will no doubt already have your own favorite variety. Any cocktail that could use some of that beerlike punch will be well served by this tincture. I like it in grapefruit-juice-based cocktails, and also in plain old seltzer water.

The bitterness of hops comes from acids. The most important bitter acids are known as alpha acids, or humulones. Hops are typically boiled for a long time to extract those alpha acids and convert them into isohumulones through a chemical process called isomerization. It is those isohumulones that do the real bittering. Remember, for bitter you need boiling. The downside: hops' aroma comes from volatile essential oils that evaporate or alter when boiled. So to add back hop aroma, brewers add more hops after the boiling process.

You have three choices when making a hops tincture in an iSi. If you want a really bitter tincture, put the pressurized whipper in simmering water. Unfortunately, boiling the tincture destroys the fresh aroma of the hops. You'd think that the aroma would be preserved because the iSi whipper is a sealed system that prevents evaporation, but it isn't. Hot hops tinctures have that characteristic beery bitterness. To maximize hop aroma, you should make the tincture cold; that is your second option. I like cold hops tinctures a lot but don't expect them to add a lot of bitterness. Your third option—the one I recommend—is to do a dual infusion: infuse once hot, vent, and infuse again cold. It is a bit of a pain, but worth it.

You may use more than one hops variety in this recipe if you like—one for bitterness, one for aroma, or a combination of different hops for both purposes. I use just Simcoe, which has a very high bittering potential and an aroma that I like.

Luckily for the tincture maker (that is, you), brewers are very particular about the amount of bitterness they add to their beer, so all hops are codified by their quantity of bitter acids. Their specs are given in terms of percentages of alpha acids (the humulones) and beta acids. When comparing bittering potential for tinctures, you can safely ignore the beta acids and look only at the alpha. Simcoe hops have an alpha acid content of 12 to 14 percent—mega-super-high. Many well-known hops are in the 6 percent range. If you are using a less bitter hop, you should increase the amount of hops you use accordingly.

One last important note: do not expose this tincture to bright light. It is possible that UV light will cause it to get skunked owing to a photochemical reaction between hop acids, riboflavin (vitamin B), and trace amino acid impurities, which forms MBT (3-methylbut-2-ene-1-thiol). MBT is some superstinky stuff that you can detect in minute quantities. If you have ever had skunky beer, you know what I'm talking about. If you haven't, consider yourself lucky. Cold-infused hops tinctures are less susceptible to skunking than hot-infused hops, but either way, I store hops tinctures in stainless flasks.

INGREDIENTS FOR HOT OR COLD HOPS TINCTURE

250 ml neutral vodka (40% alcohol by volume)

15 grams fresh Simcoe hops

EQUIPMENT

Two 7.5-gram N_2O chargers

PROCEDURE

Add the vodka and hops to the whipper, charge with one charger, and shake for several seconds. Add the second charger and shake for an additional 30 seconds.

If making a hot tincture: Place the whipper in simmering water. Allow it to simmer for 30 minutes, then place it in ice water and cool it down to room temperature. The cooling should take 5 minutes.

If making a cold tincture: Allow to rest for 30 minutes, agitating occasionally.

Vent and allow the bubbling to stop. Strain through a coffee filter and press out any extra liquid. Store in a light-tight flask.

YIELD: 85% (212 ML)

INGREDIENTS FOR DUAL HOT AND COLD HOPS TINCTURE

30 grams fresh Simcoe hops divided into two 15 gram piles

300 ml neutral vodka (40% alcohol by volume)

EQUIPMENT

Three 7.5-gram N_2O chargers

PROCEDURE

Add 15 grams of hops and the vodka to a half-liter whipper, charge with a charger, and shake for several seconds. Place the whipper in simmering water and allow to simmer for 30 minutes. Transfer the whipper to ice water and cool it down to room temperature, about 5 minutes. Vent. Return any liquid expelled from the venting procedure to the whipper, add the remaining 15-gram pile of hops, charge with a charger, and shake for several seconds. Add the third charger and shake for an additional 30 seconds. Allow to rest for 30 minutes, agitating occasionally. Vent and allow the bubbling to stop. Strain through a coffee filter and press any extra liquid from the hops. Store in a light-tight flask.

YIELD: 85% (212 ML)

15 grams of Simcoe hops, a very bitter hop. You will need twice this amount if you opt for the dual hot and cold hops tincture.

SCALING UP

Cream whippers are limited to a capacity of 1 liter. What happens if you want to make a lot of infused liquor? You can use a Cornelius soda keg. Corny kegs (which are described in the Carbonation section, page 288) can hold up to 5 gallons of product and can easily withstand up to 100 psi (6.9 bars) of pressure. Most Corny kegs are rated for 130 psi (9 bars), but the pressure-relief valve often starts venting before you get there. Unlike with a cream whipper, into which you inject all the gas right away and with which you lower the pressure by shaking, with a Corny keg you supply a constant pressure—an advantage. And because your gas pressure is constant, it doesn't matter whether you fill the keg to the same level each time. Unfortunately, most people who use Corny kegs for infusion will be limited to CO_2 for their infusion gas, because N_2O is very hard to get in tank form. Suppliers are worried that people will abuse it as a drug and potentially asphyxiate themselves.

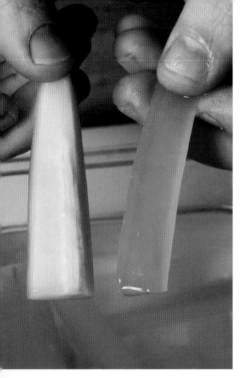

TOP: *Before infusion, on left, and after, on right. The difference is stark.*
BOTTOM: *Pineapple core garnishes: converting garbage into something great.*

Vacuum-Infused Solids: Garnish Magic

If you remember, back at the beginning of this chapter I said that infusion is really two different processes, infusing liquids into solids and getting the flavors out of solids into liquids. The latter process is what we have been talking about, using nitrous infusion. Here I introduce you to infusing liquids into solids using vacuum infusion.

When you vacuum-infuse, you surround a solid ingredient—say a slice of cucumber—with a flavorful liquid—say gin—and then use a vacuum to suck the air out of the pores of the cucumber. When you release the vacuum, the air can't rush back into those pores because they are surrounded by gin, and that gin can't get out of the way. Instead the atmosphere's pressure pushes the gin into those empty pores. This process can make for some delicious and beautiful results. Look at a slice of cucumber. It is whitish green, with a satiny surface. If you look closely, you can see that the whiteness possesses a certain granularity; those are the pores in the cucumber. As light hits those pores, it bounces around every which way, and some of it gets scattered back at you. The pores are making the cucumber opaque. When you inject those pores with liquid, they no longer scatter light, and the cucumber becomes translucent and jewel-like. For more on this effect, see the sidebar on compression, page 227.

One of the coolest things about vacuum infusion is its ability to turn a product's faults or inedible parts into assets. Watermelon rinds, a typical throwaway, become excellent cocktail garnishes when you slice the rind thin and long with a vegetable peeler and infuse it with a mixture of lime acid and simple syrup. Pineapple cores are garbage . . . but if you slice the core into long spears and vacuum-infuse them with simple syrup, they become a welcome change from the stick of sugarcane in a mojito. The firm structure of an underripe pear, a liability for eating out of hand, is a benefit for vacuum infusion, and the pear is exceedingly good when infused with a strong mixture of yuzu juice and elderflower syrup.

In the cocktail world, vacuum infusion is mostly used for garnishes, with some notable exceptions. Because you can infuse booze into solid ingredients, you can create foods that straddle the line between comestibles and cocktails. My Cucumber Martini, an "edible cocktail" of cucumber infused with gin and vermouth, was one of the first tech recipes I got attention for (see the full recipe on page 228). I came up with the idea in 2006 while teaching culinary technology at the French Culinary Institute in New York City. I wasn't allowed to serve liquor to the students, but I *was* allowed to "cook" with liquor.

VACUUM MACHINES

Vacuum infusion is preposterously simple, provided you have a decent vacuum machine, but professional chamber vacuum machines are expensive. Luckily, you don't need to break the bank to try vacuum infusion. You can also get okay (though not great) results with a hand vacuum pump or with a Vacu Vin wine saver. I've even developed a technique for infusing solids with the iSi whipper; see page 190. If you want a professional level of vacuum on the cheap and are handy, you can whomp up a good vacuum-infusion rig using a refrigeration-repair technician's vacuum pump; they are pretty inexpensive and can suck a decent vacuum.

Commercial chamber vacuum machine.

If you are fortunate enough to have a professional chamber vacuum machine, a piece of equipment designed to seal bags for storage and cooking, it is by far the best choice for vacuum infusion. The bagging function isn't key to infusion; what's vital is a good, strong vacuum pump. These machines start at about $1500. Commercial vacuum machines are expensive because they have really good vacuum pumps, and these pumps are pricey. The pumps in a commercial machine are sealed with oil and can achieve a fairly high level of vacuum (over 99 percent of the air removed, or 10 millibars) in just a few seconds, and withstand the constant use and abuse of the professional kitchen.

A new generation of sub-$1000 vacuum machines is on the market now. They all have weaker vacuum pumps than the pro models, but I'm sure some of them will do a decent infusion job. Really inexpensive electric vacuums like the FoodSaver don't have the strength to do a proper infusion.

WATERMELON RIND

MAKING A WATERMELON RIND GARNISH: 1, 2, and 3) When you cut up your watermelon, leave the rind in rings. **4)** Peel the deep green outer rind off the rings. **5)** Place your peeler on the ring and start peeling in circles.

6) Keep going—you want a long, continuous, unbroken strip. *7)* To get at the strip without cutting you will need to disassemble your peeler. *8)* Finally, vacuum infuse with All-Purpose Sweet-and-Sour. (See recipe, page 232.)

CHOOSING THE INGREDIENTS

Porosity: Just as with nitrous infusion, any candidate for vacuum infusion must be porous—it must have air holes in it. During the vacuum infusion process you will fill those pores with liquid. The larger and more numerous the pores in your solid, the more liquid you can inject into it and the more flavor your solid will acquire. Watermelon, which has boatloads of air space in it, soaks up liquids like a sponge. Apples have plenty of air holes in their flesh, but those holes are smaller and more difficult to penetrate, so they are more difficult to infuse with flavor.

Not only must your ingredient have pores, but those pores must also be available to the infusing liquid, which means it's difficult to infuse through skins, which are typically less porous than fruit flesh. Unpeeled cherry tomatoes pick up no flavor from vacuum infusion; peel them and they are infusion champions. Additionally, cutting your ingredients thinner makes it easier to infuse them, because it makes more pores available to the liquid.

Since vacuum infusion simply fills up existing pores, it never adds a large volume of liquid to your solid. The pores never constitute a large percentage of the whole. Because the volume of liquid you inject is small, the flavor must be powerful to make an impact on the flavor of the solid. The less porous your ingredient is, the more powerfully flavored your liquid needs to be. Using underflavored liquids is a big rookie mistake.

Candidates for vacuum infusion should not be too fragile. Fruits like strawberries have lots of pores and infuse well, and look awesome at first. After about 5 minutes they start to turn slimy and gross, because their structure can't handle the forces of vacuum infusion. Ripe pears, too, are heavily damaged by vacuum infusion.

Vac Infusion versus Pickling and Influence on Shelf Life: Rapid infusion is not like traditional pickling, in which an acidic, salty, or sugary brine changes an ingredient's composition slowly through osmosis and often effectively preserves the product. Because vacuum infusion doesn't radically alter the composition of your solid ingredient, it can't preserve it the way a traditional pickling process would. Remember that.

On the flip side, if you vacuum-infuse something and then let it sit around for a long time in the fridge—overnight, for a couple days, whatever—then the composition of the flesh of your solids *will* change, just as in traditional pickling, because of osmotic effects and diffusion. Sometimes that is okay. Apples do not lose their texture when stored for a long time after infusion. Sometimes, however, texture is damaged. In the Cucumber Martini recipe, I tell you to eat the cucumber slices within 2 hours. Beyond that time osmosis will cause the water in the cucumber's cells to leach out and dilute the gin. The consequence will be a loss of turgor. In other words, the cucumber will get sad and floppy.

GETTING READY: THE TEMPERATURE

Anything you plan to vacuum-infuse must be cold. *Cold.* Not lukewarm, not tepid, not cool, but cold—fridge-cold at least (34°–40°F, 1°–4.4°C). That includes the solid you are infusing and the liquid you are infusing with. Trying to infuse too warm is the most common vacuum-infusion mistake I see, and I see it a lot. The result is messy boilovers and poor infusions. Read on for the full explanation, or take my word for it and skip to the next section.

The water in this vacuum machine is at just 9 degrees Celsius, yet it is boiling, violently, because the pressure is so low.

Here's a property of vacuums that you need to understand: applying a vacuum reduces the boiling temperature of liquids. Boiling isn't just about temperature, it is about a combination of temperature and pressure. As you know from reading the instructions for a boxed cake (yes, you have), when you cook in the mountains you have to alter the recipe, because water boils at a lower temperature at high altitudes owing to lower atmospheric pressures. A commercial vacuum machine can create pressures much lower than you experience on the highest mountains, as low as the surface of Mars (6 millibars)—and we all know what happened to Arnold Schwarzenegger in *Total Recall* when he got shot onto the surface of Mars: his blood started to boil.

Why does pulling a vacuum reduce the boiling point? The molecules in a liquid are bouncing around all the time. The temperature of that liquid is a function of the average speed of those molecules. Faster speeds equal higher temperature. Aside from the attraction the liquid molecules have to each other, what keeps those bouncing molecules in the liquid phase is the pressure of

CLEANING VACUUM MACHINES

Commercial vacuum machines use pumps that are sealed with oil. As the pump is used, water and other liquids end up getting trapped in the oil, substantially reducing the performance of the pump. All oil-sealed vacuum pumps have a window that lets you see the oil's condition. Clean oil looks like . . . oil. Contaminated oil has liquid emulsified in it and looks like salad dressing. You can clean the oil by running the pump for several minutes in the open atmosphere (without building up a vacuum); all machines have some way to do this, usually by propping the lid open a bit. The pump heats up, making it easier for the contamination to boil away; the air streaming through the pump strips it out. Now you have clean oil again.

VACUUM MACHINE OIL: the oil is the lifeblood of the vacuum machine. Contaminated oil (top) is full of water and reduces the vacuum level your machine can achieve. Clean the oil by running your machine for five to ten minutes with the lid open—boiling out the water and cleaning the oil (bottom).

the atmosphere tamping them down. At any given time, some molecules have enough speed to leave the liquid, which is what causes evaporation—and evaporative cooling. As the high-speed molecules leave, the rest of the molecules have a lower average speed, so the liquid is colder. Below the boiling point, however, an average-speed molecule is more likely to stay in the liquid than leave it as a vapor molecule. Boiling happens when the pressure is so low or the temperature is so high that the average molecule of liquid is equally likely to leave the liquid as to stay in it. When we pull a vacuum and lower the pressure, we lower the boiling point of the liquid. How much we lower the boiling point depends on how good our vacuum pumps are.

The vacuum pumps we use to infuse can easily reduce the pressure enough to boil water at room temperature, and even at fridge temperature. The vacuum boiling isn't raising the water's temperature, it is *lowering* it by evaporative cooling. One of my favorite demonstrations is to boil some room-temperature water in a vacuum machine and then invite a student to put her hand in it. In any group there are always a few people who believe that the water will burn them and can be convinced otherwise only by placing their hands in the cold, just-boiled water.

What this means to you, the prospective vacuum infuser, is that all your products must be cold—really cold—if you want to get good results. If your products are not at least refrigerator temperature when you start, you will be boiling your product before you can achieve the right vacuum levels. Alcohol boils at a lower temperature than water does, so vacuum machines have an even easier time boiling alcohol—which means that alcohol should be as cold as you can get it without freezing your solids. I can't tell you how often I see people breaking this rule. Cold! Cold! Cold!

GO: THE SUCKING TIME

Vacuum infusion takes more time than you might think. Let's say you want to infuse good dark rum into ripe pineapple spears—and believe me, you do. You cover the cold pineapple with cold rum and apply a vacuum. The vacuum will suck the air out of the pineapple pores and cause the pores to open up. The air from the pores will bubble up through the rum and evacuate the system through the vacuum pump. You want to suck a serious vacuum, because you want to get as much air out of that pineapple as you possibly can. If you don't suck out all the air, you won't infuse as much liquid into the pineapple. It takes longer than you think to get all the air out of those pores. Even if your vacuum machine reaches full vacuum, you can't be sure you've gotten all the air out of the pineapple—just that you've gotten all the free air out of your chamber. Suck longer. Remember: suck longer.

When I do vacuum infusion, I typically suck a vacuum for about a minute and then turn off the machine, leaving the product under vacuum. (All commercial machines allow you to do this. Most of the time you just switch the machine off; don't use the Stop button.) Even after the vacuum pump is turned off you will see air escaping in the form of tiny bubbles. The bubbling will last for several minutes. Why turn off the pump? Two reasons: first, if your pump is good, your liquid will cold-boil no matter how cold it is, and continuous boiling strips your liquid of flavor (in essence, you are distilling); second, the vaporized liquid you produce by running the pump a long time will contaminate your pump oil, leading to poor performance.

Now comes the fun part: letting the air back in.

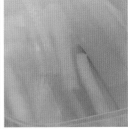

As the air rushes into the vacuum chamber watch the gin and vermouth get injected into these cucumbers.

THE MOMENT OF TRUTH

Before you sucked the air out of the pores with the vacuum, they were at the same pressure as the atmosphere around us, roughly 14.6 psi (1,013 millibars). After the vacuum is applied, the pores will have almost no pres-

MARSHMALLOWS

Whenever I demonstrate vacuum machines, I blow up marshmallows. If you have a vacuum machine, you should do this experiment as well—it will firmly fix the principles of the vacuum machine in your mind. Throw marshmallows into a vacuum chamber and they will expand to several times their original size. What is happening is that the pressure inside the vacuum chamber is being reduced, and the air that is trapped inside the marshmallows begins to expand, making the marshmallows blow up like balloons. Everybody loves this visual, but the *expansion* isn't the important part for vacuum infusion. Keep sucking a vacuum on those marshmallows after they expand and they'll start to shrink again as the trapped, expanded air slowly leaks out of them. Eventually the majority of the trapped air escapes the marshmallows and they shrink back to their nor-

mal size, but it takes a while. That's the important part. It shows that it takes longer than you think to get the air out of pores. The same is true of all the ingredients we infuse: it takes longer than you think to get the air out.

An enjoyable marshmallow-only bonus is that when you finally release the vacuum and allow air back into the chamber, the marshmallows get smashed. Just as it was difficult to suck the air out of the marshmallows, it is difficult to get it back in. When the air hits the marshmallows all at once, the squishy little guys don't have the structure to stand up to the force and they auto-compress, even without a bag. Very few things auto-compress. I like this demo: it seems to stick in people's minds and visually demonstrates the principles of the vacuum.

seconds 0   4   8   12   16   20   24   28

PUTTING A MARSHMALLOW IN A VACUUM MACHINE: *At 0 seconds the marshmallow is at rest. As the vacuum machine starts running, the marshmallow expands as the air bubbles in the marshmallow expand. The marshmallow reaches maximum size at 8 seconds. Watch what happens between 12 and 24 seconds—the marshmallow shrinks. It shrinks because the air inside the marshmallow slowly escapes. The same is true for the fruits and vegetables you put in the vacuum machine— they may not expand, but they have trapped air inside that takes a long time to escape. Unlike fruit, marshmallows will dramatically crush when the air comes back into the chamber at 28 seconds.*

sure in them. Releasing the vacuum causes air to rush back into the chamber of your vacuum machine. That air begins to compress those pores with the force of 14.6 pounds per square inch. That doesn't sound like much, but 14.6 psi of pressure on a 4-inch-square plate is 233 pounds. Instead of crushing the pores, however, the pressure injects the liquid—rum in this case—into the porous solid, the pineapple. What you're left with is vacuum-infused pineapple, which is straight-up delicious and gorgeous. You will want to look at the process as it happens. I have been doing it for almost a decade and it never gets old.

COMPRESSION

If you suck the air out of a porous food, then seal it in a bag without adding any liquid and let the air press on that bag, the full force of the atmosphere presses on those pores with a force of 14.6 psi (1013 millibars). There is no infusion, because there is no liquid. Instead the pores just get crushed flat. The result is a look similar to that of infused products—think jewelry or stained glass, but with no flavors added. This technique is called vacuum compression or texture modification, although the texture isn't changed all that much. I rarely use this technique, but some people love it. If you want to make a piece of fruit look awesome but don't want to add any flavor, compression is a good option. Remember, to do compression, you must seal your ingredient in a bag. Thicker items need multiple compression cycles to finish. No need to reopen the bag; just suck a vacuum on the sealed bag till it starts to puff up (from the vapor coming off your product), then let the air back in. This is the equivalent of lifting a gentle brick and dropping it again and again on your ingredient. The pores you crushed on the last cycle are now uncompressible, and the current cycle of letting the air back in will crush a fresh batch of pores.

USING THE VACUUM TO MAKE INFUSED LIQUIDS

If you have just run a single infusion cycle with the vacuum machine, the liquid will have picked up very little flavor from the solid ingredient. Let the infused solid rest in the liquid for a few minutes and then suck a second vacuum. Don't worry that your product isn't as cold as it was before. This time you actually want the liquid to boil a bit. Why? Because now you are using the boiling to extract the flavorful liquid from the pores inside the solid ingredient!

In addition to the liquid boiling, there will be, despite your best efforts, some residual air inside your product, and that too, when it expands, will help push flavored liquid out of the ingredient. That is how you use a vacuum machine to flavor the *liquid*. When you let the air back in, fresh liquid will be reinfused into the solid and you can repeat the process again. You can run this process several times, but I caution you against more than three cycles, or you might notice some flavor loss from distillation. If your product is sealed in a bag, however, you can run the cycles indefinitely; once the bag is sealed, no flavor can leave. The advantage of using the vacuum for this process is that you can make liters of liquid at a time without having to buy chargers. The disadvantage is that you don't have control over pressure or temperature the way you do with an iSi.

If you find that your product has not fully infused with one infusion, you can repeat the cycle. Hard ingredients like apples or thickly cut ingredients tend to require two or more cycles. If you run more than two or three cycles and your liquid is alcoholic, you risk the loss of some flavor through unintentional distillation. You'll also be adding more flavors of the solid to the liquid, making the infusion go both ways.

Making the Cucumber Martini: The Technique in Recipe Form

Here is the first alcoholic recipe I ever devised for a vacuum machine.

INGREDIENTS

6⅔ ounces (200 ml) cold gin

1⅔ ounces (50 ml) cold Dolin Blanc vermouth

⅓ ounce (10 ml) cold simple syrup

Dash of cold saline solution

2 chilled cucumbers (577 grams)

1 lime

Maldon salt

Celery seeds or caraway seeds

EQUIPMENT

Chamber vacuum machine or equivalent
vacuum system

Vacuum bag

Microplane

PROCEDURE

Everything in this recipe needs to be thoroughly chilled. Combine the gin, vermouth, simple syrup, and saline solution. (I add simple syrup to this recipe even though a martini contains no added sugar because the bitterness of the cucumber requires it.) Chill the mix till it is cold—between 32° and 40°F (0° and 4.5°C).

Refrigerate your cucumbers before you slice them. You don't want to fabricate your cucumber planks and then have them sit around in your fridge chilling, because some of the pores will collapse over time, damaging texture and lessening the infusion effect. While I'm on the subject of cucumbers, here in the United States two main varieties of cucumber are available: hothouse, also known as English cucumbers—long, skinny, and wrapped individually in plastic—and select, typically piled in bulk bins in the produce aisle. Use the select variety for this recipe. I really dislike hothouse cucumbers. People claim the skin is better on hothouse cucumbers, but we don't care about that, since this recipe calls for peeling. People say hothouse cucumbers are seedless, a lie obvious to anyone who has eaten one.

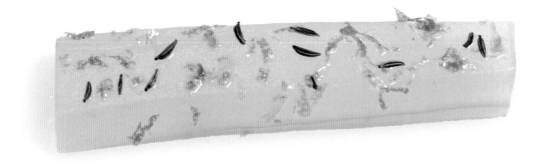

Hothouse cucumbers are also more expensive. But the real reason not to use hothouse cucumbers is that they have very little flavor, and what flavor they do have is bitter.

Slice the chilled cucumbers into 28 rectangular planks. First peel them, then slice into cylinders roughly 4 inches (100 mm) long, then cut those cylinders into 8 radial wedges. Slice the seeds off lengthwise, flip over, and slice off the curved outer surface off the cucumber. Voilà: flat planks (see the picture). You should have 210 grams.

Place the cucumbers in the vacuum bag with the liquor mixture. You only need a bag because the quantity of liquid is so small. The bag holds the liquid in close contact with the cucumbers, which need to be covered with the martini mix for the infusion to work. If you were making a lot of cucumbers, you could just put the ingredients in a pan instead, as long as the pan fit in your vacuum.

Suck a vacuum on the cucumbers and booze for at least 1 minute and then turn off the machine, allowing the ingredients to remain under vacuum. Notice that bubbles are still coming out of the cucumbers. That's good. Wait for the bubbles to stop forming, and then turn the vacuum machine back on to allow air back into the chamber. Make sure to watch what happens. As the air rushes in, the liquid will inject into the cucumbers, turning them into martinis.

Drain the cucumbers (drink the booze). Pat the cucumber martinis dry with a towel. Zest some lime peel on top and sprinkle with Maldon salt and a few celery seeds or caraway seeds. Eat the cucumber martinis within 2 hours, after which they lose some of their crunch.

FABRICATING CUCUMBER PLANKS: 1) *Peel, then cut the cucumbers into 2 cylinders.* **2)** *Cut each cylinder into 8 wedges.* **3)** *Cut the seeds off the wedges to create one flat side. Eat the seeds—they are delicious.* **4)** *Flip the wedges over and flatten the rounded outside portion.* **5)** *All the cucumber stages in one photo.*

MAKING THE CUCUMBER MARTINI WITH THE iSi

In nitrous infusion, liquid is forced into the pores of ingredients under pressure and then that same liquid violently boils out when pressure is released. When you are trying to inject the liquid into the solid, the boiling out is a problem. The liquid doesn't stay inside the solid, where you want it. The solution: put your ingredients in a Ziploc bag so the nitrous never gets into your products in the first place. The disadvantages of this technique are that you can't make too many at a time, you have to spend money on chargers every time you want to infuse, and you can only infuse objects that you can fit into the neck of the iSi. The main advantages: your products don't need to be ice-cold when you use the iSi, and—this is the biggie— you don't need to own a vacuum machine.

EQUIPMENT

1 liter iSi cream whipper

2 chargers (either CO_2 or N_2O)

3 sandwich-size Ziploc bags

Microplane

INGREDIENTS

Same as the Cucumber Martini

PROCEDURE

Combine the gin, vermouth, simple syrup, and saline solution. Prepare the cucumbers, then divide the liquid and the cucumber planks between the three Ziploc bags. Remove the air from the bags by immersing them in water. To do this, get a container of water larger than the bags. Seal each bag starting from one side, and allow only the corner to remain unsealed. Put your finger in the open spot and hold the bag up from that point so the bag looks diamond-shaped. Immerse the bag in water till the water level almost reaches the open spot by your finger while freeing any air pockets in the submerged bag with your free hand. Seal the bag. There should be almost no air in the bags.

Roll up the bags and put them in the iSi whipper. Add water to the fill line (this makes the venting procedure gentler on your product). Seal the whipper and charge with one charger. Agitate mildly for a couple of seconds and allow the product to rest for 2 minutes. Slowly vent the whipper. Go slowly. If you vent too quickly, you'll spoil the infusion. The air that was inside the cucumbers will expand again and force out some of the gin-and-vermouth mixture, reducing flavor and spoiling the look of the cucumbers.

Allow the bags to rest in the whipper for 5 minutes. During this time, air will continue to leave the cucumbers through pathways created by the initial pressure-and-vent cycle. After 5 minutes, apply the second charger, agitate mildly, and allow to rest 2 minutes. Vent slowly and remove the bag from the whipper. This time there isn't enough residual air in the cucumbers to force the liquid back out.

Drain the cucumbers. Zest some lime peel on top and sprinkle with Maldon salt and celery seeds. Use the cucumbers within 2 hours or they will lose some of their crunch.

This technique can also be used in a dry bag for compression, and can be scaled up in a Cornelius keg (see "Scaling Up," page 217).

MAKING THE CUCUMBER MARTINI IN AN ISI: *Put the martini mix and cucumbers into a Ziploc bag and seal all but one corner.* **1)** *Hold the bag as shown.* **2)** *Slowly push the bag underwater while excluding air from the bag. Just before the bag is completely submerged, seal the bag.* **3)** *It should look like this, with little internal air.* **4)** *Shove the Ziplocs into your ISI then fill to the fill line with water.* **5)** *Charge, shake, and wait 2 minutes. Slowly vent the whipper.* **6)** *If you stopped now, the cucumbers would look bad—like this.* **7)** *Wait 5 minutes, charge and shake again and allow to rest 2 minutes.* **8)** *Vent slowly and you are done.*

Here are some suggestions to start with. Use them individually as garnishes or combine several to make an alcoholic fruit salad.

All-Purpose Sweet-and-Sour

This is a good sweet-and-sour syrup useful for infusing things like watermelon rinds, underripe pears, and so on.

MAKES 1 LITER

INGREDIENTS

400 ml simple syrup (or 250 grams granulated sugar and 250 grams filtered water)

400 ml freshly strained lime juice, lemon juice, or lime acid (page 60)

200 ml filtered water

Healthy pinch of salt

PROCEDURE

Mix everything together and go.

MELONS

Watermelons infuse amazingly well. They are incredibly porous. They almost infuse themselves, which is why football fans for generations have punched a hole in a watermelon and poured a bottle of vodka into it to sneak liquor into stadiums. I don't really like the flavor of watermelon infused with liquor. If you infuse watermelon with watermelon juice, you get superwatermelon. It tastes less like a watermelon and more like honeydew. The rind is, as I have stated before, really good infused with the all-purpose sweet-and-sour mix. Crunchy!

Cantaloupe and honeydew melons infuse very well, but since these two foods are the only foods I truly dislike, I am not qualified to recommend them.

TOMATOES

Vacuum-infused cherry tomatoes make a great garnish for savory cocktails. The Jalapeño Tequila recipe on page 207 is nice served as a shot with a salted rim and a cherry tomato flash-infused with salt, vinegar, and sugar. Before you infuse, you should plunge the tomatoes a couple at a time into simmering water for 20 seconds and then put them directly in ice water. If you leave them in the water longer, they'll cook. If you add too many at once, your water temperature will drop and screw up your timing. After they have chilled, use a sharp knife to peel them. Cherry tomatoes can be quick-pickled in the fridge in a couple of hours without the vacuum, but they won't be as punchy.

FOR THE INFUSION LIQUID

100 grams granulated sugar

20 grams salt

5 grams coriander seeds

5 grams yellow mustard seeds

5 grams allspice berries

3 grams crushed red pepper (omit if garnishing Jalapeño Tequila)

100 grams filtered water

500 grams white vinegar

PROCEDURE

Combine all the dry ingredients in a small pan and add the water. Bring to a boil while stirring, making sure the salt and sugar dissolve totally. Remove from the heat and add the vinegar. Cover and chill in the fridge till cold. Vacuum-infuse as usual.

PEARS

As I mentioned, pears must be firm for you to infuse them properly. They work best when cut into thin strips or disks, and look fantastic as a garnish. They like to be infused with sweet-and-sour mixes. I have also infused pear with Poire William to good effect. Port-infused pears are not as good as you think they will be, and Asian pears do not infuse as well as Western ones.

CUCUMBERS

I have already covered these ad nauseum. In addition to gin, they can be infused with aquavit—which is delicious—or with sweeter mixtures, in which case they take on more of the qualities of a fruit and can be paired with tarter fruits in an alcoholic fruit salad. I like to infuse cucumbers right before I serve them.

APPLES

Apples don't pick up a lot of flavor, so intense liquids are called for. I typically infuse apples for culinary applications and use flavored oils such as curry oil. You might be tempted to infuse a beautiful red liquor like Campari into apples, because red apples are cool, but the color will not satisfy you. You *can* flash-infuse apples with beet juice. This makes a very pretty garnish when sliced thin.

PINEAPPLE

I've already mentioned that ripe pineapple infuses well with dark rum, and pineapple core infuses well with simple syrup. Ripe pineapple also infuses well with aperitifs such as Lillet Blanc. I don't recommend infusing pineapple itself with syrups unless the pineapple is extremely underripe. Infused pineapple makes a great garnish for built drinks on the rocks, especially old-fashioned variants. One thing to note about pineapples: they ripen from the bottom up. Next time you cut up a pineapple, taste a piece from near the base and a piece near the leaves. The one near the base will be radically sweeter. The base also ferments first. I always cut pineapple garnishes so that every piece extends the whole length of the pineapple; otherwise one person's pineapple is very different from that of the person next to them. If the top is really underripe, just cut it off and infuse it separately with some added sugar. Likewise, if the bottom is overripe, remove it. Infused pineapple lasts a long time stored in its infusion liquid in the fridge.

*TOP: **FABRICATING A PINEAPPLE:** When you cut like this, there is very little waste. Pineapples are sweeter and riper at the bottom than at the top. The spears on the lower left are cut from top to bottom so each one should taste the same as the next, with some of the less-ripe top and sweeter bottom. In the lower right is the pineapple core, cut and ready to infuse for garnish. I peel the pineapple after I cut it into quarters—you lose less good pineapple meat that way. **MIDDLE:** Pineapple spears infused with rum in a vacuum bag. If you don't want to use rum, Lillet Blanc works too. **BOTTOM:** Those same pineapples, ready for service.*

Clarification

Definition, History, Technique

WHAT IS CLARIFICATION?

Liquids can be crystal clear, completely opaque, or somewhere in between. Clear doesn't mean colorless. A glass of red wine can be both deeply colored, thanks to the pigments dissolved in it, and clear. Unclear liquids are actually suspensions, containing particles that reflect and scatter light in a random pattern that makes the liquid appear milky. Clarification removes these particles. It pays to remember the difference between dissolved substances, which don't cloud a liquid but may add color, and suspended particles, which are not truly dissolved and scatter light.

It takes a ridiculously small percentage of suspended crud—way less than 1 percent—to make a liquid cloudy. Good clarification gets rid of all, not most, of the light-scattering doodads.

From berry to clear juice.

One path to clear grapefruit juice.

WHY CLARIFY?

Why clarify? Why breathe? I love the look of crystal-clear liquids. I much prefer a beautiful limpid drink to a murky one. (Remember, clear doesn't mean colorless—think brown liquors.) But clarification's not just about looks: it is also necessary for good carbonation. Particles floating around in your drink provide places for errant bubbles to form. Those errant bubble-making particles will wreak foaming havoc in your drink and reduce the level of carbonation you can achieve. When you squeeze a lime into a gin and tonic, you witness immediate frothing and bubbling. Unacceptable!

Clarification also alters the texture of drinks by removing solids and reducing body. Mouth feel is an aspect of cocktails that is often overlooked. I don't want to chew on my drinks, so I almost never use purees without clarifying them first. (I like a traditional bloody Mary, but that's about it.)

I first started worrying about clarification in 2005, when I was obsessing over making the absolute best lime juice for gin and tonics. I didn't know it at the time, but I couldn't have chosen a more difficult clarifying problem. You need to clarify lime juice quickly because lime juice doesn't last, but many clarification techniques take time. You can't use too much heat because heat ruins the flavor of lime, but most clarification uses heat. The particles in lime juice are very small, and small particles are hard to filter. Lime juice is very acidic, and acidity thwarts some clarification tricks.

Turns out that if you can clarify lime juice, you can clarify almost everything else. My lime-juice-clarification efforts eventually unlocked the gin and tonic I sought, and a whole lot more. The journey included messing with filtration, setting gels, using equipment such as centrifuges, and searching out ingredients such as enzymes and wine-fining agents. These days I can—and do—clarify almost everything. It is a bit of a sickness.

CLARIFICATION TECHNIQUES: THE THEORY

Clarification is all about removing suspended particles, separating clear liquids from cloudy solids. It is mainly mechanical in nature. There are three primary means. Filtration blocks particles, allowing clear liquid to run through. In gelation, you trap particles in a gel and coax out the clear liquid. And with separation by density, you assist or augment gravity to allow particles to settle out of the liquid.

First, some theory and history. If you truly have no patience for learning, skip to the clarification flowcharts on pages 249–55 for the how-to's.

FILTER CLARIFICATION

Up-front confession: I hate filtration as a clarification technique. Industrially, filtration works great. In your kitchen, not so much. Clarification requires filters much finer than the ones you use for coffee, and they clog frustratingly fast. Sure, filter aids and multiple filter setups can help solve these clogging problems, and you can purchase special charged filters that clog less, but let me reiterate: filter clarification is a drag.

GEL CLARIFICATION

Gel clarification works really well. Simply put, you trap your liquid in a gel and then induce it to leak. This process is called syneresis. The particles that were making your liquid cloudy remain trapped in the gel, and the stuff that leaks out is clear. The gel acts like a massive 3-D filter that never clogs. Gel clarification doesn't require expensive equipment and can be scaled for large quantities.

There are several different ways to clarify with gels.

Old-School Gels: Egg whites and lean ground meat are the original gel-based clarifiers, traditionally used to clarify stocks into consommés. The proteins coagulate and form a gel-like raft through which you continuously ladle your simmering stock. Eventually this raft traps all the cloudy bits and leaves you with a perfectly clarified consommé. This process has many drawbacks for cocktail clarification. It is tedious and prone to error, requires prolonged heating (which can change or destroy delicate flavors), and adds meat flavors where you probably don't want them.

Freeze-Thaw Gelatin: A little over a decade ago, some European chefs, most notably Heston Blumenthal, noticed that when meat stocks containing gelatin were frozen solid and then allowed to thaw in the refrigerator, a raft of gelatin goo held on to cloudy particles and the liquid that dripped out was clear. Freeze-thaw clarification was born. In the early years of the century, on this side of the pond, New York chef Wylie Dufresne realized that this technique isn't limited to stock—you can add gelatin to almost anything, and then freeze and thaw it to get a clear liquid. This observation revolutionized the clarification world. Soon I was freeze-thaw clarifying anything I could get my hands on.

I soon discovered the major drawbacks of freeze-thaw gelatin clarification. It's a bit of a chore, and it ain't quick. You have to freeze your gelatin mixture solid—I mean solid. That takes a day. You must then slowly thaw this frozen block in the fridge, which can take another two days. If you try to cheat and thaw on the counter, the fragile gelatin raft (the thing that's holding back all the cloudy nastiness) will break, and it's all over. Also, you can't start using the stuff that drips out first while you are waiting for the rest to thaw, because the clarified liquid at the beginning is too concentrated and the stuff at the end tastes like water.

Why is this process so finicky? With freeze-thaw gelatin clarification, you use 5 grams of gelatin per liter. A 5-gram-per-liter gelatin mixture is, disconcertingly, still a liquid, not a gel like Jell-O. As the gelatin mixture freezes, the pure water starts to freeze first, and everything else—the gelatin and any color, flavor, sugar, acid, what have you—becomes more and more concentrated as the water continues to freeze. Eventually this solution concentrates enough to form a delicate gel network, the source of the raft. Then that network freezes solid as well. As the gelatin network freezes, it gets torn apart by ice crystals. As it thaws, that torn network retains just enough structure to hold back solids in a sludgelike mass but to leak clear liquid like a sieve.

If everything goes according to plan, this system works great, but because your mix is a liquid at the beginning, it's hard to tell if you've achieved the right consistency before you freeze. You'd like to make a firm gel to increase your peace of mind, but you can't. If your mix were more solid from the get-go—say the consistency of Jell-O, which is self-supporting and rigid at 14 grams of gelatin per liter—it *would* be easier to work with, but the network would not break down enough

during the thaw, the gel raft would remain too solid, and the dripping clear liquid that is the point of all this hassle would never materialize.

The upshot is, you never know if everything went right until it is too late. You must wait three days to verify whether your labors yielded a lovely clear liquid or a soupy mess. There has to be a better way.

Freeze-Thaw Agar: I don't know who first started using agar instead of gelatin for freeze-thaw clarification, but what an awesome idea. Agar is a seaweed-based gelling agent, so, unlike gelatin, it is vegetarian-friendly. Agar gels are very porous—much more so than gelatin—so they leak much more, allowing you to form self-supporting gels and still clarify. Because a real gel is formed *before* the freeze-thaw process begins, you have visual and tactile verification that your clarification

FREEZE-THAW CLARIFICATION:
The frozen grapefruit juice block at the top left thaws to the nasty looking raft at the bottom. What drips out is the beautifully clear juice at right.

*In freeze-thaw clarification the stuff that drips out at the beginning of the thaw (**LEFT**) is more concentrated in flavor and color than the stuff that drips out at the end (**RIGHT**).*

will work—a win over gelatin. Because the agar forms a gel, any unfrozen part won't cloud your product even if you don't freeze the gel thoroughly—another win. Agar rafts don't melt until they get very hot, so you can thaw an agar raft much faster than a gelatin raft—major win. One more agar win: it produces, in my estimation, a more sparklingly clear product than gelatin does.

Agar has only one disadvantage compared with gelatin. Gelatin will dissolve into a liquid with only a modicum of heat, while agar must be boiled and held there for several minutes to fully incorporate, which is too much heat for many delicate products. There is a fix; see the flowchart on pages 249–55.

What would be better than agar clarification is something faster. Even though you can freeze-thaw agar faster than you can freeze and thaw gelatin, the process still takes a couple of days. That's okay for some things, like strawberry juice, but unacceptable for lime juice, which you must use the day you make it.

Also, in any freeze-thaw clarification, what drips out changes over time. The first stuff to thaw is the last stuff that froze: sugars, acids, and other concentrated flavors. As the thawing progresses, the stuff dripping out has less and less flavor. You need to batch all the drippings over the entire process or your flavor will be unbalanced.

For years this problem of slow and uneven processing stuck in my craw. The solution ended up being very simple.

WHEN TO USE GELATIN FREEZE-THAW

When making a drink with meat stock, like a bullshot, always use gelatin freeze-thaw clarification, and not just because it makes sense to use the gelatin already present in the stock. With any other clarification method, the stock gelatin remains, and when you reduce the stock to condense the flavors, which you always do for cocktails (I reduce by a factor of 4 or more), it will get gluey and unusable. With gelatin freeze-thaw, you remove the gelatin from the stock as you clarify, so your stock can be reduced a ridiculous amount and still end up thin and limpid, even when cocktail-cold.

Trick: you'll probably need to add a little water to your stock before you free-thaw clarify. Most stocks have too much gelatin in them to freeze-thaw clarify effectively. At most, you want a stock that just barely gels when cold. Even better is a stock that just barely doesn't gel. After your clear consommé drips out, you can reduce out all the extra water.

For a nice savory shot I like concentrated gelatin-free consommé, centrifuge-clarified tomato water, iSi rapid-infused jalapeño tequila, a pickled cherry tomato, and a healthy dose of salt.

Quick Agar Clarification: In 2009 I discovered that I could clarify with agar simply by breaking the gel with a whisk to increase the surface area and letting it leak. Remember, agar gels are very porous—they want to leak. You have to work hard to get them *not* to leak. So it turned out that freeze-thaw is unnecessary. Extra equipment is unnecessary. Because there is no freeze-thaw cycle, the first drop of juice you get with this technique tastes the same as the last. Also, because there is no freezing, you can clarify otherwise unfreezable *booze* this way (remember to hydrate the agar in water first; booze can't get hot enough). With quick agar clarification, anyone can clarify anything in under an hour, including—finally!—lime juice.

Quick agar isn't perfect. A bit of skill is involved, and it takes a while to get the knack. You will see in the how-to flowchart that lots of messy hand-straining through cloth is called for, so quick agar is inconvenient for large quantities. It is rarely as perfect as freeze-thaw; some agar usually gets into the final product and can form visible wisps if left to sit overnight. For all these faults, however, quick agar remains my go-to technique when I have to clarify away from my centrifuge.

GEL CLARIFICATION RECAP

Freeze-thaw agar has some advantages over quick agar clarification: it entails less labor, is more foolproof, and produces a product that will never recloud. But remember, you'll need a lot of freezer space to make large quantities, and the technique takes a couple of days. Quick agar works great for small quantities of products that you want to clarify quickly and use the same day.

The main disadvantage of all gel clarifications is the yield: you will always lose some liquid in the gel raft that remains after clarification. Expect to sacrifice at least a quarter of your product, sometimes more.

GRAVITATIONAL CLARIFICATION: RACKING, CENTRIFUGING, FINING

Most of the time, suspended particles are denser than the liquid they are floating in. If unhindered, they will eventually fall to the bottom of the liquid. This behavior is the basis of separation based on density.

RACKING

In a liquid in which the particles are large and relatively free to move around, you can employ the simplest density-separation technique, racking. Just put the liquid in a container and let it settle over time. After all the particles settle to the bottom, decant the clear liquid off the top, and you are done.

You can't use racking alone very often in practice because many liquids settle very slowly, and some will never settle at all—at least, not in your lifetime. Sometimes settling won't occur because particles are too small, sometimes because their movement is blocked by stabilizers in the liquid. Even in liquids that settle fairly quickly, such as carrot juice, racking can be difficult because the particles don't form a compact layer at the bottom; they float around near the bottom instead. That floaty zone contains a lot of juice you won't be able to clarify, so your yield will be lousy. If you are going to clarify by racking, make sure to use round containers. Liquids moving in square containers kick up small particles and ruin the clarification.

CENTRIFUGING

To get around the problems with racking, I use a centrifuge. Centrifuges spin fluids rapidly. Anything inside the centrifuge tends to get pushed toward the outside—that's centrifugal force. Centrifuges are rated by how much force they produce relative to the force of gravity, and they can produce many thousands of times the force of gravity. They radically exaggerate the difference in density between a liquid and the particles floating in it, and drastically increase the speed at which particles settle out of a liquid. The force also tends to smash those particles into a firm cake, called a puck, so your yield is high. Sweet.

The Problem with Centrifuges: As I write, centrifuges are still esoteric, but that is changing. In 2013, a centrifuge that could handle enough product for a busy bar (3 liters) cost $8000 and was the size of two microwaves. A small unit to play with at home is the size of a toaster and costs a couple hundred bucks. (It works but holds a tiny amount of product and is therefore basically a toy.) I predict an explosion of centrifuges in the next ten years. Many more people will have them, and a centrifuge the size of a microwave with enough capacity for a professional will cost less than $1000. Centrifuges are the wave of the juice future.

A tube from the under-200 dollar centrifuge filled with crabapple juice that was treated with Pectinex Ultra SP-L. The solids are smashed into the bottom and sides of the tube. The compressed solids in a centifuge bucket, which you'll find in larger centrifuges, are called a puck, but in a tube they're referred to as a pellet. It's easy to rack the clear liquid off the top.

When I first started using centrifuges, in 2008, I was borrowing time on a superspeed centrifuge at my buddy Professor Kent Kirshenbaum's NYU lab. It could spin 500 milliliters of juice at 48,000 times the force of gravity! As you know, I was interested in the holy grail of clarification: lime juice. I hadn't yet figured out quick agar. I learned some fascinating but depressing things on that centrifuge. Lime juice doesn't start clarifying until your centrifuge gets to about 27,000 g's. A centrifuge that can pull 27,000 g's is larger, more expensive, and more dangerous than any centrifuge I'd care to have at a bar. Even worse, 27,000-g lime juice tastes metallic. At 48,000 g's, lime juice tastes great, but a centrifuge that can do 48,000 g's is even less practical for your bar—tens of thousands of dollars and the size of a washing machine.

I tried clarifying other juices as well: ginger, grapefruit, apple. Not all of them required the full 48,000 g's, but most juices and purees needed more than a reasonable centrifuge could muster. Practical bar/restaurant centrifuges max out at about 4000 g's. I needed to find a way to get the results I wanted with a fairly puny 'fuge. And I did. Here's how.

Refining the Process: In the cocktail world we mostly clarify purees and juices from fruits and vegetables. These purees and juices all have suspended chunks of cells and cell walls composed largely of the polysaccharides pectin, hemicellulose, and cellulose. These busted-up chunks of cells tend to make juices thick and gloopy, which in turn makes them difficult to clarify, because they don't flow or drain well. Pectin in

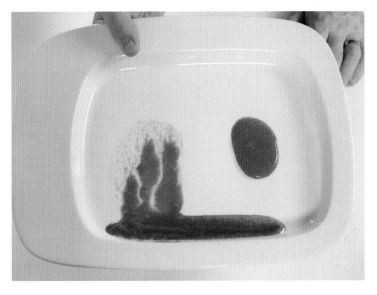

The strawberry puree on the left was treated with Pectinex Ultra SP-L. The puree on the right is untreated.

particular tends to stabilize the particles in juices and purees, making them more difficult to remove. You need to knock out these stabilizers to free the suspended particles so you can separate them out.

How do you destabilize a juice? Add enzymes.

Destabilizing: The Magic of SP-L: Ninety-nine percent of the stuff you'll want to clarify is stabilized by pectin and thickened by busted-up cell parts.

Pectinex Ultra SP-L ate the pith off of these grapefruit wedges to make these beautiful suprêmes. This trick works on all citrus I have tried—including pomelos, which are notoriously difficult to suprême—except limes.

Luckily, both of these hindrances can be handled with a single concoction of enzymes called Pectinex Ultra SP-L. Just call it SP-L. I call it my secret ingredient; about 75 percent of my drinks involve SP-L at some point. It is a mix of enzymes that are purified from *Aspergillus aculeatus,* a fungus found in soil and rotting fruit. Fungi are the world's champions at enzymatic breakdown of everything, and *A. aculeatus* is a great generalist. It produces enzymes that obliterate most things that get in clarification's way: pectin, hemicellulose, and cellulose.

SP-L maintains its activity over a wide range of temperatures, pHs, and ethanol concentrations. The ethanol bit is extremely important. Many enzymes don't work well—or at all—in concentrated alcohol solutions. SP-L does, so you can clarify booze with it. SP-L-treated juices clarify quite well in a centrifuge that operates at 4000 g's or even a bit less, so SP-L makes reasonable centrifuges worthwhile. SP-L works so well that some juices can be clarified without a centrifuge. Apple juice with SP-L will settle on its own enough to rack off clarified juice. I still use a centrifuge on apple juice to increase my yield, but you don't have to.

Using SP-L: Using SP-L couldn't be simpler. I always use 2 milliliters (roughly 2 grams) of SP-L per kilo or liter of juice. That's twice the amount used industrially, but sometimes I can't be sure how well my enzymes have been stored or how old they are, and both those factors affect potency. Remember that ratio: 2 grams per liter. Never tell someone to add 0.2% SP-L; they will almost always add 20 grams per liter, I don't know why. The really good news is that that 2 grams isn't critical—little more, little less, no problem. Don't go enzyme-crazy, though. The stuff on its own tastes fermented and weird. You don't want a perceptible amount in your juice. Even when you are gel clarifying, using SP-L is often a good idea. Knocking out the stabilizers before you clarify can increase the yield of clarified juice you get by 30 percent or more, because thinner products leak out of gels better than thick ones do.

MAKING THE CITRUS VESICLE (GROSS BUT ACCURATE WORD) GARNISH: *I use grapefruit here but the real show-stopper is blood orange:* **1)** *Cover the citrus suprêmes with liquid nitrogen.* **2)** *Make sure the suprêmes are frozen—it will take more liquid nitrogen and time than you think.* **3)** *Smash the suprêmes with a muddler.* **4)** *They should look like this.* **5)** *Allow them to thaw and* **6)** *use as an elegant garnish on any gin drink.*

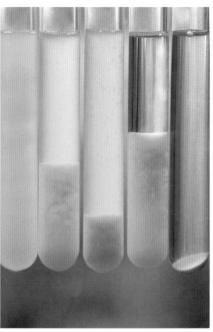

LEFT: The powerhouse wine-fining agents I use, kieselsol—suspended silica sol—and chitosan—a polysaccharide derived, in this case, from shrimp shells and prepared in an acidic aqueous solution.
RIGHT: What fining agents do to for you: At left is untreated, cloudy lime juice. The second tube has lime juice that I treated with SP-L and negatively charged kieselsol. A lot of the solids have settled, but the juice is still cloudy. Fifteen minutes later, I added positively charged chitosan to tube three. The chitosan agglomerated the negatively charged kieselsol, and the lime juice settles much more than in tube three, but is still cloudy. Fifteen minutes later I added more kieselsol to tube four. Notice this juice is crystal clear, but now the cloudiness isn't settling as much. This is a common problem, because the last particles that kept tube three cloudy don't settle as much as heavier, earlier floccing particles do. A centrifuge smashes all the solids from tube four into a tiny pellet and increases our yield to nearly 100 percent. Look at how little the pellet is. It doesn't take much to make a liquid cloudy.

I was hooked on SP-L from the moment I got my first sample. Like a drug dealer, Novozymes, the manufacturer, gave me the first sample free, but then I had to pay. Novozymes and its distributers aren't set up to sell reasonable quantities to normal folk. The smallest amount they will sell—and grudgingly at that—is a 25-liter pail. Twenty-five liters of SP-L is enough to clarify over 12,500 liters of juice and costs $570. Not a bad deal, really, but way more than you need. Luckily, online suppliers have started selling the stuff in smaller quantities with extremely fast shipping; see Sources, page 378.

When SP-L Doesn't Work: Occasionally you will run into a fruit with pulp that is resistant to SP-L; some jungle fruits from Colombia with local names I can't recall come to mind. Other fruits, like tamarind, can become SP-L-resistant if the seeds are pureed with the pulp. In these cases a hydrocolloid thickener (complex long-chain polysaccharide) other than pectin is present in the product and the SP-L can't dissolve it. In these cases, you are SOL.

SP-L won't break down starch. Starchy products such as sweet potatoes and underripe bananas will always give a cloudy result with the ~4000-g centrifuges I use.

Most important, SP-L doesn't work perfectly in very-low-pH products such as lime juice (which has a pH of 2 and change). Grapefruit juice's acidity, at roughly pH 3, is right on the edge of what SP-L can handle with aplomb on its own. With these acidic products, SP-L is still useful, but it can't work alone: you'll need further interventions. I use wine fining agents, which gave me the final key to clarified lime juice.

FINING

In winemaking, fining is the process of adding small amounts of specialized ingredients to the wine to get all the cloudy impurities to clump together—flocculation—and aggregate in large enough masses for gravity to pull them to the bottom of the vat relatively quickly. Most fining agents rely on electrical charge to do their work. You see, most of the particles that are floating in wine, or your juice, have some sort of charge on them. By adding a fining agent with the opposite charge, you can get those impurities to clump together, making it easier to get them to settle to the bottom. If the cloudy particles still aren't big enough to settle, you can use a counterfining agent. A counterfine has the opposite charge to the fining agent. It will mop up anything in the juice that the first fining agent couldn't catch, plus get the already-clumped-up stuff to clump up into even bigger particles.

Normally, in wine, you do two steps: fine, then counterfine. That doesn't work for lime juice. I figured out that to get lime juice to work, you need three steps: fine (and add SP-L), counterfine, and fine again.

The fining agents I use are kieselsol and chitosan. You can get them at any home-brew shop. Kieselsol is food-grade suspended silica; it has a negative charge. Chitosan is a positively charged hydrocolloid that comes from shrimp shells. The solution they sell at the home-brew shops is 1 percent chitosan in a weakly acidic aqueous solution. Chitosan comes from chitin, the second most common polymer on earth after cellulose. Every bug and crustacean on the planet is protected by a chitinous shell, and mushrooms and other fungi build their cell walls with chitin. The shrimp origin of chitosan is the only fly in the buttermilk of my clarification bliss. It doesn't end up in the product (it is spun out) and it is non-allergenic (I've tested chitosaned lime juice on folks with shellfish allergies), but I'm still using an animal product, and I would rather not. Luckily, non-animal-based chitosan is now being made and should be available soon.

Both kieselsol and chitosan are used in small quantities: 2 grams per liter. Unlike SP-L, the amount of fining agents you use is critical. Too much

SP-L AND TEMPERATURE

I usually warm ingredients I'm treating with SP-L, because SP-L works much more rapidly if it is warm—unless it gets too warm and gets denatured. I use body temperature as a reference, because body temperature is mostly constant, easy to judge, and warm enough to be effective but cool enough not to change my ingredients' flavors or destroy the enzyme. At body temperature, SP-L works in a few minutes. At refrigerator temperature, I have to let the enzyme work for an hour or more. If you are starting with juice or premade puree, add the enzyme directly to the juice or puree and mix thoroughly. If you are blending products to make a puree, add the SP-L directly to the blender with your fruit or veggies; it will help liquefy the puree as you blend. I use a Vita-Prep blender, which is so powerful that the friction of the blades spinning on high slowly heats my purees to just above body temp, and that is what I recommend you do if you have one. Heating with a blender never scorches anything, which is why I like to do it that way. If you don't have a Vita-Prep, soak uncut fruits in warm water for a couple of minutes to heat them to just above body temp, or else let the puree rest an hour or more for the enzyme to work before proceeding with clarification.

SP-L dissolves the white pithy part—the albedo—of citrus peels. If you make perfect citrus peels by cutting wedges into the fruit with a knife and removing perfect triangles, you can vacuum the peels in a bag with a solution of 4 grams SP-L per liter of water and let it soak for several hours. When you remove the peels from the bag, the albedo will have turned to mush. Brush it away under water with a toothbrush.

Some nice garnishes can be made this way. SP-L doesn't really work on lime peels (of course). Kumquats cut in an X shape can be soaked in SP-L to make kumquat flowers. Last, SP-L can be used to auto-suprême citrus segments. A citrus suprême is a wedge of fruit without all the connective tissue that surrounds it. Old-school kitchen technique is to make suprêmes with a knife—a good technique, but flawed in that it is wasteful of fruit and requires you to cut into the individual sacs of pulp so they leak. A better way is to peel your fruit, separate it into four parts, and soak it in a solution of 4 grams per liter SP-L and water for a couple hours. Whatever connective tissue isn't melted outright can be rubbed off or peeled away with ease. These suprêmes look beautiful. A great trick is to freeze them with liquid nitrogen and shatter them. They will shatter into their individual juice sacs without leaking. You can then do things like float blood-orange bits on top of your drink without spoiling the overall appearance of the cocktail.

is as bad as too little. If you add too much, you can actually stabilize the particles you are trying to floc together.

CLARIFICATION TECHNIQUES AND FLAVOR

Clarification changes the flavor of ingredients. That's right—clarification changes the way things taste. Usually the particles floating around in your drink making it cloudy are also contributing some kind of flavor. The flavor of the particles is often not identical to that of the liquid they are floating in, so when you strip the particles out, you alter the flavor of the remaining liquid.

Whether the flavor change is good or bad is context-dependent. Clarifying grapefruit removes some of its bitterness, making it friendlier for many cocktails and worse for others. Sometimes the SunnyD flavor that you get when you clarify OJ helps a drink out, and other times it's gross.

How much you change the flavor of your ingredients depends on the ingredients you have and the techniques you use. In general, gel clarifications strip more flavor than mechanical (centrifugal) clarifications do. Treating a juice with SP-L has very little effect on flavor—it merely obliterates tasteless pectin—but SP-L can indirectly alter the flavor of your juice by increasing your yield. For example, one of the characteristic notes of red plum juice is astringency from the skins. Astringency is leached into the juice from the skins as it sits waiting to be clarified. As the yield of clarified juice increases, extra nonastringent juice is released from the interior pulpy part of the plums, reducing the overall astringency. Life is never simple.

Wine fining agents can have major taste impacts, but I chose kieselsol and chitosan because they don't strip flavor heavily in the dosages I use. Some other fining agents are known flavor thieves.

OPPOSITE: Here I use a separatory funnel (sep funnel) to separate the solids from liquids in orange juice. A sep funnel lets you "reverse rack." Normally when racking you remove the clear liquid off the top of the solids. With a sep funnel, you drain the bottom off.

Clarification Techniques: Nitty-Gritty Flowcharts

Here is where you want to start reading again if you skipped the theory. I'm going to assume you fall into one of two camps, those with a centrifuge and those without. Choose your camp and I'll tell you how to proceed from there.

I DON'T HAVE A CENTRIFUGE: STEP 1

Look at what you want to clarify. Is it pretty thin? Is it less acidic than grapefruit juice? Does it start to settle a bit on its own? Will it last a day in the refrigerator without losing quality? Unpasteurized apple juice, pear juice, carrot juice, and even orange juice are like this. You can often clarify such juices just by adding SP-L and letting the solids settle. Settle 'n' rack doesn't provide good yield but couldn't be simpler. Expect to lose between one-third and one-quarter of your product (well, you don't lose it, but it won't be clarified).

1A: MY PRODUCT IS THIN, SEPARATES EASILY, AND IS LESS ACIDIC THAN GRAPEFRUIT. I'M GOING TO RACK.

Add 2 grams Pectinex Ultra SP-L to every liter of juice. Stir thoroughly. Put the juice in a clear round container. Your container should be clear so you can see what is happening. Your container should be round because when square containers are moved around, they kick up particles. Put the juice in the fridge overnight and let it settle. Carefully pour the clear juice off the top. **You're done.**

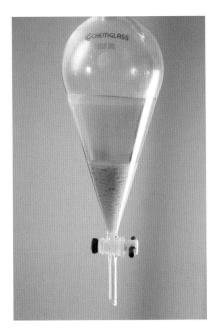
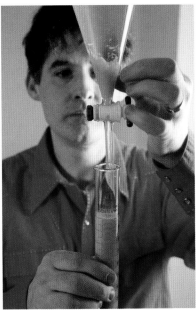

A mixture of strawberry puree and apple juice that I have treated with SP-L and am trying to clarify by racking alone. Right after treatment the juice looks like *(1)*. Even after a couple hours of sitting, the juice looks like *(2)* which has two problems: the juice is still cloudy, and the airy foam caused by juicing and blending hasn't settled. The solution? Stir in a bit of kieselsol (the wine fining agent). The stirring will help break up the bubbles and get the solids to settle and the kieselsol will mop up the last cloudy particles *(3)*. Notice that clarity has its price: the yield in the last glass is low.

Telephone brand agar-agar. The hydrocolloid so nice they named it twice.

1B: MY PRODUCT IS THICK, OR DOESN'T SETTLE, OR IS MORE ACIDIC THAN GRAPEFRUIT. I'M USING AGAR.

You are going to use agar to clarify. Buy powdered agar, because it is the easiest to use. Always buy the same brand. Different brands of agar will have slightly different properties. Get used to one brand and stick with it. I use Telephone Brand from Thailand.

Unless you are clarifying lime or lemon juice, you have the choice of freeze-thaw or quick clarification. Lime and lemon should only be quick clarified. The beginning steps are the same either way. **Go to Step 2.**

STEP 2: PRETREAT WITH SP-L?

Any thin juice that is made with a juicer does not need to be pretreated with SP-L: cucumber, citrus, and the like. If your product is like that, **go to Step 3.** Any thick juices or purees you want to clarify need to be treated with SP-L or your yield will be awful: blended tomatoes, strawberries, raspberries, or similar. To pretreat, add

2 grams of Pectinex Ultra SP-L to every liter (or kilo) of product. If you are blending, add the SP-L directly to the blender. If your product is cold, give the SP-L an hour or so to do its work. If your product is near body temperature, the SP-L will work in just a couple of minutes. **Go to Step 3**.

STEP 3: DIVIDE YOUR BATCH

Figure out the volume or weight of product you want to clarify. Weight or volume doesn't really matter here—they usually work out close enough. Do what is convenient. First, allow your product to come up to room temperature. If your product is too cold, you will have problems in Step 6. Then make the following choice:

3A: MY PRODUCT IS NOT SUPER-HEAT-SENSITIVE, AND IS THIN

Orange, grapefruit, ginger, and similar juices are like this. You are going to separate out one-quarter of your total batch and heat only that quarter with the agar. Afterwards you will recombine the agar with the rest of the batch. This is okay because you aren't heating that much of it, and it is relatively easy to hydrate agar in thin liquids. For example, if you start with 1 liter of grapefruit juice, divide it into a 750-milliliter sample and a 250-milliliters sample. You will add the agar to the 250. **Go to Step 4**.

3B: MY PRODUCT IS HEAT-SENSITIVE, OR ALCOHOLIC, OR THICK

Lime juice, strawberry puree, and gin blended with raspberries fall in this category. In these cases, you don't want to heat the products directly. Instead you will heat the agar in water and then combine the agar-water with your product. For every 750 milliliters or 750 grams of product, measure out and reserve 250 milliliters of water. In Step 5 you will add the agar to the 250 milliliters of water. Does this extra water

If the ingredient you are clarifying is alcoholic or heat sensitive you should hydrate the agar in pure water. Here we have 750 ml of lime juice 250 ml of water and two grams agar.

dilute the juice a little? Yes, but in my experience, not as much as you'd expect. Strangely, for products like lime juice, the difference is often negligible. **Go to Step 4**.

STEP 4: MEASURE YOUR AGAR

Now that you have measured and divided your batch, you are going to weigh out 2 grams of agar for every liter of total stuff you are going to clarify. So if you are doing an all-juice clarification, like grapefruit juice, you will measure 2 grams of agar per liter of juice. If you are doing a juice-water clarification, like lime juice, you will measure 2 grams of agar for every 750 milliliters of juice and 250 milliliters of water. Got it? **Go to Step 5**.

STEPS 5, 6, AND 7 OF AGAR CLARIFICATION: 1) *Remember to whisk agar into your liquid before you apply heat to so that the agar disperses properly, then heat and simmer for several minutes while stirring.* **2)** *Always add juice to the hot agar solution while whisking vigorously, not the other way around.* **3)** *Pour the tempered agar and juice into a suitable container (preferably over ice) and allow it to set undisturbed. I mean it. Leave it alone. Let it set.*

STEP 5: HYDRATE THE AGAR

Do not add agar to hot liquid. It will clump. Add agar to liquids at body temperature or below. Add the agar to the *small* batch of liquid or water you have reserved from Step 3 and then stir vigorously with a whisk to disperse the powder. When the powder is dispersed, turn on the heat. Agar needs to be heated to the boil and held there for a couple minutes to hydrate properly. (For those of you at high altitudes, I have successfully hydrated agar in Bogotá, Colombia, elevation 8612 feet/2625 meters, local boiling point 196°F/91°C—but it was a pain). Continue to stir while your product comes up to the boil, then turn the heat down and cover the pot; you don't want to boil off too much liquid. **Go to Step 6.**

STEP 6: TEMPER THE BATCH

Add the unheated liquid to the hot agar solution, not the other way around. If you add the hot stuff to the cold, the temperature will drop quickly and the agar will gel before you want it to, spoiling your clarification. Unlike gelatin, agar sets into a gel extremely quickly once it cools below its gelation temperature, around 35°C (95°F), or just below body temperature. Use a whisk to keep the agar fluid moving as you add the rest of the batch. When you are done, the entire batch should be just above body temperature. Now you can see why I had you let your product come to room temperature in Step 3! **Go to Step 7.**

STEP 7: SET THE AGAR

Pour your batch into a bowl, pan, or tray to set. In professional settings I use 2-inch-deep hotel pans; my Euro friends will know them as gastro-norms. Use whatever you like. Your product will eventually gel at room temperature, but I accelerate the process by putting my setting container in the fridge or in an ice-water bath. Do not mess with the agar while it is setting. I repeat: do not mess with the agar while it is setting. Many people I have trained have an irresistible pathological desire to stir a gel while it is setting. Wrong! When you think your agar has set, lightly touch the top. It should feel like a very, very loose gel. Tilt your container slightly—slightly! The gel should not run. You are ready to make your big decision: freeze-thaw or quick? **Go to Step 8.**

STEP 8: FREEZE-THAW OR QUICK CLARIFICATION?

If you need your stuff right away, go quick. If you have time and freezer space, freeze-thaw will pro-

vide a slightly better yield with fewer headaches. Also, quick-clarified products tend to develop agar wisps when stored longer than a day or so. You can stir them back in, but they are irritating.

8A: FREEZE-THAW

Put your container of set agar gel in your freezer and let it freeze solid. As long as the gel is 2 inches thick or less, it should freeze solid overnight. Once it is frozen, crack it out of its container. Don't use hot water or a torch. Don't shatter it into a billion pieces. There is an art to unmolding frozen gels that involves getting very angry with the containers holding them. I grab the edges on opposite sides and pull like a demon, working my way around the container.

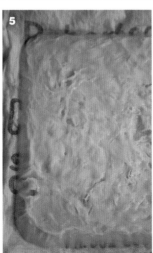

FREEZE-THAW TIPS: 1) Make sure your juice is thoroughly frozen—don't rush it. 2) To get the frozen juice out of the pan, pull on it hard, like you are drawing a strong bow, then rotate 90 degrees and repeat. 3) Put a cloth on your work surface, flip the pan and push with force. 4) Put the frozen juice in a cloth suitable for draining and place on a rack or in a perforated pan over another pan to catch the drippings. Allow it to thaw a while on the counter and finish thawing in the fridge. 5) The spent raft.

I then flip it over and punish the bottom of the vessel till the gel releases.

Once the gel is released, put it into some form of draining cloth. Normally people specify cheesecloth for this, but the average cheesecloth is absurd and close to gauze. Worthless! You are better off using an unbleached cotton tablecloth. Put the gel in the cloth into a colander set over a collection container and let it thaw. You can leave it out a couple of hours to kick-start the thawing. Once the gel starts dripping a lot, transfer it to the fridge to continue thawing. Every once and while, drain off and save what you have. When it looks like the agar is spent and what is dripping out is devoid of flavor and color, you are done. Combine all the drippings together and enjoy. **You're done.**

8B: QUICK

Get a whisk and gently break up your gel. It should resemble broken curds. Pour the curds into an unbleached cloth napkin or filter sack and let the clear juices flow into a container. Resist the urge to squeeze on the gel too hard. If you squeeze hard, you will push cloudy agar bits through your filter and into your juice. Bad. You will find that the filter clogs rather quickly. Gen-

tly pinch a portion of it in your fingers and rub the cloth against itself to free up the pores. I call this massaging the sack. Massaging the sack and applying just the right amount of pressure are the finesse skills you must develop to do this technique properly. It takes some practice. An alternate technique I show people is to tie your filter sack shut and throw it into a salad spinner. The spinner does a good job of gently extracting the liquid, and you can't spin too hard with a salad spinner. You still need to massage the sack between spins! **You're done.**

I HAVE A CENTRIFUGE: STEP 1: ADD SP-L

Add 2 grams of Pectinex Ultra SP-L to every liter or kilo of product you want to clarify. If you are blending fruits like strawberries, blueberries, peaches, plums, apricots—whatever—

*QUICK AGAR: **1)** Once the agar has set, gently break the gel with a whisk so it looks like curds. **2)** Pour the curds into a draining cloth and let it drip. **3)** As the cloth clogs, gently "massage the sack". If you squeeze hard you will push agar through the cloth and ruin your product. **4)** Quick agar is not quite as clear as freeze-thaw. This juice could be passed through a coffee filter to catch the errant bits of agar in it.*

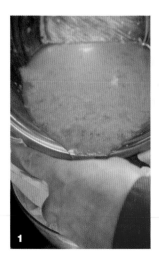

add the SP-L directly to the blender and blend till the product is just above body temperature. Remember, this process won't clarify starchy things completely. **Go to Step 2.**

STEP 2: ASSESS ACIDITY

2A: MY PRODUCT IS LESS ACIDIC THAN GRAPEFRUIT JUICE

If your product is body temp when you add the SP-L, you are ready to spin. If your product is refrigerator temperature, give the SP-L an hour to work. **Go to Step 3.**

2B: MY PRODUCT IS SIMILAR TO OR MORE ACIDIC THAN GRAPEFRUIT JUICE

2B1 When you add the SP-L to the juice (Step 1), also add 2 grams per liter of kieselsol (suspended silica) and stir. This measurement needs to be fairly accurate. I use a micropipette

to measure it, because micropipettes are fast and I do this a lot.

2B2 Wait 15 minutes.

2B3 Add 2 grams per liter of chitosan (1% chitosan solution) and stir well. Again, this measurement must also be accurate.

2B4 Wait 15 minutes.

2B5 Add 2 more grams per liter of kieselsol (measure accurately) and stir. **Go to Step 3.**

STEP 3: PREP AND SPIN

Air bubbles won't necessarily pop in the centrifuge when you spin. If the bubbles don't pop, you will get floaty doodles on top of your product when you spin. I hate floaty doodles. Bananas don't make them, but tomatoes do; it is hard to predict. If you have a chamber vacuum machine, you can use it to de-aerate your product before you spin, which will obliterate floaties, but this step is optional. When you load your centrifuge, make sure to balance it properly, and then spin for 10 to 15 minutes at 4000 times the force of gravity. If you don't have a refrigerated centrifuge, make sure your buckets are extremely cold before you spin so your product doesn't heat up. **Go to Step 4.**

STEP 4: POUR OFF THE PRODUCT

When you pour your product out of the centrifuge tubes or buckets, it is good practice to pour through a coffee filter or fine strainer to catch anything that might be floating on top or in case the puck breaks free and falls toward your clarified product. **You are done.**

AN ALTERNATIVE TO MASSAGING THE SACK: tying the top of the sack closed and putting it in a salad spinner—the poor cook's centrifuge. In between spins you still need to do some massaging to unclog the cloth, but this method is almost foolproof.

REALLY ADVANCED CENTRIFUGED GRAPEFRUIT

Centrifuge-clarified grapefruit juice tastes more bitter than agar-clarified grapefruit juice does. The agar gel traps and holds on to a portion of the bitter molecule in grapefruit, naringin. In my carbonated cocktail Gin and Juice (page 329), I think the less bitter juice is better, but centrifuging is a far more convenient technique for me to use at the bar: better yield and much faster. In a technique similar to washing, you can use agar to strip some of the naringin from grapefruit juice that is clarified in a centrifuge.

You do this by making an agar fluid gel. A fluid gel is something that acts like a gel when it is standing still but acts like a liquid when you stir it or agitate it. Chefs use fluid gels to make sauces that plate like purees but feel very liquid in the mouth. Thinner fluid gels are used to suspend objects in drinks or soups. You aren't using any of those properties here. An agar fluid gel is a bunch of tiny gel particles suspended in a liquid. All those tiny particles have a huge surface area that will help soak up naringin and are easy to spin out of suspension with a centrifuge.

To make the fluid gel you first make a regular grapefruit gel with 1 percent agar—10 grams of agar per kilo of juice—and set it solid. This is a much firmer gel than you use in gel clarification. Put the gel in a blender and blend until totally smooth. With this step you have created a fluid gel.

Add 100 grams of the grapefruit fluid gel to every 900 grams of regular grapefruit juice right when you add the SP-L and kieselsol as part of your normal centrifuge routine. Proceed as usual with the rest of clarification and you will have a less bitter grapefruit juice.

On the upper left is grapefruit juice gelled solid with 10 grams of agar per liter of juice. On the lower right is that same gel blended into a fluid gel. Fluid gels are fantastic in culinary applications because they plate like a puree but eat like a sauce. Here we use it to strip naringin from grapefruit juice in centrifugal clarification.

Clarifying Booze in the 'Fuge: The Justino

Bad news: you need a centrifuge to try the following technique. Good news: once you have a centrifuge, this technique will change your drink-making life.

Turns out you can make a beautifully clear spirit from a straight liquor and the fruit or vegetable or spice of your choice. You blend the liquor and the other ingredients, add the enzyme Pectinex Ulta SP-L, and use a centrifuge to spin this mix into a clear spirit that I call a Justino (pronounced *whoo-stee-no*). (If the process in that last sentence holds no meaning for you, read the earlier part of this clarification section). Justinos rely on the fact that Pectinex Ultra SP-L, an enzyme that destroys the structure of blended fruit and allows effective clarification, works quite well in highly alcoholic solutions. Many enzymes don't.

The birth of Justino went something like this: I was interested in making a banana cocktail that wasn't thick and gloopy like a smoothie. While there are plenty of banana-flavored liquors, I wanted to use just the straight juice, and I wasn't having much success making it. My yields were poor, and the taste was off. I knew I needed to add more liquid to the banana to up the yield, but I didn't want to add a non-alcoholic liquid, so I just blended the bananas with the booze and centrifuged it clear. The result was rum with pure banana flavor. Beautiful! When a reporter asked me what the liquor was called, I came up with Justino—and the name stuck.

Although you can blend almost anything into booze to make a Justino, I like to use low-water products so the alcohol content stays high. I find that higher-proof Justinos mix better and last a lot longer than lower-proof ones. Lower-proof Justinos tend to be less versatile and have unstable flavors. If you want to Justino with high-water products like honeydew melon, you should put those products in the dehydrator before you blend them with the liquor. Commercially produced dried fruits are great choices for Justinos.

Starchy products don't work well. The SP-L enzyme doesn't break starches down and the liquor will not clarify in the low-speed centrifuges that are commonly used in restaurants and bars. For instance, you can't make a good Justino with unripe bananas—they have too much starch.

THE JUSTINO PROCESS: 1) Add fruit to liquor, **2)** then Pectinex Ultra SP-L, **3)** then blend at high speed until the friction of the blender heats the mix up to body temperature. I use the back of my hand on the pitcher to judge the temperature. **4)** Pour the mix into centrifuge buckets, making sure to balance the buckets, and **5)** place them in the centrifuge and spin for 10–15 minutes at 4000 times the force of gravity. **6)** Pour off the clear Justino.

The basic starting recipe for Justino is 250 grams of fruit or vegetable for every liter of liquor, a ratio of 1:4. This is a good starting point. If you are using an ingredient that has very low water content and the mix seems to be more of a paste than a puree, lower the ratio to 200 grams of fruit or vegetable per liter of liquor—1:5. Sometimes you might even do 1:6. If your Justino is too thick before you spin it, your yield will be low. Sometimes if your yield is too low and your Justino ratio can't be reduced further without damaging flavor, you can fix the problem *after* you spin by adding some water to the puck and spinning again (see the Apricots Justino recipe on page 260 for the technique).

After you spin the product, taste the Justino. If it is too weak and your yield is high, increase the ratio of solids to liquor. If the ratio is already high and the Justino tastes watery, dehydrate your solids more before making the Justino. If the Justino is too strong in flavor (usually too sweet), start adding straight liquor till you like it. Once you find a ratio you like, you can try using that ratio from the beginning the next time you make it, or you can just keep making the Justino the same way and add fresh liquor at the end. Strangely, the two techniques produce different-tasting liquors. Sometimes one is better than the other. You will just have to test, taste, and see for yourself. Here is an example.

BANANAS JUSTINO

Let's say you are making a Justino of Medjool dates in bourbon and you test my recommended ratio of 1:4 dates to liquor. You will find that the yield is okay but the Justino is too sweet. Most likely, you'll like it best when you add 250 milliliters of fresh bourbon to every 750 milliliters Justino, an equivalent Justino ratio of 1:5.3. Strangely, you will also find that if you make the same Justino with an initial ratio of 1:5.3, it doesn't taste as good as the 1:4 Justino with the fresh liquor mixed in. I don't know why.

SOME OF MY FAVORITE JUSTINOS

BANANAS JUSTINO: 3 peeled ripe bananas (250 grams) per 750 ml of liquor. Works well with aged rums (not rums that have caramel color added; the Justino process strips it out), bourbons, even Hendrick's gin. Remember that the bananas must be ripe: brown but not black. If they are not ripe, the starch in them will create a cloudy drink with a starchy taste. This liquor is fantastic poured over a rock and served with a wedge of lime and a pinch of salt. For a real treat, keep the lime but pour the Justino over a large ice cube made from high-quality coconut water, and float a star anise pod on top.

DATES JUSTINO: 187 grams Medjool dates per 750 ml liquor. Justino, then add 250 ml of fresh liquor. Works well with bourbon, Scotch, and Japanese whiskey. Serve over a rock with a dash of bitters.

RED CABBAGE JUSTINO: Dehydrate 400 grams of red cabbage to 100 grams and Justino with 500 ml Plymouth gin. If you skip the dehydrating step, your drink will smell like an ill wind from the nethers. Works well in shaken drinks.

APRICOTS JUSTINO: Use dehydrated apricots for this. I prefer Blenheim apricots from California as the base of this Justino. In my opinion, they are the royalty of dried apricots. Blenheims are full of flavor and have a lot of bright acidity. If you use any other apricot, the results will be totally different. Make sure not to get apricots that are untreated for oxidation. The most common treatment is sulfuring, but there are others. It is easy to tell if an apricot has not been treated—it will be brown and taste oxidized.

Blenheim Justinos are one of the few that I actually think are low in sugar. You can either add a bit of sugar to the finished product or

substitute garden-variety dried apricots (which have less acidity) for a portion of the Blenheims. Apricots absorb a lot of liquor during the Justino process, so yield is low.

Here is how to proceed: Make a Justino with 200 grams of dehydrated Blenheim apricots and 1 liter of liquor. When you drain the Justino, save the pucks of apricot solids from the centrifuge buckets, place them in a blender with 250 ml filtered water and an additional 1 to 2 grams SP-L, and blend. I call this a remouillage (rewetting), or remmi, after the French technique of adding a second batch of water to extract used stock ingredients a second time. Spin the remouillage in the centrifuge and add the resulting clear liquid to the original Justino. This process pulls out a good bit of the trapped alcohol and flavor from the pucks, so add it to the first-spin Justino.

This recipe works well with genever, gin, rye, vodka . . . I'm hard-pressed to think of something it wouldn't work with.

PINEAPPLES JUSTINO: Use 200 grams of dried pineapple per liter of liquor. As with Apricots Justino, do not use the "natural"-type dried pineapples, which are brown and sad-looking/ tasting. This recipe works well with dark or white rum—but you already knew that. Try it with whiskey or brandy for more of a pineapple-upside-down-cake feeling.

OPPOSITE: To make Justinos out of cabbage, **1)** dehydrate the cabbage until **2)** it has lost ¾ of its initial weight. **3)** Red-cabbage Justino.

Washing

In 2012, ESPN asked me to make an alcoholic version of the Arnold Palmer, a mixture of iced tea and lemonade named for the famous golfer. I replied that the Arnold Palmer is clearly best as a nonalcoholic drink; diluting and chilling an alcoholic tea cocktail makes the tea's astringency too dominant. Think of how a big, tannic red wine tastes when it is too cold, then multiply that effect in your mind, and you get the idea. I made the film crew several alcoholic Arnold Palmer variants. We all agreed they weren't very good.

After the crew left, I got to thinking. Many people, especially in the U.K., take their tea with milk. Milk proteins—casein in particular—bind with the tannic, astringent compounds in tea and mellow the brew. I decided to make some tea-infused vodka, add milk, and then curdle the milk-vodka to clarify the liquor and remove the astringency. I steeped tea in vodka till I had a very strong decoction, then added that vodka to milk and stirred in a little bit of citric acid solution. It worked beautifully. The milk broke and the solids settled to the bottom. I spun my vodka in a centrifuge (because I have one, but you can just strain the liquid through a fine cloth). I reduced the astringency of the tea so much that the cocktail tasted balanced even when cold, though the tea flavor was still very strong. When I added simple syrup and lemon juice to the tea vodka and shook it with ice, I reaped an unintended side benefit: booze washed with milk this way gets a silky texture and makes an incredibly rich head when shaken. Although the casein in the milk curdles and is removed in the washing process, the whey proteins remain—and they are fantastic whipping agents.

Milk washing, therefore, has two purposes: it reduces astringency and harshness, and it augments the texture of shaken drinks.

The concept of washing liquids takes a little getting used to. You wash clothes to remove dirt; you wash ingredients to remove flavors. You can use washing in your cocktail ventures in two different ways. You can *booze-wash*, as I did in the Arnold Palmer, by adding a "detergent"—usually milk, gelatin, hydrocolloids, or eggs—to bind with unwanted compounds in the liquor so that you can remove them. You can also *fat-wash* to wring good flavors out of a fat and into a liquor, and then use that liquor to make something delicious. In the first example you're washing a liquor; in the second you're washing a fat.

Good news: all the techniques in this section can be pulled off without fancy equipment, although having a centrifuge is nice for some. Let's tackle booze washing first.

POLYPHENOL ASTRINGENCY AND PROLINE-RICH PROTEINS

The astringency of a polyphenol correlates well with how strongly it binds to a specific group of salivary proteins called proline-rich proteins (PRPs). These proteins contain large amounts of the amino acid proline. The prolines give the proteins more affinity for polyphenols. If you remember, plants produce astringent polyphenols to make themselves less digestible and therefore less likely to be eaten. PRPs in saliva bind with the polyphenols and mitigate their antidigestive properties—an herbivore's countermeasure to the plants' defense mechanism. Tellingly, carnivore spit doesn't have PRPs. Tigers eat only meat, not leaves and bark, so they don't need the PRPs. Herbivores have lots of PRPs in their spit. We, as omnivores, are in between.

Many other proline-rich proteins besides saliva PRPs also bind to polyphenols, including milk protein (casein), egg whites, and gelatin. These are the proteins that we use in booze washing.

Booze Washing

To remove flaws from poorly distilled spirits you typically use hard-core fla-vor- and color-stripping media such as activated charcoal. We are not talking about that kind of removal here. We are talking about selectively stripping flavors from perfectly good booze. Why would we do this? Because certain flavors, while nice on their own, can overwhelm the other ingredients in a cocktail. Bourbon, for example, is delicious. Bourbon cocktails are delicious. Carbonated bourbon cocktails? Often not delicious. The pleasing woodiness in bourbon becomes harsh and overpowering when carbonated. You *could* just add less bourbon, or perhaps cut the bourbon with a neutral spirit like vodka, but isn't it better just to moderate that harshness and leave the rest of the bourbon flavors intact? Another example: Black tea tastes good. Black tea vodka tastes good. Cocktails made with black tea vodka are usually not good—they are harsh, astringent, and difficult to balance properly. You *could* just dial back the tea in the cocktail till the astringency didn't bother you anymore . . . or you could moderate the harshness by washing the booze, leaving the rest of the tea flavor intact. Choose washing!

BOOZE-WASHING SCIENCE

The flavors we target with booze washing are called polyphenols, a group of chemicals produced by plants, often as a defense against predators or dam-age. Polyphenols are decent defense mechanisms because they typically have bactericidal, insecticidal, and antidigestive properties that animals leave alone. Many polyphenols are astringent. Tannins, for instance, are polyphenols, and their presence in grape seeds and skins is responsible for the tannic flavors of red wine. Ditto the tannic flavors in cranberry, cassis, and certain apple varieties. Polyphenols in oak wood give whis-key and brandy their trademark woodiness. You might remember from the nitro-muddling section of this book that a damaged herb leaf causes enzymes called polyphenol oxidases to link small phenolic molecules together into large, dark-colored polyphenols. In tea, those polyphenols are desirable: they create the characteristic astringency of dark tea.

To attack polyphenols, I steal from the winemaker's playbook. The winemaker combats the problems of excess astringency, protein haze, off flavors, and turbidity through a process called fining: adding small quantities of ingredients to wine to correct for perceived problems. I use these same ingredients for booze washing. They all rely on some combination of three basic principles: protein binding, charge, and adsorption.

PROTEIN BINDING

Protein-rich agents—egg white, blood (yep), gelatin, casein (milk protein), and isinglass (fish gelatin)—bind to impurities in a complex way. Proteins are very good at stripping tannins and other polyphenols. They also strip color and some flavors, which can be good and bad.

CHARGE

Some agents rely solely on electric charge to do their work. Kieselsol (suspended silica) and chitosan (a polysaccharide found in all arthropod skeletons and in some fungi) are two such agents. Kieselsol has a negative charge and will therefore attract positively charged impurities, while chitosan is positive and will attract negatively charged impurities (for more on these two agents, refer to the Clarification section, page 235) The polyphenols that we seek to reduce when we booze-wash are negatively charged, so chitosan, with its positive charge, is our best electrostatic weapon.

ADSORPTION

Other agents rely on adsorption, the process of a liquid or gas sticking to a surface. Adsorptive agents, such as activated charcoal, have an immense surface area of tiny pores that trap impurities. Adsorptive agents tend to be rather broad-spectrum flavor muters—too blunt an instrument. I don't use them much.

Every fining/booze-washing agent removes different flavors in different amounts, and some bring textures and flavors of their own. While wine fining is still a bit of a dark art, there's plentiful information to guide the booze washer; I have called out some sources in the bibliography. My booze washing differs from wine fining in some key ways. I focus on quick results, while wine fining is typically a slow process. I use large amounts of fining agents to strip flavors, while wine producers are usually looking for more subtle effects. Let's look at how I developed three different booze-washing techniques—milk washing, egg washing, and chitosan/gellan washing—and how they can help you.

Milk Washing: What's Old Is What's New

I must point out that adding milk to alcohol and then clarifying, as I did with the Arnold Palmer, is nothing new. Milk punch has been around since the seventeenth century. The difference between milk punch and my milk washing: milk washing is practiced on straight booze, not a cocktail, and it anticipates shaking to produce fantastic foam. Milk punches were not typically shaken. Over time, the whey in milk-washed booze degrades and loses its foaming power. It doesn't go bad, it just loses its awesomeness. Use milk-washed boozes within a week or so.

My tea drink from the beginning of the section came out so nicely that I put it on the menu at the bar. Here it is, on the following page.

MILK PUNCH

A traditional milk punch contains booze, milk, and other flavors. The milk is induced to curdle and the curds are strained out; after the straining you are left with a clear, stable beverage. Here is Benjamin Franklin's recipe from a letter he wrote in 1763:

> Take 6 quarts of Brandy, and the Rinds of 44 Lemons pared very thin; Steep the Rinds in the Brandy 24 hours; then strain it off. Put to it 4 Quarts of Water, 4 large Nutmegs grated, 2 quarts of Lemon Juice, 2 pound of double refined Sugar. When the Sugar is dissolv'd, boil 3 Quarts of Milk and put to the rest hot as you take it off the Fire, and stir it about. Let it stand two Hours; then run it thro' a Jelly-bag till it is clear; then bottle it off.*

Why make milk punch? Milk punches are known for their soft, round flavors. That softness isn't caused just by the presence of milk but by the removal of phenolic compounds from the brandy via the casein-rich curds. Old Ben Franklin might have been dealing with some pretty rough brandy back in 1763, and the milk would have stripped away a lot of its harshness. Ben didn't mention the awesome foaming properties, because at that time no one was shaking cocktails with ice. Pity!

*Courtesy Bowdin and Temple Papers, in the Winthrop Family Papers, Massachusetts Historical Society

Tea Time (an alcoholic Arnold Palmer variant)

I use a Selimbong second-flush Darjeeling tea for this drink. Darjeeling is a famous tea-growing district in the mountainous northeast of India. In March the teas of Darjeeling grow their first crop of leaves—the first-flush Darjeelings. The first-flush teas are the most expensive, but they aren't my favorite. Several months later there's a second flush of new leaves. These second-flush Darjeelings are unique in the tea world for their fruity aroma, known in the biz as muscatel. Selimbong Estate is particularly well known for the quality of its second-flush teas. I infuse it into vodka because I want to highlight the tea rather than the intrinsic flavor of a spirit. Originally I made this cocktail with lemon juice and simple syrup. Piper Kristensen, who works with me at Booker and Dax, suggested that I use honey syrup instead, and he was right, not only because tea, lemon, and honey are a classic combination but because proteins in honey augment the foaminess of the milk-washed tea vodka. The honey syrup recipe is easy: add 200 grams of water to every 300 grams of honey (for ounce people: 10¼ ounces of water per pound of honey). Note that this recipe is by weight, not volume.

INGREDIENTS FOR THE TEA VODKA

32 grams Selimbong second-flush Darjeeling tea

1 liter vodka (40% alcohol by volume)

250 ml whole milk

15 grams 15% citric acid solution or a fat 1 ounce (33 ml) freshly strained lemon juice

When you make your tea infusion, make it dark. Don't worry about over-steeping, you will strip out the astringency later.

PROCEDURE

Add the tea to the vodka in a closed container and shake it up. Let the tea infuse for 20 to 40 minutes, shaking occasionally. The time will change based on the size of the leaves you use and the type of tea you use if you don't use the Selimbong; what's important is the color, which provides a decent indicator of brew strength in tea. Go dark. When the tea is dark enough, strain it from the vodka.

Put the milk into a container and stir the tea vodka into the milk (note that if you add the milk to the tea vodka instead, the milk will instantly curdle and reduce the effectiveness of the wash). Let the mix rest for a couple of minutes, then stir in the citric acid solution. If you don't want to buy citric acid, use lemon juice, but don't add all the lemon juice at once; do it by thirds. When the milk breaks, stop adding. Don't stir too violently after you add the acid. Once the milk breaks, you don't want to reemulsify or break up the curds at all, or you'll make straining more difficult.

After the milk breaks, you will see small clouds of tan curds floating in a sea of almost clear tea-colored vodka. If you look closely, you'll see that the vodka is still faintly cloudy. It still has some casein in it that hasn't agglomerated onto the curds. Take a spoon and gently move the curds around to mop up the extra casein. You should see the vodka get noticeably clearer, and the curds will get noticeably more distinct. Do the gentle curd-mopping several times, then let the vodka sit undisturbed for several hours to settle out before you strain the curds with a fine filter and a coffee filter (or just spin the stuff in a centrifuge right away, as I do).

INGREDIENTS FOR TEA TIME

MAKES 1 4⅗ OUNCE (137 ML) DRINK AT 14.9% ALCOHOL BY VOLUME, 6.9 G/100 ML SUGAR, 0.66% ACID

- 2 ounces (60 ml) milk-washed tea-infused vodka
- ½ ounce (15 ml) honey syrup
- ½ ounce (15 ml) freshly strained lemon juice
- 2 drops saline solution or a pinch of salt

PROCEDURE

Combine all the ingredients, shake with ice, and serve in a chilled coupe glass. Garnish with pride in a job well done.

All milk washing follows this procedure, but some liquors will curdle the milk even if you don't add any acid. In coffee infusions, the combination of alcohol and coffee is usually enough to curdle the milk on its own; ditto with cranberry-infused liquor. Whenever you milk wash, follow the tips above, like making sure to add the liquor to the milk, not the other way around, and gently stirring the curds to mop up all the free casein, and you'll get good consistent results.

Most of the time when I'm milk washing, I'm focused on very astringent ingredients such as tea. I have tested milk washing on less astringent polyphenol-rich aged spirits such as bourbon and rye and brandy. In these spirits, the oak in concert with the alcohol typically curdles the milk by itself. The oak in aged spirits not only destabilizes milk by adding polyphenols that bind to casein, it also reduces the pH to around 4 or 4.5, making the milk break more easily. Unfortunately, the milk washing really, really strips the oak flavor and color. Way too much, in my opinion.

You can milk-wash spirits without polyphenols, not to remove flavors but just to get the textural effects. Robby Nelson, a former bar manager at Booker and Dax, made daiquiris with milk-washed white rum, and were they good! I use milk-washed white rum to make an orange Julius variant called the Dr. J, which makes use of typical daiquiri specs (2 ounces rum, ¾ ounce lime juice, flat ¾ ounce simple syrup, pinch salt) but replaces the lime juice with lime-strength orange juice (add 32 grams of citric acid and 20 grams of malic acid to a liter of orange juice; see the Ingredients section, page 50, for details) and adds a drop of vanilla extract.

OPPOSITE: MILK WASHING: 1) Always add the liquor to the milk, not the other way around, or the milk will instantly curdle. Some liquors—like coffee infusions—will break on their own. Others—like the tea shown here—will not. 2) Stir the liquor to get it moving then add a bit of citric acid solution or lemon juice. In (3) the milk has just started to break. Gently agitate the liquor with a spoon to allow the curds to mop up any errant cloudy particles and 4) allow the liquor to settle before coffee-filtering or centrifuging.

Egg Washing

After I tackled the problem of creating tea cocktails with milk washing, I remembered an unwashed tea cocktail that I really liked: the Earl Grey MarTEAni that my friend Audrey Saunders developed at Pegu Club, one of my favorite NYC bars. It's a mixture of Earl Grey–infused gin (Earl Grey is a blended tea flavored with bergamot citrus rind), lemon juice, simple syrup, and an egg white, with a lemon twist. I had always figured that the egg white in the MarTEAni was there for texture, but now I realized that the cocktail *needed* the egg white to bind with the tea's polyphenols and mellow the astringency. I asked Audrey about it and she said, "Of course—that's why the egg white is there." Duh! Then I got to thinking some more. Why is a whiskey sour shaken with an egg white while other shaken sours with similar sugar:acid:booze ratios (like the margarita, the daiquiri, and the fresh-lime gimlet) are not? It's the whiskey! The egg white in a whiskey sour is mellowing out what would otherwise be too astringent a drink at the temperatures and dilutions a whiskey sour achieves. I decided to give egg washing a try on its own, without making a cocktail.

A whole egg white represents a substantial portion of the liquid in the undiluted cocktail. A large egg white, roughly 30 milliliters, can represent a quarter of the total liquid before dilution—a 3:1 ratio of cocktail mix to egg white. I figured I'd try slightly less egg white, aiming for an egg wash at a ratio of 4 parts booze to 1 part egg white. I chose bourbon whiskey for my test. I mixed the egg white with a fork and stirred the whiskey into the egg, the same way that I stir booze into milk for milk washing. The egg quickly coagulated hard and was easy to filter out, but all that egg obliterated the flavor and color of the booze, rendering it like weakly flavored vodka. In an egg-white cocktail, the egg white stays in the drink, so the flavor isn't completely stripped. In egg washing, all that protein is removed, so the stripping effect is starker. I tried 8:1. Better, but still too stripped and a bit out of balance—the spiciness of the bourbon was gone. I then tried 20:1 and 40:1. The winner: 20:1 booze to egg white! This ratio left the character of the whiskey unharmed and allowed for a balanced drink. The procedure was simple and foolproof.

Note that this 20:1 ratio is designed to wash pure aged booze. If you want a harsher stripping effect, as you might for coffee or tea liquor, you'll be better off with a 8:1 ratio, or even a little higher. Those higher egg ratios will give you stripping power similar to milk washing. But remember that milk washing adds foaming power; egg washing does not. For a stirred tea cocktail, use an egg wash. For a shaken drink, go with milk and take advantage of the awesome texture.

Egg-Washing Technique

INGREDIENTS

1 extra-large egg white (after removing the yolk and losing the stuff stuck in the shell, this is about 32 grams, a little less than the 37 needed for a true 20:1)

1 ounce (30 ml) filtered water

750 ml liquor of your choice at 40% alcohol by volume or higher (lower-proof liquors don't coagulate the egg as hard, and once the alcohol content goes below around 22% alcohol by volume the egg won't coagulate at all)

PROCEDURE

Mix the egg white and water together to combine, and then add the liquor to the mixture while stirring. Note that the water is just there to increase the volume of the egg mix. A single egg white would be difficult to add liquor to without insta-curdle effects. If you are using an 8:1 or higher booze-to-egg ratio, you can omit the water. After you add the liquor, the egg white should break pretty quickly. Let the mixture sit for a couple of minutes and gently stir it to swab any stray proteins into the coagulum. Let the mixture stand for an hour, then strain it through a coffee filter. It should be clear.

EGG WASHING, IN THIS CASE WITH BLENDED SCOTCH: *(1) Add liquor to the egg-white and water mix while stirring. In (2) the egg is just beginning to coagulate. With some gentle stirring it should look like (3) and should eventually settle like (4). (5) Strain through a coffee filter. In (6) you see untreated liquor on the left and egg-washed liquor on the right.*

One of the fantastic things about egg washing is that you need no equipment, and unless you are vegan, you probably have eggs in your fridge right now. Also, because egg washing leaves very little residual protein in the liquor, you can use it to mellow drinks that are too harsh to carbonate on their own, and you can carbonate them without too much foaming. Milk washing will never work for a carbonated drink because it exacerbates the foaming issue.

Here is a recipe for an egg-washed red wine and cognac carbonated cocktail! That is a drink that needs some mellowing.

Cognac and Cabernet

If you set out to mix a red wine with a spirit like Cognac, you might choose a sweet wine—one that would mask the tannins present in both. But here we will stay dry, and strip the tannins out instead. This cocktail is deep pink, dry, raisiny, and satisfying. It drinks less like a cocktail and more like a wine. To do the booze washing you'll use 1 part egg to 6 parts booze—a fairly large proportion.

MAKES TWO 4⅘-OUNCE (145-ML) DRINKS AT 14.5% ALCOHOL BY VOLUME, 3.4 G/100 ML SUGAR, 0.54% ACID

INGREDIENTS

1 large egg white (1 ounce/ 30 ml)

2 ounces (60 ml) Cognac (41% alcohol by volume)

4 ounces (120 ml) cabernet sauvignon (14.5% alcohol by volume)

2 ounces (60 ml) filtered water

½ ounce (15 ml) clarified lemon juice or 6% citric acid solution

½ ounce (15 ml) simple syrup

4 drops saline solution or a generous pinch of salt

PROCEDURE

Add the egg white to a small mixing container and thoroughly break it up with a bar spoon. Combine the cognac and the cabernet and add them, stirring, to the egg white. The mix should go cloudy. Continue to stir the mix slowly, making sure that all of the egg white comes into contact with the liquor. You should see wisps of denatured egg white throughout the mix. If you have a centrifuge, you can spin the mix now and collect the clear washed booze. Alternatively, let the mix sit for several minutes and then stir again. Allow the mix to settle for several hours and strain it first through a cloth napkin and then through a coffee filter.

CONTINUES

When you have a clear liquid, add the water, lemon juice or citric acid, simple syrup and saline solution or salt. Chill the mix down to −6° C (20° F) and carbonate with the method of your choice (see the Carbonation section, page 288). Note: you'll have some residual egg protein in this drink, so it will foam quite a bit when you carbonate.

I've figured out an even better technique for mellowing aged liquors that don't require too much stripping prior to carbonation, but it calls for some specialty ingredients: chitosan and gellan. Keep reading, or skip ahead to the Carbonated Whiskey Sour, page 282, and use egg-washed whiskey to make it.

OPPOSITE: EGG WASHING THE CABERNET AND COGNAC: This recipe uses much more egg than regular egg washing; 1) add no water to the egg. 2) Pour the cabernet and cognac into the egg while stirring. 3) It will look momentarily chalky, 4) then break. Stir to mop up the cloudy bits and allow to settle before coffee filtering.

Chitosan/Gellan Washing

When I began clarifying apple juice years ago, I really wanted to make a refreshing carbonated apple cocktail with bourbon, but the oak flavors dominated the apple and became unpleasantly harsh. My solution at the time was to redistill the bourbon with a rotary evaporator. Unlike a typical still, the rotary evaporator can distill at low temperatures and recover almost 100 percent of the volatile flavors that get boiled off during the distillation process. I would split the bourbon into two parts: a clear liquid containing all the alcohol plus the aromas from the oak and the base spirit, and a dark opaque liquid with all the nonvolatile oak extractives that were messing with my carbonated drink. Since unaged clear bourbon is known as white dog, I called my redistilled clear bourbon gray dog—and it is pretty darn good, even on its own. I used the leftover oak extractives to make a great ice cream. Efficiency!

Now, redistillation has some problems. First, the equipment is expensive and there's a pretty steep learning curve. Second, and more important, it is illegal to distill liquor without a license here in the United States, and a bar can't ever get that license. Distilling at the bar puts the liquor license in jeopardy, which, once I owned a bar, I could not abide. After I began booze washing, I experimented with combinations of washing ingredients that would produce an aged whiskey or brandy that could be carbonated well, was foolproof, and removed only the flavors that scratch the back of my throat when I drink carbonated unmodified whiskey. I finally settled on a two-step (and fully legal) process using chitosan and gellan gum.

If you've read the section on carbonation, you've met chitosan, one of the magic wine fining agents that I use to clarify lime juice in the centrifuge. Chitosan is a long-chain polysaccharide (sugar) with a positive charge, commercially produced from the shells of shrimp, but it does not cause reactions in people with shellfish allergies. You can get it in liquid-solution form at any home winemaking shop. Unfortunately, the chitosan on the market today is not vegetarian, but that will change. As I write, vegan-friendly fungal chitosan is available in Europe. Chitosan, as a positively charged molecule, is going to attract the negatively charged oak polyphenols in whiskeys and brandies. The problem is, now you have to remove the chitosan. For that I use gellan.

Gellan is a gelling agent derived from microbial fermentation and is primarily used in cooking. Gellan has a lot of interesting properties. It can form fluids that, when standing still, can act like gels to make drinks with suspended solid particles (note: I don't like this application for cocktails). It can form any texture, from soft and elastic to hard and brittle. It is heat-proof. But none of these cool properties are important for us here. Despite what the manufacturers planned, we aren't going to use gellan to make a gel at all. For booze washing, all we are concerned with is gellan's negative charge, which makes it attract the chitosan; and its insolubility in booze, which makes it easy to filter out. There are two types of gellan: low-acyl gellan, aka Kelcogel F, and high-acyl gellan, aka Kelcogel LT100, both made by the CP Kelco corporation. Typically the Kelcogel F makes hard, brittle gels and the high-acyl gellan makes soft, elastic gels. For booze washing you should use the Kelcogel F, because unlike the high-acyl variety, it doesn't swell in water at all and cause us to lose precious, precious booze. The names are random and confusing, but there's nothing we can do about that. Just remember: Kelcogel F low-acyl gellan.

Original blended Scotch on left, chitosan-gellan washed Scotch in the middle and egg-washed Scotch on right. Note the different stripping intensity of the two techniques.

Chitosan/Gellan Washing Technique

15 grams liquid chitosan solution (2% of the booze amount)

750 ml booze to wash

15 grams Kelcogel F low-acyl gellan (2% of the booze amount)

PROCEDURE

Add the chitosan to the booze and shake or stir to combine. Allow to rest for an hour, agitating periodically. Add the gellan to the booze and shake or stir to suspend the gellan. Resuspend the gellan in the booze every 15 to 30 minutes and allow the gellan to remain with the booze for 2 hours, then strain the liquor through a coffee filter. You are done.

NOTES: This recipe uses 2 percent chitosan. That's a lot of chitosan. For comparison, I use 0.2 percent in clarification—an order of magnitude less. I tested the recipe using less chitosan, and strangely, using less chitosan seemed to strip the liquor *more*. Yeah, I don't get it either, but it is hard to argue with empirical data. Two percent gellan is also *a lot* of gellan to use, much more than is used to make a gellan gel (0.5 percent low-acyl gellan makes a hard gel). The reason I use so much is that I'm using only the surface of the gellan powder. The vast majority of the gellan is on the inside of the granules and is useless to us. I could use less gellan if I used a finer particle size, but finer gellan isn't readily available.

Speaking of surface area, remember from the wine fining discussion that adsorptive flavor-stripping ingredients such as activated charcoal work by having tremendous surface area in which flavor molecules can get trapped. I wondered how much flavor-stripping gellan would do on its own owing to surface area effects, so I tested just gellan, without the chitosan. It did strip out some flavor, but not too much.

OPPOSITE: CHITOSAN-GELLAN WASHING: 1) Once you have the ingredients, this technique is preposterously easy and the yield is high. 2) Add chitosan to liquor. Stir and let sit for an hour. 3 and 4) Add the gellan and stir. Stir and let settle a couple more times over the course of 2 hours, then 5) strain through a coffee filter. 6) Look at the leftover gellan. All that color is the stuff you stripped out.

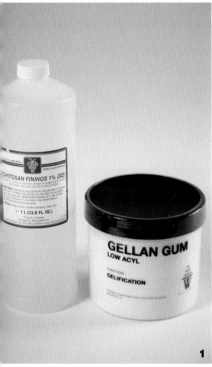

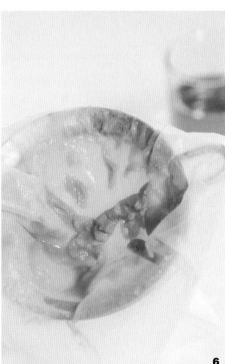

Carbonated Whiskey Sour

Here is a simple carbonated drink to test booze washing. If you don't want to use chitosan/gellan washing, you can use egg washing instead. If you haven't already, read the Carbonation section for carbonation technique (page 288).

**MAKES ONE 5⅖-OUNCE (162.5-ML) DRINK AT 15.2%
ALCOHOL BY VOLUME, 7.2 G/100 ML SUGAR, 0.44% ACID**

INGREDIENTS

2⅝ ounces (79 ml) filtered water

1¾ ounces (52.5 ml) chitosan/gellan washed bourbon (47% alcohol by volume)

5/8 ounce (19 ml) simple syrup

2 drops saline solution or a pinch of salt

Short ½ ounce (12 ml) clarified lemon juice (or add the same amount of unclarified lemon juice after you carbonate)

PROCEDURE

Combine everything (except the lemon juice if it isn't clarified), chill to 14°F (−10°C), and carbonate with whatever system you choose.

What I've presented here on booze washing only scratches the surface of the possible. There's lots more for you to discover on your own. But, as I promised at the beginning of the section, I will move on to a bit on fat washing, a technique for washing flavor *into* (instead of out of) booze.

A Short Word on Fat Washing

Fat washing is simple. Anyone can do it. Pick a flavorful fat or oil. Common choices are butter, bacon fat, olive oil, peanut butter, sesame oil—whatever. Whatever fat you choose, make sure it tastes good. Just because you like bacon doesn't mean all bacon fat is good—properly rendered bacon fat is delicious; fat from overcooked bacon is gross. Butter that has been unwrapped in your fridge is gross. Your fat needs to be fresh and delicious. Next, ask yourself how strong your fat is. Smoky bacon fat is strong in flavor; butter is more delicate. With stronger-flavored fats, use a ratio of around 120 grams (4 ounces) of fat per 750 ml of liquor. With butter I use closer to 240 grams (8 ounces) per 750 ml.

If your fat is a solid at room temperature, melt it. If not, move on.

Add the fat to your liquor in a wide-mouth container, close it, and shake the container (you shake to increase the surface area where the liquor contacts the fat). The wide-mouth container will make it easier for you to get the liquor separated from the fat later on. Let the liquor rest an hour or so, agitating every once in a while for the first half hour, and then proceed to the next step.

Most of the fatty stuff should have floated to the top by now. Place the container of liquor and fat in the freezer. Most fats will form a nice puck at the top after a couple of hours in the freezer; just poke a hole in the fat puck, pour the clear, flavored liquor through a coffee filter into a bottle, and you are done. If your fat will not solidify (olive oil won't, for example), you can use a gravy separator or a separatory funnel (that's what I use) to separate the fat from the liquor.

Fat washing is a great technique, but I don't use it very often, because some of my good friends—Sam Mason and Eben Freeman, formerly of Tailor; Tona Palomino, formerly of wd~50; and Don Lee, formerly of PDT—were its real pioneers, so I leave it to them.

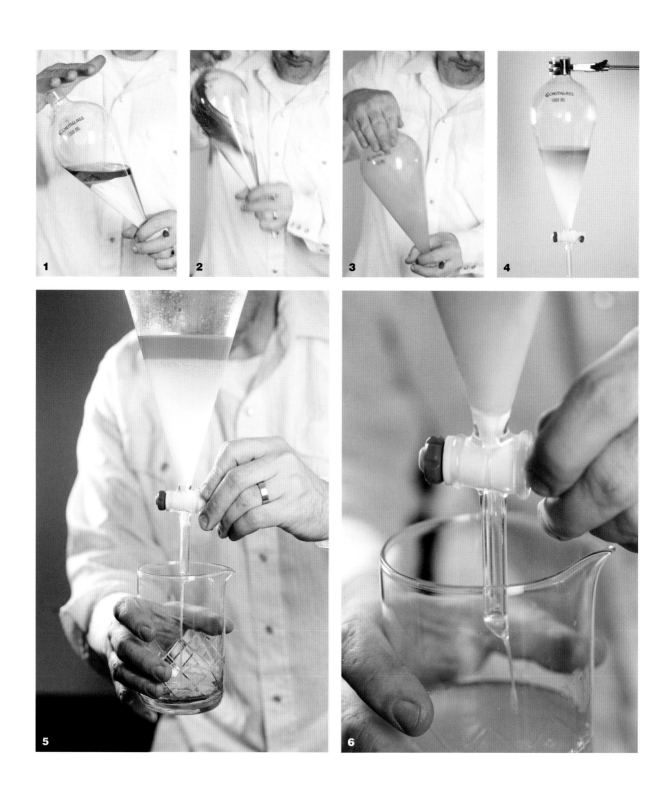

Peanut Butter and Jelly with a Baseball Bat

In 2007 Tona Palomino, who ran the bar program at the famed restaurant wd~50 at the time, made a carbonated peanut butter and jelly cocktail called the Old-School. Because he was carbonating, Tona went to great pains to keep the liquor clear. We don't have to worry about that because we are going to make a shaken cocktail. If you want clear liquor that you can carbonate, you'll have to do what Tona did: spread a thin layer of peanut butter on the bottom of a hotel pan and then pour a thin layer of booze over the top and let it sit covered in a fridge for days.

PEANUT BUTTER AND JELLY VODKA INGREDIENTS

25 ounces (750 ml) vodka (40% alcohol by volume)

120 grams creamy peanut butter

125–200 grams Concord grape jelly

PROCEDURE

Thoroughly mix the vodka and peanut butter. Put the mix in a covered container and place it in the freezer for several hours to settle out. If you have a centrifuge, spin the mix; you should have a yield of around 85 percent (635 ml). If you don't have a 'fuge, pour the mix through a cloth napkin to remove the big particles and then pass the strained liquid through a coffee filter. The coffee filter will clog frequently, so you'll need to use several. The booze should be clearish. Using the napkin-and-coffee filter method, you should get a yield between 60 and 70 percent (450–525 ml). Don't use expensive vodka! Add a shade over 30 grams (a heavy ounce) of grape jelly to every 100 ml of peanut butter vodka and shake or blend to combine. Strain the mix to get rid of any stray jelly particles, and you're done.

OPPOSITE: FAT WASHING GIN WITH OLIVE OIL IN A SEPARATORY FUNNEL (SEP FUNNEL): 1) Combine gin and olive oil in the funnel and cover tightly. 2 and 3) Shake violently to combine. Do this several times, a few minutes apart, and 4) allow to settle. The beauty of the sep funnel: 5) it allows you to drain the heavier liquid booze off the bottom without disturbing the liquid fat on top. 6) The steep conical shape of the sep funnel encourages settling and allows you to get almost every drop of gin separated from the oil.

**MAKES ONE 4⁷/₁₀ OUNCE (140 ML) DRINK AT 17.3% ALCOHOL
BY VOLUME, 9.0 G/100 ML SUGAR, 0.77% ACID**

PB&J WITH A BASEBALL BAT INGREDIENTS

2½ ounces (75 ml) Peanut Butter and Jelly
Vodka (32.5% alcohol by volume)

½ ounce (15 ml) freshly strained lime juice

2 drops saline solution of a pinch of salt

PROCEDURE

Combine the ingredients in a cocktail shaker with
copious ice and shake for just 6 seconds. Don't
overshake; this drink is not good when overdiluted.
Strain into a chilled coupe glass and enjoy.

MAKING THE PEANUT BUTTER VODKA: 1) *The vodka and peanut butter have settled in the freezer for several hours. Either
centrifuge them, or* **2)** *strain first through a cloth,* **3)** *then through a coffee filter.*

PEANUT BUTTER AND JELLY WITH A BASEBALL BAT

Carbonation

Carbon dioxide gas (CO_2) gives bubbly drinks their distinctive taste. The taste of carbonation is difficult to describe. I call it prickly, but that doesn't get it quite right. Carbonation is hard to describe because, as was recently discovered, we have specific elements in our mouths that sense CO_2. In other words, carbonation is an actual taste, like salty and sour. You'd be hard-pressed to describe those tastes as well. Current research indicates that our carbonation sense is related to our sense of sour, but carbonation doesn't taste acidic. People used to think that the sensation of carbonation was due to pain in the mouth caused by the acidity of carbonic acid (formed when CO_2 dissolves in water) combined with the mechanical action of the bursting bubbles. Clearly not true. You can make bubbly drinks with nitrous oxide (N_2O), aka laughing gas, but because N_2O tastes sweet instead of prickly, the drink won't taste carbonated, even if you add acid to the drink. Carbonation is an ingredient, like salt or sugar.

Carbonated drinks are supersaturated with CO_2, meaning that they have more gas in them than they can permanently hold—that's why they bubble. The more CO_2 a drink contains, the sharper the carbonation tastes. Controlling that sharpness is the art of carbonation.

Using premade bubbly mixers cedes cocktail bubble control to the mixer, which isn't a good thing. Most commercial mixers are low quality. Even good mixers can deliver only meager bubbles to a finished cocktail once they've been diluted by alcohol and melted ice. Put simply, if you want bubbles done right, you'll have to carbonate your cocktails yourself.

BUBBLE PHILOSOPHY

I grew up drinking bubbly water, and it has been my primary form of hydration for decades. Flat water is, well, flat by comparison. When I ask for sparkling water, I want the throat-ripping experience of copious carbonation. I believe this is a very American preference, and I'm suspicious of any native son or daughter who prefers lightly carbonated water. But not all drinks benefit from maximum carbonation. Some effervescent cocktails taste good with puny bubbles. Carbonation is an ingredient, and too much is as bad as too little. Many American sparkling wines, for instance, are overcarbonated and taste better after they sit for a while. Overcarbonation can destroy the nuance of fruit flavors, overemphasize oaky and tannic flavors, and add a harsh carbon dioxide bite.

The goal of carbonation is to *control* the bubbles in your drink. You should be able to conjure up the exact level of carbonation that you desire. While there has been a recent proliferation of relatively inexpensive and easy-to-use equipment offering wide access to carbonation, most carbonation technique at the bar remains poor. Carbonating water is fairly easy; even subpar equipment and lax technique can produce decent sparkling water, so people are lulled into thinking the same carelessness will suffice when they move on to carbonated cocktails. Not so. Only great care and perfect technique can achieve good carbonation in an alcoholic beverage. If you understand how carbonation works, you can hone your carbonation technique and achieve good results even with suboptimal equipment. So before we get into specific carbonation techniques, a word on how bubbles work. Expect detail and minutiae. If you have no patience, you can skip to Carbonation in a Nutshell, page 315.

BUBBLES 101

The amount of CO_2 that a bottled drink contains is primarily a function of two parameters, temperature and pressure (well, okay, it is also a function of the ratio of headspace to liquid, but we can safely ignore that variable). The higher the pressure in the bottle, the more CO_2 the drink will contain. More accurately, the higher the pressure of CO_2 in the headspace above the drink, the more CO_2 the drink will contain (in chemistry this fact is known as Henry's law). If the pressure in the headspace comes from regular air as well as CO_2, there will be less CO_2 dissolved in the drink—one of the reasons that air is an enemy of good carbonation.

At a given pressure, the amount of CO_2 a liquid can hold increases as it gets colder. In a sealed bottle the amount of CO_2 can't change, so as the temperature goes up, so does the pressure, and as the temperature goes down, so does the pressure. The two are tied together.

How fast the CO_2 gets into your drink is a different story. Simply applying CO_2 to the empty space above a drink doesn't carbonate it very quickly at all, because the CO_2 can diffuse into the drink only through the relatively small surface area of the liquid at rest. Just as you chill a drink with ice by stirring or shaking to bring fresh cocktail in contact with fresh ice, you carbonate by radically increasing the fresh surface area that the drink shares with the CO_2. You can achieve this by shaking the drink under pressure, or by injecting large amounts of tiny CO_2 bubbles into the drink, or by spraying the drink as a mist into a pressurized container of CO_2. I don't care how you do it—just up the surface area.

TINY BUBBLES

Many people think they want tiny bubbles. They think tiny bubbles are a hallmark of quality. The singer Don Ho immortalized those "Tiny Bubbles in the Wine" and set the consuming public on a seemingly inexorable path toward bubble prejudice. But let's look at how bubble size is determined. In a highly carbonated drink, more CO_2 rushes into the bubbles at any given time, making bigger bubbles. Similarly, the warmer a drink is, the more CO_2 tries to leave, making larger bubbles. Last, the taller you pour a drink, the larger the bubbles become, because they have longer to live and grow from where they form till they pop at the surface of the drink. None of these bubble-size factors are markers of quality.

The composition of the drink also influences bubble size. Different drinks—let's say a glass of champagne and a gin and tonic—carbonated to the same level and served at the same temperature in the same kind of glass will have different bubble sizes. The ingredients of a drink affect how easily bubbles are formed, how fast they grow, and how much CO_2 wants to leave. Even among fairly similar liquids, like different white wines, the grape variety and the amount of yeast breakdown products radically affect bubble size. Different bubble sizes in these cases are not hallmarks of quality but a function of drink composition.

So how did the tiny-equals-good mythical equation arise? An analysis of champagne aging helps.

LE CHÂTELIER

Strangely, the reason the solubility of CO_2 goes up as temperature goes down is that CO_2 releases energy when it dissolves in liquid; the reaction is exothermic. In fact, the heat from dissolving CO_2 in a highly carbonated drink can raise its temperature by over 5°C!

My assumptions: dissolution enthalpy of CO_2 = 563 calories per gram. carbonation level = 10 grams per liter, beverage = pure water

The temperature rise is a rarely considered consequence of carbonation, but it is the reason we can apply one of the fundamental principles of chemistry, Le Châtelier's principle. Named after the French chemist Henry Louis Le Châtelier, Le Châtelier's principle states that if you push a chemical system away from equilibrium, that system will tend to push back toward equilibrium. If you have bubbly water and CO_2 at a given temperature and you remove heat from the system by making the bubbly water colder, Le Châtelier's principle says that the system will react by trying to make more heat. How? By making more CO_2 dissolve. True? Yes. Intuitive? No.

Young champagne has a lot of carbonation and fairly large bubbles. As champagne ages, CO_2 gradually diffuses through the cork, which is not a hermetic seal, and lowers the carbonation, producing tinier bubbles. So in champagne, weaker carbonation correlates with greater age. Since only fine vintage champagnes are typically aged very long, age and the tiny bubbles it brings denote quality. Additionally, recent research shows that as champagne ages, its composition changes, so that for a given level of CO_2, old champagne will have smaller bubbles. So in champagne, tiny bubbles correlate with age, which can often correlate with quality (we tend to age only fine champagne). Certainly the tiny bubbles aren't the cause of the quality. If I take a lousy sparkling wine and partially decarbonate it, I will end up with an equally lousy sparkling wine with tiny bubbles. If I take great vintage champagne and up the carbonation, it will still be great, but with bigger bubbles.

SABERING CHAMPAGNE AND OTHER BUBBLY WINES

While not really a carbonation technique, a section on bubbles seems the most apt place to describe sabering a bottle of champagne.

Many people feel that sabering sparkling wine is useless and wasteful. I disagree. Sabering expensive champagne is wasteful (if you make a mistake). Sabering a $7 cava is an exhilarating, awesome party trick. Whether a bottle will saber depends only on the bottle, not the price of the wine, so stick with the inexpensive.

Sabering is the art of cleanly knocking the top off a bottle of sparkling wine. You hit the lower lip of the top of the champagne bottle and snap off the top of the neck. Yes, you break the glass. No, the glass doesn't get into the drink, because the momentum carries it away from the neck. Sabering works because there is a ledge where the lip meets the neck. Stress concentrates at this ledge, making the bottle want to snap cleanly. You will get small shards of glass on the floor, so be careful. If you have young children, be doubly careful; I know from bitter experience that glass shards will find tiny feet the next morning. Obviously, saber away from any people and don't aim toward mirrors or closed windows. And don't saber over food, unless you're a glass-eating circus freak.

THE PROCEDURE

Select a bottle that looks pretty standard. Don't pick one with a funky neck—sabering might not work. Superimportant tip: if you are sabering in front of a crowd you're looking to impress, select a bottle you *know* will saber. If you sabered a particular brand before (Paul Chenaux cava, for instance, or Gruet sparkling), odds are it will work again. If you have failed with a bottle before (Cristalino Cava), you will probably fail again. Embarrassing! A corollary to this tip: practice first to figure out which bottles will work.

Chill the bottle. Let it rest upright for a while before you saber it, and be gentle with the bottle before you saber. Warmer bottles are easier to saber but tend to gush. The best saber jobs don't gush at all

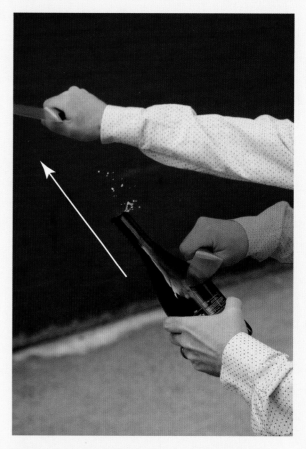

When you saber a bottle of bubbly, avoid swinging your knife in an arc. With confidence, swiftly and smoothly run the back of your knife squarely up the side seam of the bottle, strike the glass lip, and "punch" the glass collar and cork off of the bottle. This trick takes no strength—except strength of will. Don't hesitate or pull back.

(take that, anti-saber snobs). Don't take off the wire cage until you are ready, lest the cork come out on its own. Some people saber with the cage on, but I think it's more difficult.

Get a knife. It doesn't need to be heavy. It doesn't need to be sharp. In fact, it doesn't even need to be a knife—I made a stainless steel ring for sabering at parties. You'll use the back (dull) side of the knife, not the sharp one. I saw a friend forget this rule one night and ruin her host's good chef's knife.

Find the seam running up the side of the bottle; this seam is a weak point and further concentrates the stress when you hit the lip. Angle the bottle away from you, your friends, any glass, and any food. Place the knife on the bottle's seam at the bottom of the neck, making sure you keep the knife flat against the bottle. If you don't, the knife has a tendency to pop over the lip of the bottle.

The moment of truth: slide the knife smoothly, surely, and SQUARELY up the neck of the bottle and sever the top. It doesn't take force, just confidence. The biggest and most common mistake: swinging the knife in an arc. If you swing in an arc, even a small one, you won't hit the glass in the right place and you won't sever the neck. Mortifiying.

If it doesn't work, try one more time, maybe two. Don't try five or six times on the same bottle. Whacking away seems desperate. If the bottle doesn't want to saber and you force it to, you might get a bad break, shattering the bottle completely.

Remember that the momentum carries all the glass shards away from the neck and your drink (that's why I told you to hold it at an angle). Pour with confidence and enjoy.

Turns out I always blink during the money moment.

LEFT: My Carbon Dioxide/Nitrous mixing rig. I took apart and modified a Smith gas mixer designed for mixing welding gases, converting it from a flow-based device to a pressure-based one that allows me to carbonate at any pressure up to 60 psi with any ratio of CO_2 to N_2O. RIGHT: An alternative to the mixing rig: make pre-mixed gas cylinders.

The quantity of CO_2 in a drink, not bubble size, is the most important carbonation characteristic. The amount of CO_2 escaping from the drink determines how rough and biting that drink will be and how the volatile aromas will be punched into the air above the glass to greet your nose. Sometimes I want to make extremely lively drinks with huge bubbles that effervesce massively on the tongue and eject a lot of aroma compounds toward your nose. If I were to create those drinks using only CO_2, they would sting your nose painfully as you tried to drink. Instead I make super-bubbles with a gas mix. Remember that nitrous oxide (N_2O) is highly soluble in water and tastes not sharp but sweet. When I add a percentage of nitrous to my carbonated drinks, the result is big lively bubbles that aren't painful. The gas-mixing system is a bit involved, and unless you are a dentist, tanks of N_2O are hard to come by (people sometimes abuse it as a drug and die when they fall asleep with a nitrous mask on), so I won't tell you how to do it, but if you can decode what is going on from a picture of my rig, go for it.

HOW MUCH CO$_2$ IS IN MY DRINK?

When I quote carbonation levels, I use grams of CO$_2$ per liter of beverage (g/l). If you read technical soda literature, you'll notice that CO$_2$ is measured in an arcane unit called "volumes of CO$_2$." A volume of CO$_2$ equals 2 g/l of CO$_2$ (see the sidebar on volumes of CO$_2$, page 297). Your average cola contains about 7 g/l. Orange soda and root beer have less carbonation, typically around 5 g/l. Mixers such as tonic water, which are supposed to be watered down with booze, and seltzer water, which is supposed to rip your throat out with bubbly goodness, have much more carbonation—around 8 or 9 g/l. In nonalcoholic drinks, levels much above 9 g/l become painful. But with alcoholic drinks, it's a different story.

IMPORTANT CARBONATION PROPERTY OF ALCOHOL

CO$_2$ is more soluble in alcohol than in water, so less CO$_2$ escapes a boozy drink to hit your tongue. Because the sensation of carbonation depends on CO$_2$ leaving your drink and hitting your tongue, you need to put more CO$_2$ into an alcoholic drink to have it taste as carbonated as its nonalcoholic cousins. According to the literature, young champagne, which is 12.5 percent alcohol by volume, can have as much as 11.5 to 12 g/l of CO$_2$ dissolved in it. That much CO$_2$ in seltzer would be horrific. Frankly, it's too much CO$_2$ for champagne as well, but it won't rip your face off.

In general, **the more alcohol a drink contains, the more CO$_2$ you will need to add when carbonating.**

Measuring carbonation is a tricky business, and in practice I don't do it. I just carbonate my drinks at a constant temperature and pressure. If I don't like the results, I maintain the temperature and dial the pressure up or down. It is sometimes useful, however, to measure how much CO$_2$ you are putting into your drink. The easiest way to measure CO$_2$ is to weigh your drink before and after the carbonation proce-

ANOTHER IMPORTANT CARBONATION PROPERTY OF ALCOHOL: CARBONATING RESPONSIBLY

Carbonated cocktails can get you majorly shellacked. CO$_2$ really helps to get ethanol into your bloodstream. That old saw about champagne going to your head is not a myth. Alcohol is absorbed mainly in your small intestine, not your stomach. The faster alcohol gets to your intestine, the faster you can absorb it and the faster, and higher, your blood alcohol content (BAC) can spike. The current wisdom is that the CO$_2$ in drinks helps speed gastric emptying into your small intestines, causing a rapid rise in BAC. Compounding the issue, people are accustomed to drinking carbonated drinks rather quickly, so they do, regardless of alcohol content.

The takeaway message here is to keep the alcohol levels of your carbonated drinks down. When I first started carbonating drinks, I regularly carbonated cocktails at shaken and stirred drink dilutions: 20 to 26 percent alcohol. People were on the floor in short order. This outcome was not my goal, and I'm sure it's not yours. Nowadays my carbonated cocktails average between 14 and 17 percent alcohol by volume. These drinks are more civilized and taste better. As you will see, drinks with less alcohol have better carbonation, last longer in the glass, and are less cloying than higher-alcohol versions.

 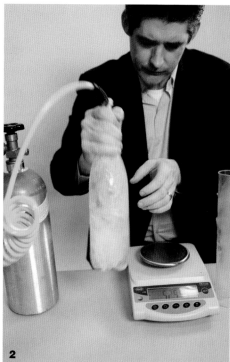 

1

2

3

**HOW TO MEASURE THE AMOUNT OF
CO₂ YOU ADD TO A DRINK: 1)** *Weigh the
un-carbonated liquid on a scale in the carbonator bottle without the cap.* **2)** *Carbonate.* **3)** *Weigh the bottle again with the cap
off. The weight on the scale is the weight
of added CO_2—in this case 4.3 grams into
500 ml. It's important that you remove the
cap so that the scale does not register the
weight of compressed CO_2 in the bottle.*

dure (with the cap off). The extra weight is the CO_2 you've dissolved into
your drink. You should convert your answer to grams per liter.

THE PROBLEM OF FOAMING AND THE THREE C'S
OF CARBONATION

CO_2 in the bottle is great, but it doesn't amount to a hill of beans if it never
makes it to your tongue. What's important is the CO_2 your drink has in the
glass. When you open a bottle of bubbly and pour a drink, you lose carbonation—maybe a lot of it. Gerard Liger-Belair, the champagne physicist (yes,
that is a real job, and you are too late to get it), has studied the phenomenon
of CO_2 loss in champagne quite closely. His studies show that uncorking a
bottle of fridge-temperature champagne and pouring it carefully down the
side of an inclined glass—the best-case pouring scenario—can lower the
amount of CO_2 in the champagne from 11.5 grams per liter to 9.8 grams per
liter. That loss is probably a good thing; 11.5 grams per liter is too intensely
carbonated. Carelessly pouring that same champagne into the same glass
reduces the carbonation to 8.4 grams per liter, my lower limit for good car-

bonation in a fresh glass of young champagne. If you open the same champagne when it's room temperature, you can knock a couple more grams per liter off those numbers. If you open a bottle carelessly and it gushes out the neck and over the side, the problem is much, much worse. Gushing, or fobbing as it is called in the soda business, can destroy precious carbonation lickety-split. In fact, foaming is carbonation's biggest enemy. Nine-tenths of good carbonation is foam control.

CO_2 leaves a drink in two ways: directly from the surface of the drink and through bubbles. The first path is controlled by the glass. Roughly, the less surface area of drink is exposed per unit of volume, the less CO_2 you lose per unit of time. This is why champagne flutes and other tall glasses are good for carbonated drinks, and why you should fill them high if they are going to sit around for a while (more volume for the same exposed surface area).

Controlling how much CO_2 leaves in bubble form is more difficult. If you pour a perfectly clear carbonated liquid into a perfectly clean, perfectly smooth glass, you get no bubbles—none—because in a pure liquid it is extremely hard to form a bubble. Bubbles in bubbly drinks inflate the same way a balloon does. The pressure of CO_2 in the drink is higher than it is in the bubble, so the bubble blows up like a balloon. The surface tension of the liquid surrounding the bubble acts like the rubber in a balloon to resist inflation. Bubbles are also like balloons in that the smaller they are, the harder they are to blow up. There is a critical bubble radius—dependent on the liquid's surface tension and the pressure of CO_2 in the drink—below which any bubble formed will get crushed back to nothing. Any bubble larger than that critical radius will tend to grow. Surprisingly, you can't just raise the pressure to initiate bubbles. You'd need hundreds of times more CO_2 than you get in ordinary bubbly drinks to form spontaneous bubbles.

What you need to form bubbles are **nucleation sites**: anything in or touching a liquid that enables bubbles to form.

VOLUMES OF CO_2 FOR IDEAL GAS LAW FANS

In the 1800s an Italian dude named Avogadro figured out that at equal temperatures and pressures, equal volumes of gases contain the same number of molecules regardless of how much those molecules weigh: a deep insight. Remember that. In a related story, the chemistry community named an important number after Avogadro—Avogadro's number, which equals 6.022×10^{23}, which is mind-bogglingly huge. Avogadro's number is important because it helps us to shift between weights on an atomic scale and a macroscopic scale. Different kinds of atoms and molecules have different weights. Those weights are expressed in a relatively useless unit, the atomic mass unit (amu). Avogadro's number is the conversion factor between atomic mass units and grams. If you know that the mass of a single molecule of CO_2 is 44.01 amu, you know that 6.022×10^{23} CO_2 molecules weigh exactly 44.01 grams. Unfortunately, you also have to remember another unit, because a pile of CO_2 containing Avogadro's number of molecules isn't called an Avogadro of CO_2, it is called a mole. The mole is one of the most important units in chemistry—it allows you to work with ratios of actual molecules using grams.

Okay, so one mole of CO_2 has a mass of 44.01 grams and contains 6.023×10^{23} molecules. Back to Avogadro's original hypothesis: one mole of gas at standard temperature and pressure (hereinafter referred to as STP) will fill up the same volume regardless of how much that mole weighs. Ideal gas law fans will remember that the volume of one mole of gas at STP is 22.4 liters. So one mole of CO_2 at STP will occupy 22.4 liters and weigh 44.01 grams.

Finally we can come back to the arcane unit of carbonation: volumes of CO_2. One volume of CO_2 is defined as the mass of CO_2 gas, in grams, that would fill a particular volume, in liters, at STP. That is 44.01 grams divided by 22.4 liters: roughly 2 g/l. Sufficiently arcane?

The water on the left is highly carbonated. It is not bubbling at all because the glass has no nucleation sites—I analytically cleaned the glass so it was free of all residue. Drop a rock of sugar into the drink and it immediately starts bubbling.

Bubbles like to form on anything that is discontinuous or already contains trapped gas. In a glass, those nucleation sites can take the form of scratches or marks in the glass, or leftover minerals or schmutz from washing, or, most famously, leftover fibers from the towels used to dry and polish. In the drink itself, nucleation sites come from suspended particles, dissolved and trapped gas, or any currently existing bubbles. If you have too many nucleation sites, there are too many places where CO_2 can form and rush into bubbles, and you get gushing. Gushing loses tremendous amounts of CO_2. Even worse, nucleation sites keep nucleating for the life of the drink. Drinks with lots of nucleation sites lose lots of gas right away and continue to lose it faster than drinks with fewer nucleation sites.

You can't eliminate nucleation sites, but you can reduce them or limit the effect they have by following the three C's of carbonation: clarity, coldness, and composition.

C #1: CLARITY

If your drink isn't clear, it won't be easy to carbonate. I clarify everything I'm going to carbonate. Some commercially carbonated drinks are cloudy, such as Orangina, but notice that they have very light carbonation and no alcohol. Even adding small amounts of unclarified juice to a highly carbonated cocktail can cause major foaming. Notice that the flip side of clear isn't colored or dark, it's cloudy. Colored drinks are okay. Drinks that have so much color they appear opaque, like red wine and cola, are still okay to carbonate if they aren't cloudy when you hold them up to the light. The first duty of the carbonation maven is to clarify. To learn how, see the section on clarification (page 235). If you don't want to clarify, choose ingredients that are already clear. If you absolutely must use cloudy ingredients, use small amounts and add them at the end, after you have carbonated the bulk of your drink.

C #2: COLDNESS

Warm drinks bubble and foam violently when they are opened. Some people try to fix the foaming problem by jacking the pressure to add more gas. This won't work. Once a drink foams violently when it is opened, you have exceeded the level of CO_2 that the drink can viably hold. Increasing the pressure beyond that point will actually reduce the amount of CO_2 you get in the glass, because the foaming will get worse and worse. I have seen many people boost carbonation pressures higher and higher, only to find their drinks tasting flatter and flatter. Better to chill your drink properly first.

How cold should you chill? My rule of thumb is, as cold as possible: at, or just above, the freezing point of the beverage. The colder you get your drinks, the more CO_2 they can hold without foaming and the less carbonation you'll lose when you serve them. I carbonate water-based drinks at 0° Celsius. I carbonate most of my cocktails between −6°C and −10°C (21°F and 14°F). Very highly alcoholic drinks, such as carbonated straight shots, I carbonate at −20°C (−4°F).

If you chill your drink too much, it will start to freeze and form zillions of tiny ice crystals. Tiny ice crystals are fabulous nucleation sites. If you open a bottle with those crystals, you'll produce huge quantities of bubble-robbing foam. As long as the bottle is sealed, don't worry—

you won't lose carbonation. Don't open the bottle until the ice crystals have melted. Once they melt, your drink will be in top form.

A last note on temperature: your carbonation will be more consistent if you maintain a constant temperature when you carbonate. If you try to carbonate at any old temperature, you will never get consistent results. Remember, as the temperature goes up or down, so does the pressure required to reach a particular level of carbonation.

C #3: COMPOSITION

A cold clarified drink can still foam because of its composition. We'll break the composition problem into three parts: suds, air, and alcohol.

Suds:

Some ingredients, even when totally clarified, foam like demons. Clarified milk whey, which I use in a lot of uncarbonated drinks, is full of protein, and protein loves to make foams. I always suspected that whey would pose problems, but other ingredients, like cucumber juice, surprise me with their foamy intransigence. In general, if ingredients have a large amount of protein, emulsifiers, or surfactants, or are viscous, you can plan on dealing with foam. A good way to know if you'll have problems is to shake some of your test ingredient in a clear container and observe the foam. Does it make a layer of foam on top? If it does, does the foam break down rapidly? The foamier and more persistent the suds are, the harder it will be to carbonate your ingredient. Other than avoiding such ingredients, your only option for controlling foam is to cover your other bases properly: clarify, chill, and carbonate carefully. I have tried using antifoaming agents like polydimethylsiloxane to help fix the foam problem. No luck. I'm kind of glad. Who wants to say they add polydimethylsiloxane to their cocktails?

Air:

As I've mentioned before, air is an enemy of carbonation. Besides taking up space in the bottle that is better used by CO_2, air causes foaming. Small amounts of air trapped on microscopic dust specks are major nucleation sites. In addition, the small amounts of air dissolved in the drink itself, and the small air bubbles formed when a drink is agitated, inflate when you open a carbonated drink and produce loads of annoying foam. Luckily, the solution is simple: carbonate more than once.

I'll say that again: to get good carbonation you must carbonate more than once. An illustration of the procedure will show why. Carbonate your drink. Open it right away and let it foam. Don't let it spray everywhere—that's messy and wasteful—but let it foam. At this point you *want* the drink to foam. As the drink foams, it forms large, quickly expanding CO_2 bubbles that carry with them a lot of the trapped air and other nonsense that was stuck in your drink. The CO_2 rushing out of your drink also pushes air out of the headspace. After the foam starts to settle, carbonate again and let it foam again. Notice that this time it foams a bit less (you hope). Carbonate again. This time let the drink settle before you ever so carefully and slowly release the pressure. After three rounds of carbonation you should have strong, stable, foam-resistant carbonation that lasts.

Alcohol:
Adding alcohol makes drinks more difficult to carbonate. Alcohol lowers surface tension and simultaneously increases viscosity—a double whammy. Lower surface tension means bubbles can form more easily. Higher viscosity means that bubbles don't dissipate quickly on the surface. The upshot is that alcohol makes drinks foam, even if you do everything else properly.

It gets worse. Remember that alcohol requires more CO_2 than water to get the same feeling of carbonation on your tongue, because CO_2 is more soluble in alcohol than in water. So when you add alcohol, you are in a real pickle. It automatically foams more than water, and you have to add more CO_2 to it than to water, increasing the foaming even more. Ouch. This is one reason I keep my carbonated drinks between 14 and 17 percent alcohol by volume. For another reason to keep the alcohol by volume down, see the sidebar on carbonating responsibly, page 295.

Last Word on Foam
Remember, no matter how good you are at carbonation, your drink will foam. If you don't want your drink spraying everywhere when

Here I break all the rules of foam prevention. I'm quickly opening a recently agitated, poorly clarified, too-warm alcoholic drink—with predictable results.

you open it, you need to leave enough headspace in your carbonation vessel. Learn from my mistakes.

FINALLY! HOW TO ACTUALLY CARBONATE SOMETHING: EQUIPMENT AND TECHNIQUES

There are a million ways to skin the carbonation cat. I'll describe three. Hopefully you can adapt one of these techniques to whatever equipment you have. The systems I describe will certainly become outmoded, but the principles of carbonation and how they are applied are immutable.

TECHNIQUE ONE: BOTTLE, CAP, AND TANK

The easiest of the techniques: just use a special adapter to hook up a plastic soda bottle to a large-format CO_2 tank. This is the system I use at home and at the bar. It is simple, relatively cheap, and produces amazing results very predictably for pennies per drink.

THE HARDWARE

CO₂ TANK: I like this system because of the large tank and the regulator. Large CO_2 tanks, as opposed to cartridges or minibottles, are superior, because the gas is much cheaper and you'll make a whole lot of drinks before you run out. CO_2 tank sizes are measured by how many pounds of CO_2 the tank can safely hold. I use 20-pound tanks at home and at the bar, and 5-pounders when I do demonstrations on the go. A 20-pounder will just fit in a standard under-counter cabinet and will carbonate hundreds of gallons of water, wine, and cocktails.

You can rent a CO_2 tank from your local welding supply shop, but just buy one. As I am writing, a brand-new empty 20-pound CO_2 tank can be had for less than $100. The folks at your local welding shop (you have one; you just need to find it) will swap your empty tank for a full one and they'll just charge you for the gas. Depending on where you live, you will pay between $1 and $2 a pound for the gas. You need to chain or secure the tank in an upright position and keep it away from extreme heat. Before you use a tank you must familiarize yourself with basic CO_2 safety principles—not rocket science, but you need to study them. Check the Internet—a number of companies post advice.

CARBONATOR RIG: *On right, a standard soda bottle. In the center, a 5-pound CO_2 cylinder connected to a pressure regulator that can supply 120 psi of pressure. A gas hose comes out of the regulator and terminates in the gray ball-lock connector. In use, the red carbonator caps at left screw on the soda bottle and connect with the gray ball-lock connector.*

REGULATOR: Your system will need a regulator to drop the high pressure of the tank—roughly 850 psi at room temperature—down to the lower pressures you'll use for carbonation—30 to 45 psi. This regulator enables you to effortlessly change the pressure with which you are carbonating. Prices vary widely. If you are using the system at home, go for inexpensive regulators like the 0–60 psi regulators from Taprite, which will set you back only about $50. At the bar, where our regulators get regular abuse, I use heavy-duty models with protective cages over their pressure gauges (the most fragile part of the regulator). CO_2 regulators have two gauges, one to show the pressure of the tank and one to show the pressure coming out of the regulator. You cannot use the tank pressure to figure out how much CO_2 you have left. The majority of the CO_2 in your tank is a liquid. As you take gas out of the tank, some of the liquid CO_2 turns to gas and maintains a constant pressure inside the tank, so the gauge will still read full pressure until the tank is almost empty. The weight of the tank and the sound of liquid sloshing about inside it are the only reliable measures of CO_2.

BALL-LOCK CONNECTOR: A length of reinforced hose will connect your regulator to a special fitting called a ball-lock gas connector. Ball-lock connectors are special doodads that were designed in the 1950s to pressurize and dispense from 5-gallon soda kegs (one of two such systems; the other is known as pin-lock). Soda suppliers would ship the soda, known as premix soda, to bars and restaurants in those kegs. The competing system, also designed in the 1950s, is called bag-in-box. With a bag-in-box system, syrup is added to carbonated water at the last minute, just before the drink enters the glass. The two systems existed side-by-side for a while, but bag-in-box eventually won out and rendered soda kegs and their connectors obsolete. In the 1990s and early years of this century, while bag-in-box was consolidating its hegemony over the fountain soda world, the market was flooded with surplus premix soda kegs and connectors. Home brewers snatched them up. Buying surplus, home brewers (yep, I was one) could brew, keg, and tap their beers in convenient 5-gallon quantities, almost for free. Those were good times. Today the surplus is gone, but the home brewer's love is not, and the ball-lock soda fittings and kegs live on, at a higher but still reasonable price.

Some crazy home brewers loved ball-lock fittings so much that they made a cap—the carbonator cap—to attach their beloved ball-lock fittings directly to standard soda bottles. They weren't trying to make a great cocktail system; they just wanted an easy way to keep their beer carbonated in small quantities. I started using them as soon as I found out about them, and they were quickly embraced by tech-minded cocktail folk. They cost around $15 apiece and last for many, many hundreds of carbonation cycles, as do the soda bottles you'll use with them.

This entire carbonation rig—tank with gas, regulator, hoses, fittings, caps, and bottles—should cost you less than $200. Here is how to use it.

CARBONATING WITH BOTTLE, CAP, AND TANK: Turn on the CO_2 gas cylinder and adjust your pressure (for most cocktails I use 42 psi.) Fill a soda bottle two-thirds to three-quarters full with very, very cold beverage.

At the bar I have a special freezer, the Randall FX, to chill my drinks to exactly the temperature I want—typically 7°C (20°F). At home you can just leave your drinks in the freezer till they get syrupy (high alcohol) or just start forming ice crystals (lower alcohol). The ice crystals will melt as you carbonate. You can also chill drinks with dry ice or liquid nitrogen, but make sure neither gets into the bottle: explosion hazard!

Now squeeze all the extra air out of the bottle—remember that air is your enemy—and screw on the carbonator cap. Use your fingers to pull up on the ring around the ball-lock connector, press it down hard on the carbonator cap, and release the ring. Many people have problems managing the connector the first few times. You will get the hang of it pretty quickly. As soon as the connector is on, the bottle will inflate quickly, which is fun to watch.

Now, with the gas connected, shake the drink like your life depends on it. You will hear and feel the gas leaving the tank and going into your drink. Sounds like the creaking hull of a ship. Don't hold the bottle upside-down when you shake lest cocktail get into your gas line, which is icky. After you shake the drink, pop off the connector and crack open the carbonator cap to let the drink foam up. Don't unscrew the cap all the way, or you and everyone around you will be covered in cocktail. After the foam has died down, carbonate again and foam again. Carbonate a third time and you should be done.

Let the bottle rest for at least 30 seconds—longer is better, up to a minute or two—before you open to pour. A little residual foam on the surface of the drink is acceptable. The problem is bubbles in the drink itself. You want to wait for all those bubbles to rise to the top and pop. You'll know when you are ready to open because the drink will appear clear. When you open to pour, crack the cap very gently. At this point you want to prevent foam. After you pour your drink, squeeze a little excess gas out of the bottle (in case some air got in), screw on the carbonator cap, and hit the bottle with CO_2 again—a process I call fluffing. If you keep the temperature right and fluff every time you pour, the last cocktail you pour out of the bottle will be as good as the first.

I find that my results are best when my bottles are filled two-thirds to three-quarters full. Any fuller and it is hard to defoam the drink. Any less full and it is hard to get the air out. I also feel in my heart that the carbonation I get in an underfilled bottle isn't as good, but my head doesn't know why. To make sure I'm always using optimum

Carbonating a vermouth cocktail with a carbonator cap. **1)** *Squeeze all the air out of your bottle.* **2)** *Screw on the carbonator cap.* **3)** *Pull up on the ring of ball-lock connector, push it down on the carbonator cap and release the ring to apply gas. Sometimes it takes people a while to get the hang of this.* **4)** *When the pressure of the gas is applied, the bottle inflates rapidly.* **5)** *Shake.*

6) *Vent and allow foaming.* 7) *Pressurize and shake again.* 8) *Vent and allow foaming.* 9) *Pressurize and shake again.* 10) *Wait until all bubbles subside and then gently open the bottle.* 11) *Pour at an angle.* 12) *The drink.*

fill levels, I stock a variety of soda bottles: 2 liters, 1.5 liters, 1 liter, 20 ounces (591.5 ml), 16.9 ounces (500 ml—Poland Spring makes these), and 12 ounces (354.9 ml—hard to find, but Coke makes some). I use the 2 liters for large events where I'm pouring a lot in a short time, although the sloshing you get when you pour out of a 2-liter bottle makes me wince. I use the 1-liter bottles at the bar. The small 16.9- and 12-ounce bottles are superhandy to keep around for recipe testing. They can carbonate a single drink.

There are two downsides to this system: the recycled plastic bottles don't look sexy behind the bar, and the carbonation cycle takes a bit too long to make individual drinks to order. Regardless, it's the best system out there.

TECHNIQUE TWO: SODASTREAM

The Sodastream carbonator has revolutionized home carbonation. The manufacturer warns you not to carbonate anything other than pure water, but you can, provided you are careful. With water, Sodastreams are supersimple to use: just fill the special bottle up to the fill line with cold water, screw it into the machine, and press the carbonator button till the machine breaks wind. The Sodastream doesn't require any shaking and doesn't require you to purge the headspace of the bottle. Instead, it submerges a carbonator wand just below the surface of the water. When you push the button, the wand injects CO_2 into the water, creating zillions of tiny bubbles with their attendant surface area and obviating the need for shaking. At the same time, CO_2 rushes out of the drink and pushes up into the headspace. When the pressure in the bottle exceeds the factory-set carbonation pressure, a relief valve opens up and lets the CO_2 push the air out of the headspace with a rather impolite sound. Smart system. You can't adjust the pressure in a Sodastream, but it can produce adequate carbonation in most cocktails.

The problem with carbonating cocktails is the pesky foam. If you carbonate something that foams, you can clog the relief valve. If you clog the relief valve, the pressure that builds up can cause damage to the Sodastream itself and, if the bottle ruptures or flies out of the machine, to anything nearby.

The trick to carbonating wine or cocktails in the Sodastream: make sure you never let the foam get all the way to the top of the bot-

PLASTIC SODA BOTTLES

Plastic soda bottles are made from polyethylene terephthalate, aka PET (or PETE), the most common form of polyester. It's the same stuff from which fine leisure suits are crafted. PET soda bottles are cheap and flexible and can easily handle the pressures involved in carbonation. PET bottles for still beverages can't handle the pressure; I've had them blow up. There isn't any real danger when one of those bottles blows while you're carbonating, but it sure is embarrassing: imagine a small chef's office with a 360-degree line of carbonated iced coffee sprayed across the wall at chest level. Even the carbonation bottles aren't indestructible, however. Don't stick chunks of dry ice in them; they will blow up. Ditto with liquid nitrogen. Don't put boiling hot liquids in them either, or they'll deform. Over long periods of time, PET bottles can pick up aromas and colors from the sodas they hold, so stay away from root-beer bottles and orange-soda bottles.

Carbonator caps are supposed to fit on any standard soda bottle up to 2 liters. Unfortunately, the soda industry introduced a new cap style several years ago. The new caps are shorter. Shorter caps are good because they save plastic, but bad because carbonator caps don't fit their bottles as well. You *can* use the carbonator caps with new-style bottles, but get the old style if you can.

Unlike glass, which is a perfect gas barrier, PET soda bottles slowly leak gas. The majority of the leakage is straight through the walls of the bottle; they are semipermeable to gases. This is the reason you shouldn't stock up on soda in plastic bottles and leave them in your pantry for months. The lifetime of a PET bottle of soda is measured in weeks before it starts tasting flat. The smaller the bottle, the more surface area the bottle has per unit volume, so the faster it goes flat. This is why bars always buy tiny glass bottles of soda—plastic bottles that size would go flat really fast. If you want to store a home-carbonated product for a long time, make sure to fluff the bottle with CO_2 every week or so to maintain carbonation.

A newer version of soda bottle is made of layers of PET and polyvinyl alcohol (PVA). PVA is a much better gas barrier than PET and degrades into harmless products when the bottles are recycled, so these bottles are still recyclable and they maintain carbonation much better than plain PET bottles. Unfortunately, the bottles are just labeled with the PET recycling code, so you can't know whether you have one.

From an environmental and health perspective, it might interest you to know that in 2011, only 29 percent of all PET bottles were actually recycled (according to EPA data). Frown. But according to the National Association for PET Container Resources (NAPCOR), the rate of recycling is increasing. NAPCOR also vehemently asserts that it is safe to reuse, freeze, and store products in PET, that nothing is ever leached into your beverage from PET, that PET contains no evil bisphenol A, and that even though the word *phthalate* is in the name of the plastic, the phthalates in PET aren't the ones we should be worried about. Of course, whether you choose to believe data provided by a trade group whose sole mission is to support the use of PET containers is up to you. From my perspective, I have drunk out of PET my whole life. I currently use PET, and serve drinks made in it to my family. That doesn't mean I can state with apodictic certainty that there is nothing wrong with PET, but I will continue to use it till I hear credible rumblings of safety concerns.

tle. You will not be able to fill up the bottle to the fill line. In fact, don't fill the bottle more than one-third of the way to the fill line—about 11 ounces (330 ml), which equals two drinks. You need to leave plenty of room for the drink to foam. At one-third full the carbonator wand will not be submerged. That's okay. The force from the spraying CO_2 is strong enough to create tiny carbonating bubbles even if the wand isn't submerged. You'll just need to press the button a few more times and waste more CO_2 than you would when carbonating water. You can't make less than two drinks, however, because if your fill level is too low, the CO_2 won't get injected well.

Therefore, in one sentence, the secret of the Sodastream is **always carbonate exactly two 5½-ounce (165-ml) drinks**.

Just as with other all carbonation techniques, you should vent the Sodastream to atmospheric pressure and recarbonate several times. Again, you must be careful when releasing pressure. Your cocktail will foam. Make sure you can stop the releasing procedure quickly. Practice releasing the pressure and sealing it again several times while carbonating pure water to get the hang of it. The foam can shoot into the relief valve with surprising rapidity. Do not let the foam get up and into the relief valve. Never put any pulp into this system—you will create messy, messy problems. Let the drink settle down a bit after your last carbonation cycle before you do your final venting. Make sure your product is supercold, almost frozen cold, to keep foaming down and increase bubble quality.

Sodastream has released many different models. These instructions should work with most of them. Don't get the glass one for carbonating cocktails—it operates a bit differently and has to be hacked. It also makes me nervous.

The advantages of the Sodastream system are its small footprint, ease of use, and ubiquity. But you'll spend a lot more money on CO_2 than you will with a proper tank. Carbonating wine and cocktails burns through CO_2 much faster than carbonating water.

Never allow a Sodastream to foam into the area above the top of the bottle— you could clog the safety valve. When unscrewing the bottle only unscrew a tiny bit at a time.

OPPOSITE: CARBONATING A GIN AND JUICE IN A SODASTREAM: 1) Make sure your liquid is cold—this one is almost crystallizing. 2) Charge with gas until the pressure release vents. 3) Carefully vent. Be careful—you must be able to instantly re-tighten and seal the bottle. Do not let foam hit the top of the bottle. 4) Charge with gas again until the pressure release vents. 5) Carefully vent. 6) Charge with gas again until the pressure release vents. 7) Wait for bubbles to subside. 8) Notice there is no foaming when I unscrew the bottle this last time. 9) The drink.

WATER FOR CARBONATION

Good seltzer starts with good water. If your water doesn't taste good, your seltzer will taste even worse. If your municipality has bad water, filter it or use jugs of spring water. My favorite seltzer base? New York City tap water, which is soft and has no off flavors for the bubbles to magnify. Many people like water with a lot of dissolved minerals. The minerals add taste and mediate the way your tongue perceives CO_2, usually by making things taste less carbonated than they really are. While I enjoy mineralized waters such as Apollinaris, Gerolsteiner, and Vichy on occasion, for maximum refreshment give me seltzer made from soft water any day.

Even if your water tastes fine, make sure that it contains no chlorine. Chlorine tastes poisonous in seltzer. If your water contains chlorine and your pipes are lead-free, you can use heat to drive off the chlorine. Start with hot water that you allow to cool, or heat cold water on the stove and then cool it down.

OPPOSITE: CARBONATING A RED CABBAGE JUSTINO COCKTAIL WITH AN ISI: 1) Chill your drink and add it to the pre-chilled iSi whipper. 2) Add a couple extra ice cubes. 3) Charge with CO_2. 4) Shake. 5) Vent. 6) Charge with CO_2 again. 7) Shake. Wait a minute for the bubbles to subside, and 8) slowly vent. 9) The drink.

TECHNIQUE THREE: ISI WHIPPER

Using whipped-cream makers and cartridges of CO_2 to carbonate is my least-favorite common technique, and I resort to it only when there is no alternative. Whipped-cream makers are not the same as seltzer-water makers, which also use cartridges. Strangely, the whipped-cream maker makes better bubbles than the seltzer version, which I never, ever use.

Cream whippers use 7.5-gram cartridges of gas: CO_2 for soda, N_2O for whipped cream. Compared to any other form of buying CO_2, these cartridges are superexpensive—up to a dollar apiece. You'll need at least two of them every time you carbonate. Even worse, you cannot control the pressure inside; your only choice is whether or not to add another 8-gram cartridge.

If you must carbonate using a cream whipper, keep it in the freezer for a while before you carbonate, because the vessel has enough steel in it to significantly warm your drink. I also recommend throwing an ice cube or two into the whipper along with your chilled beverage. Scale your recipe to make enough mix to fill your whipper approximately one-third full. Don't fill the whipper more than halfway. Try to keep your fill levels the same all the time: the pressure in the bottle—and therefore the level of carbonation you get—is dependent on the fill line. Before you screw the top on, make sure the valve is clean—it easily gets filled with schmutz if you use the whipper for infusions—and that its main gasket is in place. If either of those two things is messed up, the whipper won't pressurize and you will have wasted a cartridge.

Now put in your first CO_2 cartridge and shake the whipper violently. After you shake it, hold it upright and use the dispensing handle to vent out the headspace air and to blast the air off potential nucleation sites in the cocktail. If cocktail starts spraying out the top, release the handle and wait a second before resuming your vent. After the cocktail is vented, add another CO_2 cartridge and shake again. Let the whipper sit undisturbed for a while so the drink settles before you slowly vent the gas. To serve, unscrew the top and pour.

PUTTING BUBBLES ON DRAFT

At home I have seltzer water on draft. Given the quantities I consume, any other system is cost-prohibitive for me. I use a 5.3-liter Big Mac McCann commercial carbonator plumbed to my water supply. The Big Mac is the type of unit a restaurant uses to make draft sodas at a bar. It works quite well. The reason most bars have crappy soda isn't a result of their carbonators; it's a result of poor filtration, poor chilling, and poor dispensing—the same factors you need to control for good seltzer.

Filtration first: before the water gets to the carbonator I run it through a dedicated filter to remove chlorine and sediment. In your area you might need a more sophisticated filtration or water-treatment system to make good seltzer. Just remember, if it tastes bad flat, it will taste worse with bubbles.

Next, chilling: I use a cold plate, as bars do for soda, but a little differently. A cold plate is a block of aluminum filled with stainless steel tubing, kept in ice to chill drinks on the fly. You see, carbonators carbonate water at room temperature at very high pressure, like 100 psi. As the water passes through the cold plate it is chilled and the liquid is slowed down, causing a pressure drop that prevents the soda from spraying everywhere when you dispense it. Cold-plate circuits aren't long enough to chill the seltzer

or to slow it down adequately, so I run the seltzer through two circuits in a series. Works perfectly, and everyone I've shown this technique to swears by it. Make sure to buy a cold plate with more than one circuit in it.

Last, dispensing: most people ruin their soda with poor dispensing equipment. A CMBecker premix soda tap is the only tap you should ever use to dispense highly carbonated beverages (they are scarce; see Sources, page 378). Never attempt to use beer taps, aka picnic taps, for this purpose unless you hate quality. The Becker premix valves have a special pressure-compensation system that gently eases carbonated beverages from the world of high pressure to our world of low pressure with a minimum of bubble loss. They also have a tunable flow rate. At home I have a Becker valve that has been in continuous service, pumping out seltzer for over fourteen years with nary a problem. My seltzer is good.

As your mania for bubbles grows, you might decide to make some draft sodas or even draft cocktails. From a quality standpoint, these are fraught with peril. You have all the problems of seltzer, magnified. If you really want to draft cocktails, I can only recommend drinks that want minimal carbonation and can be served on the warmer side (like at 36°F/2°C).

cold-plate

ice machine

filtered water

filtered water into carbonator
100 psi CO2 into carbonator
carbonated water out of carbonator, into cold plate, and out to seltzer tap

McCann carbonator
carbonator pump

CO2

The only carbonation advantage of the whipper: it lets you experiment with nitrous oxide, the standard gas for making whipped cream. Whipped cream is always made with nitrous so that the cream doesn't taste carbonated and spoiled. Nitrous doesn't add much flavor to whipped cream because it is expanded out, but what flavor it does have is sweet. Used as a bubble source in drinks, it is noticeably sweet. For this reason, it works well in coffee- and chocolate-flavored drinks. It adds body and liveliness with no carbonation taste. Avoid milk in these drinks, as it foams way too much.

If you want to try mixed gas, carbonate normally using two CO_2 cartridges and then add a nitrous cartridge to finish. Done in super-cold ice water, a two-CO_2 one-N_2O carbonation job approximates my favorite bubbly water, which I make with my souped-up mixed-gas system (80 percent CO_2, 20 percent N_2O at 45 psi in ice water).

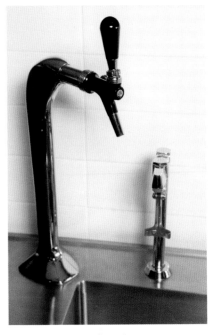

My seltzer tap, a pre-mix soda valve made by CMBecker, fitted to an Ibis tower.

CARBONATION IN A NUTSHELL

Everything that came before, distilled to a few sentences for easy reference.

Always remember the three C's of carbonation:

- **Clarity**—the drinks you carbonate should be clear.
- **Coldness**—the drinks should be cold, usually as close to freezing as possible.
- **Composition**—get rid of bubble nucleation sites. Try to avoid ingredients that foam too much, try keep your alcohol levels on the low side most of the time, and remove as much air as possible from your drink.

When you carbonate, remember these tips:

- **Carbonate each drink several times** and allow it to foam between each carbonation cycle. This is one of the key tricks to getting rid of nucleation sites and getting long-lasting bubbles.

SIDEBAR OPPOSITE: MY DRAFT SELTZER RIG: *Regular tap water is filtered and runs to my McCann carbonator (blue lines). A pump on the carbonator forces the water under pressure into the stainless-steel carbonator tank. CO_2 is fed into the stainless tank at 100–110 psi to supply gas and pressure (yellow lines). Room-temperature seltzer exits the tank and heads for a cold plate in my ice machine (green lines). The seltzer runs through the cold plate two times (one of my secrets) then exits and heads above the cabinet to my seltzer tap.*

- **Don't overfill your carbonation vessels**, or you will find it impossible to control foam and to carbonate properly. Proper headspace will depend on your application, but bottles that must be shaken should never be filled more than three-quarters full, and Sodastream bottles should never be filled to more than one-third of their normal capacity when carbonating foamy drinks (like cocktails).
- **Don't underfill your carbonation vessels.** For each system there is an optimum recipe size. When you go too low, you risk inconsistent results.
- **Don't use too much pressure**. If your drink is foaming excessively and tastes flat in the glass, your carbonation pressure is too high—you are trying to add more CO_2 than your drink can hold without gushing. Try lowering the pressure. If that doesn't work, go back and address the three C's—clarity, coldness, and composition. See the recipe section below for specific pressure recommendations.
- **Up the surface area**. Carbonation requires agitation to get CO_2 into your drink, typically by shaking under pressure or by injecting bubbles. Any effective carbonation scheme has a mechanism for increasing surface area and agitation. Pressure alone won't work.

Three important things to remember about carbonation theory:

- **Pressure**—as the pressure goes up, so does the amount of CO_2 you put into a drink.
- **Temperature**—as the temperature goes down, you increase the amount of CO_2 a liquid can hold.
- **Alcohol**—the higher the alcohol content of your drink, the more CO_2 you need to add for the same feeling of carbonation in your mouth.

RECIPES: WHAT TO CARBONATE

In the following recipes I give you the usual bartender-style measurements—a fat ¾ ounce of lime juice and such—and also the much more precise milliliter measures. The standard bartender measures will work fine from drink to drink, and they convey the meaning of recipes to within the accuracy of most folks' jiggering skills. When you are batching multiple drinks at a time, you will be much better served doing as I do: using milliliter measurements and a graduated cylinder.

NONALCOHOLIC DRINKS

Recipes for carbonated sodas are fairly straightforward. The main ingredient of soda is usually water, so use water that tastes good. Make sure it doesn't smell of chlorine.

Most sodas contain between 9 and 12 grams of sugar per 100 milliliters of beverage, which is pretty sweet for something you might pound in large quantities. Unfortunately, in my experience, reducing the sugar quantities much lower than 8 grams per 100 milliliters doesn't make a soda seem drier, it just makes it seem underflavored and watery. I suspect the reason for this is the lack of alcohol. Alcoholic drinks have a built-in flavor structure that exists without sugar, so sweetening an alcoholic drink doesn't require as much sugar. Water tastes like... water. It is possible to make highly flavored nonalcoholic drinks that are low in sugar, be they bitter or sour, but people don't perceive these as legitimate sodas; they perceive them as flavored seltzers—not the same thing. People expect sodas to be relatively sweet and, most of the time, tart as well.

When making sodas, you have two chilling options: mix your ingredients and put them in the fridge for several hours, or use ice. Ice is faster, and because it gets all the way down to 0°C (32°F), it makes sodas colder and easier to carbonate. To use ice for soda you need a container marked for the full volume of your finished drink. Get some ice water—not cold water, but water with pieces of ice in it. Add all of your nonwater ingredients to the marked container. Take some ice from the ice water, add it to your ingredients, and stir till the ingredients are chilled to 0°C (32°F). If you still have a bunch of ice left in the container, pull some out and just leave a chip or two. If all the ice melts, add a bit more.

When you are satisfied that your flavor base is chilled, give your ice water a stir (to get the whole shebang down to freezing again) and fill your mixing container with water to the fill line. Leave a small piece or two of ice in the mix—it will melt as you carbonate and help keep your drink cold during the process. Don't worry about the extra nucleation sites the ice produces. One or two pieces won't cause the same problems as scads of tiny ice crystals.

Simple Lime Soda

For this recipe you may use either clarified lime juice or lime acid. Lime acid is a mixture of malic and citric acids that imitates regular lime juice (4 grams citric acid and 2 grams malic acid in 94 grams water). Use this recipe as the basis of all your sour-type soda recipes. The ounce of simple syrup this recipe contains supplies 18.5 grams of sugar, making the whole 6-ounce (180-ml) recipe a hair over 10 percent sugar (w/v).

MAKES 6 OUNCES (180 ML) AT 10.5 G/100 ML SUGAR, 0.75% ACID

INGREDIENTS

1 ounce (30 ml) simple syrup

¾ ounce (22.5 ml) clarified lime juice or lime acid base

4¼ ounces (127.5 ml) filtered water

2 drops saline solution or a pinch of salt

PROCEDURE

Combine all the ingredients and chill in the refrigerator for several hours before carbonating at 35–40 psi, or follow the ice-water procedure above.

Strawbunkle Soda

I made this recipe a couple of years ago when my son Dax was reading *The Big Friendly Giant*, by Roald Dahl. The BFG mispronounces many things, including *strawberries*. Dax pointed out that I share this intentional mispronunciation trait—hence the soda name. Clarified strawberry juice is about 8% sugar and 1.5% acid, a bit too acidic to make directly into a soda. Even if the acid level was right, straight strawberry soda would also be too dang juicy and overpowering. I add simple syrup to balance the acid, and water to mellow the flavor.

MAKES 6 OUNCES (180 ML) AT 10.1 G/100 ML SUGAR, 0.94% ACID

INGREDIENTS

½ ounce (15 ml) simple syrup (this contains 9.2 grams of sugar)

3¾ ounces (112.5 ml) clarified strawberry juice

1¾ ounces (52.5 ml) filtered water

2 drops saline solution or a pinch of salt

PROCEDURE

Combine all the ingredients and chill in the refrigerator for several hours before carbonating at 35–40 psi, or follow the ice-water procedure above.

WINES AND SAKE

Traditional carbonated wines, like champagne, aren't force-carbonated. They are naturally carbonated by yeast's action on the sugar that is added after the wine's primary fermentation and before the bottles are sealed. The flavor and body resulting from this extended second fermentation are completely different from the results you get with force carbonation. Not better, just different. When you choose a wine to carbonate, ask yourself if the wine will be good cold—not slightly chilled like a nice glass of Beaujolais in the summer, but cold. Put a bottle in the fridge and leave it there. Taste it. If a cold wine is out of balance or overly tannic when flat, carbonation won't help it.

Unlike cocktails, sparkling wines don't taste best when chilled below refrigerator temperature. I overchill the wines I plan to carbonate by submerging the bottles in ice water, since the cold facilitates carbonation and guarantees consistency. But I let them come up a couple of degrees before serving.

The pressures you'll use to carbonate depend on the alcohol content of the wine. At 0°C (32°F) a low-alcohol white—in the range of 10 to 12 percent—should be carbonated with 30 to 35 psi of CO_2. Any more and you'll lose the fruit. Wines in the 14–15 percent alcohol range should be carbonated to 40 psi at 0°C. Sake, which could be up to 18 percent alcohol, should be carbonated to 42–45 psi at 0°C.

When I first began carbonating wines, I thought all my experiments were delicious. I was a carbonation genius! But then I conducted a rigorous series of tastings, sampling the original still wine against my carbonated version. You know what? Most of the time, even if the carbonated wine was good, the still version was better. Only occasionally was the carbonated version a revelation compared to the original.

The moral? Don't become enamored of your capabilities. As doctors of our ingredients, our first mission is to do them no harm—the Hippocratic Oath of cooking.

FLUFFING EXISTING CARBONATED COCKTAILS

If you have no patience for making your own mixers but you do have the ability to carbonate, it is simple to up your bubbly cocktail game by simply making a basic drink—whiskey soda or vodka tonic, for instance—and then force-carbonating the whole kit and caboodle. If you keep your straight booze in the freezer (usually −20°C/−4°F) and your mixer in the fridge, you can mix a typical cocktail and carbonate right away with very little fuss. I recommend a ratio of 2 ounces (60 ml) of freezer booze to 3.5 ounces (105 ml) refrigerated mixer.

MIXING YOUR OWN

I shoot for between 14 and 16 percent alcohol in my carbonated drinks. I push some drinks with stronger flavors to 17 or 18 percent alcohol. To get the correct alcohol range, use my magic ratio: every 5½-ounce (165-ml) carbonated drink contains between 1¾ and 2 ounces (52.5 and 60 ml) of booze between 40 and 50 percent alcohol. I find that this ratio almost always works. It is almost always right. Tasted flat, these drinks are absurdly weak. They taste watery. Believe me, when you carbonate them, you'll like them better than stronger versions.

In fact, all the flavors in a carbonated beverage should taste a bit weak when the drink is flat. Most acidic shaken drinks will have about 0.8 to 0.9 percent acidity—the equivalent of ¾ ounce (22.5 ml) of lime juice in a 5¼-ounce (160-ml) finished drink. Most acidic carbonated drinks are between 0.4 and 0.5 percent acidity—the equivalent of a short ½ ounce (12 ml) of clarified lime juice in a 5½-ounce (165-ml) finished drink. Shaken drinks will usually be between 6.5 and 9.25 percent sugar. Carbonated drinks are mostly between 5 and 7.7 percent sugar. Notice that the acid is typically scaled back more than the sugar is. Also notice that these sugar levels are well below those found in most sodas. I'm not sure why we like our cocktails less sweet than our sodas, but we do.

Learn to taste flat drinks and know how they will taste carbonated. As a further experiment, make a stronger carbonated drink, then add a splash of seltzer and see if it doesn't taste better.

In the following recipes, all the ingredients are clarified. Some recipes, like Gin and Juice, can't be made without clarification. Others, like Gin and Tonic, will let you add some basic lime juice at the end. For the purposes of the recipes, however, I assume you'll be clarifying.

Fluffing a gin and tonic in a Sodastream.

At the bar we batch and carbonate all our drinks prior to service. If we didn't prebatch, we'd never get the drinks out on time. Because we premake our carbonated drinks, freshness becomes an issue. We never add clarified lime or lemon juice to our drinks before we carbonate them. We make fresh clarified lemon and lime juice daily. Day-old is garbage. If we added lemon or lime before we carbonated, we'd have to throw away whatever was left in the bottles at the end of the night, which would make me sad, or serve a day-two drink that I wasn't proud of, which would make me supersad. Instead we add a small amount of clarified lemon or lime juice to the glass right before we pour the drink and serve it to our guest. The ¼ to ½ ounce (7.5 to 15 ml) of noncarbonated stuff we add to the drink doesn't hurt the bubbles. If you are carbonating and serving right away, don't worry about adding highly perishable ingredients like lime juice. If you want your drink to last a couple of days, add the fragile stuff later.

I serve carbonated drinks in champagne flutes chilled with liquid nitrogen. Serve this way and you'll never go back. Looks great, tastes great. Take the time to pour your drink gently down the side of the flute. You've worked so hard to make the drink—don't ruin it at the last minute with a careless pour.

WHICH BOOZE TO CHOOSE?

For my palate, gin is the easiest spirit to carbonate. Gin drinks just taste good carbonated. Vodka can be carbonated to good effect when you have fruits or spices that you want to highlight without an assertive spirit competing for attention. Tequila can be difficult. Strongly flavored tequilas tend to get even stronger when they are carbonated, and they can overpower. Very light tequilas carbonate just fine. Strong tequilas can be cut with vodka to mellow their tone a bit. White rums can be carbonated, but I like carbonated rum drinks much less than I thought I would, considering how much I love rum; somehow, most carbonated rum drinks taste cheap and fake to me. I've had—and made—decent carbonated rum drinks, but I find them very challenging.

Oaked spirits intensify their oakiness when they are carbonated. If you like a whiskey soda, you'll probably like carbonated whiskey drinks too. I find they are difficult to balance properly. Many liqueurs are champions in carbonated cocktails. Campari and its cousin Aperol scream to be carbonated. Just remember when using liqueurs not to make the batch too sweet, or the drink will be cloying.

SOME UNITS TO REMEMBER WHEN YOU MAKE AND ADJUST RECIPES

- One standard carbonated drink is 5½ ounces (165 ml). Scale the recipes as you wish.
- Every ½ ounce (15 ml) of 1:1 simple syrup will add 9.2 grams of sugar and will increase the sugar in a standard size (5.5-oz, 165-ml) carbonated drink by 5.6 percent.
- Every ½ ounce (15 ml) of clarified lime juice will add 0.9 grams of acid (0.6 grams citric and 0.3 grams malic) and will increase the acidity in a standard size (5.5-oz, 165-ml) carbonated drink by 0.55 percent.

X, Y, OR Z AND SODA

Anytime you have to make an "-and-soda" drink, your choices are fairly simple: decide on the ratio of booze to water. With stronger-flavored liquors, like whiskey, I like to use a short 2 ounces (57 ml) of spirit to a fat 3½ ounces (108 ml) of water. Use good, filtered water and don't pour the drink over the rocks—it doesn't need any more water. For fairly neutral spirits like vodka, I up the alcohol percentage slightly, maybe up to a full 2 ounces. I don't go over that percentage unless someone specifically requests it. My only secret with these -and-soda drinks: add a drop or two of saline solution (or a couple grains of salt).

CARBONATING A CLASSIC

The first trick to carbonating a classic is to choose the right cocktail. Some, like the Manhattan, are an abomination when carbonated. I carbonate Manhattans as object lessons in what can go wrong when you carbonate. They are wretched and unbalanced. Luckily, many classics, like the margarita and the Negroni, carbonate quite well. Here's how to modify those two recipes to make them work with bubbles. See how I modified them and then modify your own.

Carbonated Margarita

Classic margaritas are a mix of agave-based spirits, orange liqueur, lime juice, and sugar. In this recipe I omit the orange liqueur; I think it muddies the taste. Instead I twist an orange rind over the surface of the drink before I serve it. The ratios in this recipe work for many sour drinks, so use it as a master recipe.

MAKES ONE 5½ OUNCES (165 ML) DRINK AT 14.2% ALCOHOL BY VOLUME, 7.1 G/100 ML SUGAR, 0.44% ACID

INGREDIENTS

Short 2 ounces (58.5 ml) light-bodied, clean tequila like Espolón Blanco (40% alcohol by volume)

Fat 2½ ounces (76 ml) filtered water

Short ½ ounce (12 ml) clarified lime juice

Short ¾ ounce (18.75 ml) simple syrup

2–5 drops saline solution or a generous pinch of salt

1 orange twist

PROCEDURE

Combine the first five ingredients and chill till almost frozen. Carbonate at 42 psi. Pour into a chilled flute. Express the orange twist over the top of the drink and discard. If you want the drink to keep for several days, carbonate without the lime juice and add the lime juice before serving.

Carbonated Negroni

The classic Negroni is

- 1 ounce (30 ml) gin (47% alcohol by volume)

- 1 ounce (30 ml) Campari (24% alcohol by volume, 24% sugar)

- 1 ounce (30 ml) vermouth (16.5% alcohol by volume, 16% sugar, 0.6% acid)

The volume of the recipe as written is 3 ounces (90 ml). If you add 2½ ounces (75 ml) water to make the drink volume proper for a single carbonated drink, the result would have an alcohol by volume of 16% with 7.3% sugar. Both those numbers are pretty good. The acid level is a bit low at 0.18%, but the Negroni is not an acidic drink. If you need a refresher on how I got these numbers, check the cocktail math section of the book, page 18.

If you want to make it more refreshing, you can sub out ¼ ounce (7.5 ml) of the water with clarified lime juice or an acid of your choice. Finish it off with a twist of grapefruit-peel oil.

MAKES ONE 5½-OUNCE (165-ML) DRINK AT 16% ALCOHOL BY VOLUME, 7.3 G/100 ML SUGAR, 0.38% ACID

INGREDIENTS

1 ounce (30 ml) gin

1 ounce (30 ml) Campari

1 ounce (30 ml) sweet vermouth

¼ ounce (7.5 ml) clarified lime juice

2¼ ounces (67.5 ml) filtered water

1–2 drops saline solution or a pinch of salt

1 grapefruit twist

PROCEDURE

Combine everything but the twist and chill till almost frozen. Carbonate at 42 psi. Pour into a chilled flute. Twist the grapefruit peel over the drink and discard the peel. If you want the drink to keep for several days, carbonate without the lime juice and add the lime juice before serving.

Champari Spritz

Campari and soda is a classic low-alcohol summer favorite, good for times when the carbonated Negroni is a bit too much but you still want that Campari hit. Once carbonated with a squeeze of lime, Campari is the epitome of bitter, bracing, and refreshing. Instead of adding a twist of lime, I add champagne acid; hence the name. Champagne acid is the same mix of acids that is naturally present in champagne diluted to lime-juice strength (30 grams tartaric acid and 30 grams lactic acid in 940 grams water); see the Acids section, page 58, for details. Champagne acid shifts the drink into a winy, champagney area that I really like. If you make this recipe more than a couple hours in advance, don't add the champagne acid until you are ready to serve. Although champagne acid doesn't spoil like lime juice, over time it makes the Campari taste progressively and unpleasantly bitter. I don't know why. This cocktail will last a week in the fridge. Notice I said fridge, not freezer. This drink is low in alcohol (7.2% alcohol by volume) and freezes easily.

MAKES ONE 5½-OUNCE (165-ML) DRINK AT 7.2% ALCOHOL BY VOLUME, 7.2 G/100 ML SUGAR, 0.44% ACID

INGREDIENTS

Fat 1½ ounces (48 ml) Campari (24% alcohol by volume, 24% sugar)

⅜ ounce (11 ml) champagne acid (6% acid)

3⅛ ounces (94 ml) filtered water

1–2 drops saline solution or a pinch of salt

PROCEDURE

Combine all the ingredients and chill in ice water to 0°C (32°F). Carbonate at 42 psi. Serve in a chilled flute.

Gin and Tonic

I explain the derivation of my gin and tonic recipe in depth elsewhere, but here is my recipe. This version is extremely dry and austere—the way I like it.

MAKES ONE 5½-OUNCES (165-ML) DRINK AT 15.4% ALCOHOL BY VOLUME, 4.9 G/100 ML SUGAR, 0.41% ACID

INGREDIENTS

Full 1¾ ounces (53.5 ml) Tanqueray gin (47% alcohol by volume)

Short ½ ounce (12.5 ml) Quinine Simple Syrup (61.5% sugar; see page 367)

Short 3 ounces (87 ml) filtered water

1–2 drops saline solution or a pinch of salt

⅜ ounce (11.25 ml) clarified lime juice (6% acid)

PROCEDURE

Combine all the ingredients (except the lime juice, if desired) and chill between –5° and –10° Celsius (14°–23°F). Carbonate at 42 psi. If you carbonate with the lime juice, serve the drink that day. If you carbonate without the juice, add it as the drink is poured into a chilled flute; your G&T will keep indefinitely.

Chartruth

This drink is for those of you who love Green Chartreuse as much as I do. If you aren't yet a fan, Green Chartreuse is an herbal smack in the face produced by Carthusian monks in France who've taken a vow of silence. This liquor is so badass that they named a color after it. This is the simplest expression possible: Chartreuse and water with a hint of lime. I call it the Chartruth. This drink is higher in both alcohol (18% alcohol by volume) and sugar (8.3%) than most good carbonated drinks. Because of the intensity of flavor, I like to pour a single drink into two small glasses and serve it to two people as a refreshing minidrink.

MAKES ONE 5½-OUNCE (165-ML) DRINK AT 18.0% ALCOHOL BY VOLUME, 8.3 G/100 ML SUGAR, 0.51% ACID

INGREDIENTS

Fat 1¾ ounces (54 ml) Green Chartreuse (55% alcohol by volume, 25% sugar)

Short 3¼ ounces (97 ml) filtered water

1–2 drops saline solution or a pinch of salt

Short ½ ounce (14 ml) clarified lime juice (6% acid)

PROCEDURE

Combine all the ingredients except the clarified lime juice and chill to between −5° and −10°C (14°–23°F). Carbonate at 42 psi. Add the clarified lime juice as the drink is poured into a chilled flute. If you are going to serve the drink immediately, you may carbonate with the lime juice.

Adding the clarified lime juice to a Chartruth.

Gin and Juice

If I'm remembered for anything, I hope it is my Gin and Juice. Essentially just gin and clarified grapefruit juice, it is simple and satisfying.

Grapefruit juice is characteristically bitter, mainly owing to the compound naringin. The bitterness of the naringin in grapefruit is counteracted by sugar (10.4%) and high acidity (2.40%). To my taste, grapefruit juice balances perfectly in still drinks. In carbonated drinks, especially with gin, the bitterness becomes too much. Luckily, clarification removes some of grapefruit's bitterness. How much bitterness remains in the juice depends on the clarification technique you use. Most people making this recipe will clarify their juice using the seaweed gel agar (see the section on clarification, page 235). Agar strips a lot of bitterness from the juice and is a fantastic clarification technique for this drink.

At the bar I use a centrifuge to clarify grapefruit juice, because the yield on centrifugal clarification is so much higher than with agar—I waste almost no juice. Unfortunately, centrifugal clarification strips much less of grapefruit's bitterness. Simple syrup and a tiny amount of champagne acid are added to counteract the bitterness. I give you recipes for both agar-clarified and centrifuge-clarified Gin and Juice.

Grapefruit juice is too juicy-tasting to use without adding some water to the drink. How much water you add depends on how juicy you want your drink.

The recipes below balance with the grapefruits I get most of the time at the bar (California-grown Ruby Reds). Remember that grapefruits taste different depending on variety, orchard locale, and season, so you might have to tweak these recipes a bit depending on the fruit you have.

Last, I have used many different gins to make Gin and Juice, but nothing makes me as happy as Tanqueray. It just has an affinity for grapefruit juice.

The ingredients for Gin and Juice.

Pouring a Gin and Juice.

GIN AND JUICE: AGAR-CLARIFIED

MAKES ONE 5½-OUNCE (165-ML) DRINK AT 16.9% ALCOHOL
BY VOLUME, 5 G/100 ML SUGAR 1.16% ACID

INGREDIENTS

Scant 2 ounces (59 ml) Tanqueray gin
(47% alcohol by volume)

Short 2¾ ounces (80 ml) agar-clarified
grapefruit juice

Fat ¾ ounce (26 ml) filtered water (If a slightly
sweeter drink is desired, replace a bar spoon
(4 ml) of the water with simple syrup; that will
make the drink 6.3% sugar, 1.10% acid)

1–2 drops saline solution or a pinch of salt

GIN AND JUICE: CENTRIFUGALLY CLARIFIED

MAKES ONE 5½-OUNCE (165-ML) DRINK AT 15.8% ALCOHOL
BY VOLUME, 7.2 G/100 ML SUGAR, 0.91% ACID

INGREDIENTS

Fat 1¾ ounces (55 ml) Tanqueray gin
(47% alcohol by volume)

Fat 1¾ ounces (55 ml) centrifuge-clarified
grapefruit juice

Short 1½ ounces (42 ml) filtered water

Fat ¼ ounce (10 ml) simple syrup

Scant bar spoon (3 ml) champagne acid (30
grams lactic acid and 30 grams tartaric acid
in 940 grams water)

1–2 drops saline solution or a pinch of salt

PROCEDURE

Combine all the ingredients and chill between –5°
and –10°C (14°–23°F). Carbonate at 42 psi. Serve in
a chilled flute.

TECHNO-VARIANT

My other favorite clarified grapefruit juice recipe is
Habanero-n-Juice. The recipe is the same, except instead
of gin you use redistilled habanero vodka. Blend 200
grams of red habanero peppers with a liter of 40-proof
vodka and distill it till you recover 650 ml of product in a
rotary evaporator with a condenser temperature of –20°C
and a bath temperature of 50°C. The habaneros must be
red for the drink to work. Habaneros are one of the hottest
peppers known, but also have fantastic flavor and aroma.
Capsaicin, the compound responsible for the spiciness
of the pepper, is too heavy to distill, so none of the heat
is present in the distillate. Roto-habanero is one of my
long-standing and favorite distillations. It goes amazingly
well with grapefruit. If you make it, be aware that the distillate has a short shelf life. After a month or so, the "red"
flavor of the distillate will fade and it will begin to taste
"green," more like a jalapeño.

LEFT: Habanero gin being vacuum distilled at low temperature
to produce non-spicy habanero gin.
RIGHT: Flying the rotovap.

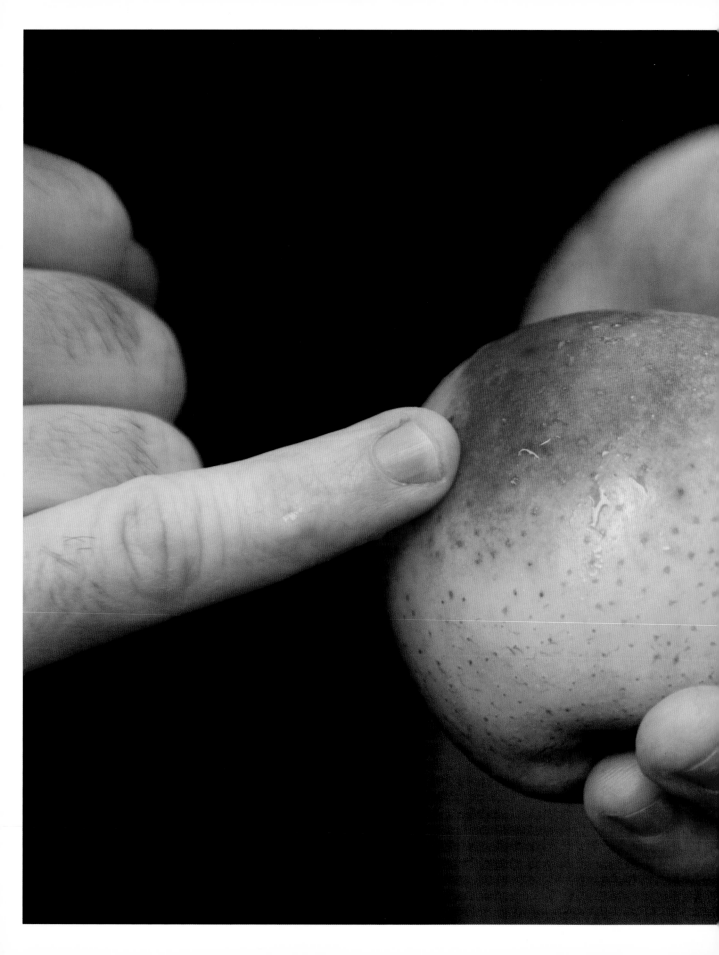

PART 4

Little JOURNEYS

In this last section I'll riff on three different subjects and see where they take me: apples, coffee, and the gin and tonic. I hope these journeys will give you insight into how I approach cocktail development. I usually start with a concept, a flavor, a fruit, an idea, or a memory. I develop a goal and try to get there. This approach to cocktails is the hardest thing to teach. So often, people think of cocktail development as the mere rearranging of spirits and juices.

I'll assume some familiarity with the techniques and concepts presented earlier in the book, so explanations here will be more laconic.

Apples

I love apples. I grew up in New York State, and apples are one of the things we do really, really well. Apples have not received their due as a cocktail ingredient. This is partly because apple juice isn't very concentrated, meaning that you must use a lot in a cocktail recipe to get the right flavor—and then it no longer works well in standard shaken, stirred, and built drinks. But I think the main reason is a lack of education.

Many of us grew up thinking there were two flavors of apple, red and green. Of course, we were wrong. There are thousands of apple varieties with startlingly different flavors, including notes of quince, orange, rose, anise, and wine. Some have preposterously high levels of acid and sugar. Some are delicate and aromatic. Some are austere. Each one of the thousands of varieties you can find today was named and propagated by someone, at some time, for some reason or another. Every apple was at one time loved by somebody. The trick is figuring out why—and then whether it is useful in a cocktail.

Some apples were loved for reasons that are irrelevant to us today. The Flower of Kent was saved for posterity because it was the apple that fell on Newton's head; it tastes bad. Some apples were loved because they performed well in a particular area. (Can you guess where Arkansas Black apples were first grown?) Other apples were loved because of their timing: early-season apples allowed cooks to make fresh apple pies in late June and early July, after their winter store of apples was exhausted. The need for early-season apples has been obviated by modern storing and shipping technology, so these apples now must stand on their own merits to survive. Sometimes they do, sometimes they don't.

My serious study of apple flavors began in 2007, when the eminent food writer and thinker Harold McGee and I visited the United States' apple collection in Geneva, New York. Yes, the United States maintains a collection of apple trees—thousands of them, two per variety, as in Noah's ark—just in case one of those trees holds some genetic material that will someday be useful to agribusiness. We tasted a couple hundred varieties over a two-day period, and what a treasure. Strangely, the keepers of the collection hadn't expected us to *taste* the apples. Apparently their usual visiting temperate-fruit pomologists just want to look at trees. When the keepers figured out our real purpose, they gave us a bemused look and allowed us free run. From the hundreds of apples we tasted, I was able to bring home and juice about twenty. The juices of those apples were the basis of my first serious apple cocktails, and one of them, Ashmead's Kernel, became my all-time favorite cocktail apple.

Since then I've been buying apples from local growers at my well-stocked New York City greenmarket and from other growers from all over the United States, as well as sampling many dozens of varieties as they grow in England. I have learned that my experience of an apple variety might not match yours. Apples are extremely dependent on where they are grown, when they are harvested, and the vagaries of the year's weather. Apples that grow well in mild climates might not ripen well in colder areas. Varieties meant for cold climates might taste insipid grown too warm. To make the best apple cocktails, you have to make the best juice. To make the best juice, you just have to taste a lot of apples, remember who grew them, and go back to those same purveyors year after year.

APPLE JUICE, THE COMMODITY

Commercial apple juice is a commodity product, and I advise that you leave it out of your cocktails. Oceans of apple juice are made every year from fruit that wasn't quite good enough to sell for eating. It's clarified, pasteurized, then often concentrated, shipped, reconstituted, blended with other reconstituted juice from who knows where, and sold. While this sort of product is fine for juice boxes, it's not good enough to partner with booze. Supermarket American sweet cider (sweet, as opposed to hard: the United States is the only country that refers to a nonalcoholic apple juice as cider) is more robust and versatile than regular juice, but I still can't recommend it for your cocktail work.

You can find some delicious single-variety and carefully blended apple juices and sweet ciders at specialty stores and farmers' markets, but they are typically marred by pasteurization. Everyone with whom I have blind-tasted has chosen unpasteurized apple juice and sweet cider over pasteurized. Some cold pasteurization methods use ultraviolet light that doesn't damage flavor nearly as much as heat, but you're unlikely to find products made this way in your market.

Last, almost without exception, store-bought apple juices and ciders are overoxidized and therefore brown. Fresh apple juice may be green, yellow, red, orange, or pink—it's never brown. The juice turns brown very quickly when exposed to oxygen, the same way a cut apple does. This oxidation destroys nuanced varietal flavors.

MAKING APPLE JUICE THE RIGHT WAY

For your cocktail work you'll need to make your own apple juice.

Wash your apples before you juice them, and look them over for signs of worms, rot, mold, and excessive bruising. The washing is important, because we will not be using heat. If the apple isn't clean enough to eat, it isn't clean enough to drink. Cut away any moldy or off-looking parts. A moldy apple can spoil a lot of juice and, even worse, can contain patulin, a possible carcinogen.

Never, ever peel an apple before you juice it. Most of the varietal-specific aromas, flavors, tannins, and apple pigments are concentrated in the flesh very close to the peel. Cut your apples up, but leave the skin on and throw

JUICING APPLES: I use a Champion juicer and make sure the juice pours into a container that already contains ascorbic acid (vitamin C) so the juice never has a chance to oxidize (LEFT). After I juice, I skim the majority of foam off the top, strain the juice through a fine strainer (the pulp is delicious) and, if I am going to clarify, add Pectinex ultra SP-L (RIGHT).

A SPECTRUM OF APPLE JUICE COLORS (LEFT TO RIGHT): *Unnamed crabapple; Stayman Winesap; Honeycrisp; Suncrisp; Granny Smith. The crabapple juice is redder than the Stayman Winesap juice, even though the Winesap is a darker apple, because the crabapple is smaller, with a larger surface-area-to-volume ratio. There is more skin to leach color into the juice.*

THOSE SAME APPLE JUICES TREATED WITH PEXTINEX ULTRA SP-L: *The solids are floating on top because of trapped air. Gently rapping the glasses on the table and stirring would allow that stuff to settle to the bottom. Notice the juices contain different amounts of solids and clarify differently. Granny Smith on the right is almost completely clear, while the crabapple on the left hasn't settled at all.*

The crabapple juice from the previous photo after it has been spun in a centrifuge—a gorgeous pink. Some apple juices lose their richness when clarified. This crabapple, on the other hand, wasn't so great to eat and its juice wasn't much better, but once clarified it was spectacular.

them into a masticating juicer. These juicers, like my Champion, do a good job of extracting the flavor and color of the flesh near the skin. They have tiny teeth that pulp and shred the apple to bits and then smash those bits against a screen to get at the juice. Once upon a time I removed the cores and seeds prior to juicing, since the cores have little flavor and the seeds contain cyanide. Not anymore. I did a side-by-side taste test of apples juiced with and without cores and seeds, and I couldn't tell the difference. The Champion doesn't seem to extract any bitter flavor from the seeds; it leaves them mostly intact or simply broken in half, not pulped (hence you are not consuming them—no cyanide risk), and the lack of flavor in the core is insignificant since the core doesn't yield much juice anyway.

You will prevent oxidation by using vitamin C (ascorbic acid). Remember from the Ingredients section that vitamin C and citric acid are *not* the same (see page 59). Citric acid is the primary flavor acid in lemons. It does not directly prevent browning. Ascorbic acid doesn't change the tartness of lemons much, but is responsible for nearly all of their antibrowning power. Either toss the cut apples in ascorbic acid powder before you juice them, or put some ascorbic acid in the container you are juicing into. Make sure to stir the ascorbic acid into the juice after you juice your first apple or two. I use about 2.5 grams (1 teaspoon) of ascorbic acid per liter of juice, much more than is used commercially.

You will now have real, fresh apple juice that tastes just like the apples it came from. If you have never tasted this kind of apple juice before, you will be angry that you have lived this long without it. The juice, however, is not cocktail-perfect: it is cloudy. You now must decide whether to clarify.

Apple juice flavor is greatly affected by clarification. Think of the difference between commercial apple juice and sweet cider. It isn't just that the cider has more viscosity and body because of the suspended particles; those particles bring their own flavor, and that flavor is often good. So why clarify? If you plan to carbonate, you have no choice. If you plan to make a stirred drink, you *should* clarify—who wants a soupy-looking stirred drink? Choose an apple that retains its awesomeness once clarified. If you're working toward a shaken drink, forget clarifying. Just strain the juice thoroughly to avoid unsightly pulp particles against the side of your glass and you're good to go.

Before we get into some special cocktail-worthy apples, let's look at some uses for supermarket varieties, both clarified and not.

Clarified: Granny Smith Soda

I'll go nonalcoholic with this one. Granny Smiths, Australia's most famous apple, are the go-to apple for cooks because they are tart, brisk, consistent, and easy to get. Their juice clarifies very well even without a centrifuge: just juice them, add some Pectinex Ultra SP-L, and let the juice sit overnight before you pour the clear stuff off the top (see the Clarification section, page 235; truth be told, if you let Granny juice sit long enough, you can get by without the enzymes). Grannies have a sugar-to-acid ratio that is perfect for soda (roughly 13 grams per 100 ml sugar and 0.93% acidity). Unfortunately, they just aren't that interesting. But while monotonic flavor is a flaw in a cocktail, where the juice must stand up to and blend with liquor, it can work well in a soda.

MAKES 6 OUNCES (180 ML) AT 10.8 G/100 ML SUGAR, 0.77% ACID

INGREDIENTS

5 ounces (150 ml) clarified Granny Smith juice

1 ounce (30 ml) filtered water

2 drops saline solution or a pinch of salt

PROCEDURE

Combine the ingredients, chill, and carbonate with your method of choice.

Unclarified: Honeycrisp Rum Shake

Honeycrisps are one of the better new commercially available apple varieties. They are a wee bit low in acid, so this recipe requires extra acidity, in the form of either lime juice or straight malic acid. Apple juice is too diluted to work in a drink shaken with ice, so for this recipe you will do a juice shake (see the Alternative Chilling section, page 140) using frozen juice.

**MAKES ONE 5.4-OUNCE (162-ML) DRINK AT 14.8%
ALCOHOL BY VOLUME, 7.8 G/100 ML SUGAR, 0.81% ACID**

INGREDIENTS

2 ounces (60 ml) clean-tasting white rum (40% alcohol by volume)

Short ½ ounce (12 ml) lime juice,

or 0.7 grams malic acid dissolved in 10 ml water and 2 drops saline solution or a pinch of salt

3 ounces (90 ml) unclarified Honeycrisp apple juice frozen into three 1-ounce (30-ml) cubes

PROCEDURE

Combine the rum and the lime juice or malic acid and salt, then shake with the apple juice ice cubes in a cocktail shaker till the ice has completely melted to a slush (you can hear this happening). Serve in a chilled coupe.

Now let's take a trip beyond the supermarket.

LEARNING ABOUT APPLES

We are living in a great time for apples, and interesting varieties are within everyone's reach. If you don't live near good growers, you can buy specialty apples directly from growers online. If you live in apple country, farmers' markets have more varieties on display every year as more and more people express interest in fruit variety. You've got to get out and taste. You'll find a list of critically important apple references in the bibliography, but remember that they are no substitute for your mouth.

CHOOSING AND TASTING APPLES

Whether you are choosing apples at the greenmarket or off the trees, keep a few pointers in mind. Bring a knife: it is much easier to get a good sense of how an apple really tastes by slicing off a piece of flesh and skin instead of sinking your teeth into a whole fruit, especially if you are tasting dozens

HOW APPLE TEXTURE MATTERS FOR COCKTAILS

You wouldn't think an apple's texture matters for cocktails—after all, we are just going to juice them—but it does, indirectly. America's texture preferences have all but removed whole swaths of potentially good cocktail apples from contention. Let me explain.

We Americans are prejudiced. We accept *only* crunchy apples. This is not a good thing. Why must an apple be crunchy? Other textures, like crumbly, can be good too. We have lost the ability to distinguish between crumbly and mealy. A mealy apple is a formerly crunchy apple that has been stored too long and is losing quality. It is bad. A crumbly apple, however, is entirely different: one whose texture was never crunchy when ripe. Accept the beauty of difference.

How does this affect the quality of your juice? American growers planting heirloom varieties know you will buy only crunchy apples, so they often pick their fruit

viciously underripe. Mature fruits would be unacceptably soft and wouldn't sell. These underripe apples are sad. They have very little of their characteristic varietal flavors. They are flat. They haven't developed their sugar yet, so they also taste like acid bombs and are full of unconverted starch. Picking underripe yields a crunchy apple that is flavorless, acidic, and starchy. Early-season apples are the biggest losers in this category, because as a group they are light in flavor to begin with, have a very short season, and turn mushy faster than any other apples do. It is almost impossible to get a grower to pick them for you when they are ripe.

As cocktail folks, we don't even care what the texture is—we are just making juice—but we still pay the price. Do your part and tell your local growers that if they pick apples when they are ripe, you'll buy them! Lots of them!

The blush side of an apple tastes different.

of apples and you want to avoid sore gums. Remember that when you taste an apple for cocktails, the texture does not matter! Erase your perception of texture and focus only on how the juice tastes. This task is difficult at first. Practice.

When you cut into an apple, feel how the knife moves through the flesh. You can sense the starchiness of an underripe apple even before you taste it: it feels like you are slicing a hard potato. You can see starch in the juice on the cut face of an underripe apple before you taste it. Avoid these apples. When you look at a bin of apples or a tree full of apples, look for two apples with different coloration. Even in apples that stay green forever, the color will change as the fruit ripens, usually getting darker or developing a blush. Even on a particular tree, apples will be in varying stages of ripeness depending on where they are. And beyond simple ripeness, the flavor of an apple can change depending on how much sun it gets compared to others on the tree and how far down the branch it grows. Tasting the two most differently colored apples of a variety will show you the range of flavors a particular apple is likely to have. On an individual fruit, look for one side that is more colored than another or that has a blush on it. First taste the side farthest from the blush. Hopefully it is good. Now taste the side with the blush—it will be richer, sweeter, more sun-kissed. The juice from that apple will be in between the two tastes.

After you perform these tests, take only the apples that fit your requirements. In an orchard this is easy, and in a market it isn't too difficult. Just buy two different apples from each variety that interests you, taste them on the spot, and then buy however many you need for cocktails. Remember that apple juice freezes well: if you find an apple that you love, buy a bunch, then process and freeze.

SUGAR AND ACID

Apples that do well in cocktails tend to be high in both sugar and acid. Low-sugar apples are rarely good because they are mostly underripe, so they will also have little flavor, not just low sugar. High-acid juice is good, because it allows you to make a cocktail without any other added acid. Low-acid apples can be greatly improved by adding a little acidity, in the form of straight malic acid, lemon or lime juice, or even acidic alcoholic ingredients like vermouth. But I prefer to use apples that don't require too much adjusting, allowing you to present the pure flavors of the apple without distractions.

Commercially, especially in hard-cider production, people will quote apple varieties based on the sugar-to-acid ratio. This enables you to judge the balance of a juice based on one number. You can then look at the sugar level (usually quoted in Brix) to determine the overall taste strength of the juice. Most people won't have refractometers to measure their juice, and fewer still can adequately measure the acidity of their apple juice (pH meters won't work for this), but many benchmarks for particular varieties can be found online and are useful guidelines for choosing which apples to test if you can't taste them before you buy.

For cocktails, I like apple juice with a sugar-to-acid ratio between about 13 and 15; 13 is tart, 15 pleasantly acidic. For reference, Granny Smiths are about a

OVERRIPE AND OVER-THE-HILL APPLES

You want to avoid apples that have languished. In a market, you can tell which apples are past their prime because they no longer feel hard in the hand; maybe they have even shriveled a bit. In an orchard, feel the apples. If they feel greasy, this is a sign that they are overripe; the naturally occurring epicuticular wax coating on many apples gets greasier as they ripen. (In the grocery store this test is not so useful: the grower may have removed the natural coating during washing and put a different one on top.) If you suspect an apple is overripe, cut it open. Many overripe apples develop water core, a wet-looking inside that appears as though it has been frozen and thawed. Usually, but not always, overripe apples will be softer than they should be.

Overripe or overstored apples are no good for eating, but mildly overripe apples can make for an interesting cocktail ingredient. As they overripen, apples lose acidity, so overripe apples will always need more acid correction than properly ripe ones will. They can also develop very intriguing perfumey floral aromas as the ethylene gas in the apples kicks into overdrive and produces volatile esters in the fruit. In small amounts these esters can be fabulous. In overabundance they make the apples smell of solvent. If you want to capture these flavors, beware: they are very fugitive.

14. Gala apples, which are low in acid and fairly sweet, clock in around 21. As for sugar levels, apple juice should be above 11 Brix to be useful, and is better around 14 or 15—as sweet as or sweeter than soda.

Let's look at some cocktail experimentation with two high-sugar, high-acid apples: Ashmead's Kernel and the Wickson crabapple.

ASHMEAD'S KERNEL

Ashmead's Kernel is a russeted, yellowish-skinned English apple dating back to the early eighteenth century. I get mine from a grower in New Hampshire. When these apples are good, they are really, really good. The juice just tastes rich. Many russeted apples share a pearlike note, but with more acid; Ashmead's has a bit of that but is more full-bodied than usual. The Brix can approach 18, which is very high, and it has a high acid to match. I don't have any hard data on the sugar-to-acid ratio, but I'd guess it's around 14.

Ashmead's Kernel desperately wants to be paired with whiskey and carbonated, but for one sticking point: the oak in whiskey swamps the flavor of the apple. This problem eventually got me started on washing whiskeys (see page 265). But long before I tried to soften with washing, I solved the problem through simple redistillation in a rotary evaporator. This redistilled whiskey was clear and colorless but still very plainly whiskey. I simply mixed it with the clarified Ashmead's, softened the drink with a bit of water and a pinch of salt, chilled it, and carbonated it. The Kentucky Kernel, as I called it, is exactly the kind of drink I love to make: just two ingredients, manipulated and combined to create a flavor that people hadn't experienced before.

I first made that drink in 2007, with the very first batch of Ashmead's Kernel that I swiped from the U.S. apple collection in Geneva. Now that I have a bar, I can't distill anymore (pesky legal issues), so I can't make my Kentucky Kernel the way I used to. The answer, as you know by now, is washing the whiskey. Here is the recipe:

Kentucky Kernel

MAKES ONE 5.25-OUNCE (157.5-ML) CARBONATED DRINK
AT 15% ALCOHOL BY VOLUME, 8.6 G/100 ML SUGAR,
APPROXIMATELY 0.6% ACID (ASSUMING YOUR BATCH OF
APPLES HAS AN ACID-TO-SUGAR RATIO OF 14; YOU'LL
HAVE TO TASTE YOUR APPLES TO SEE IF YOU NEED TO
ADJUST MY RECIPE)

INGREDIENTS

1¾ ounces (52.5 ml) chitosan/gellan-washed
   Makers Mark bourbon (45% alcohol by
   volume)

2½ ounces (75 ml) clarified Ashmead's Kernel
   juice

1 ounce (30 ml) filtered water

2 drops saline solution or a pinch of salt

PROCEDURE

Combine the ingredients, chill, and carbonate. Serve
in a chilled flute.

That recipe is equally delicious made with de-oaked
Cognac instead of bourbon. If the oak is still too
strong for you, you can try egg washing (see the
Booze Washing section, page 265).

BOTTLED CARAMEL APPLETINI TWO WAYS, AND THE AUTO-JUSTINO

The Wickson crabapple is not really a crabapple at all. It has an illustrious and definitely uncrabby parentage, and is called a crab simply because it is very small. From what I can gather, it is a cross between the Newtown Pippin, the first apple America exported to Europe in colonial times (discovered in New York), and the Esopus Spitzenburg, another fabulous and famous colonial-era American apple (also from New York State), with which I have made many good cocktails. Discovered in California in 1944, the Wickson packs a wallop in a tiny package. It can reach over 20 Brix, although the ones I get are closer to 15. The acid level can reach 1.25 percent. These are great levels for cocktail work, and it has a great flavor as well, rich and round. I decided to make appletinis.

The appletini, as you are no doubt aware, has a well-deserved bad rep attributable to the fake sour-green apple schnapps of which it is typically constructed. The Wickson has the acidity and sugar to pull off a beautifully refined appletini, one you can be proud to order. You could make this drink with a mixture of Plymouth gin and vodka, but I stick with the pure vodka here, and finish with Dolin Blanc sweet white vermouth. I also mess with the Wickson a little: I decided to produce a caramel-apple flavor for a nice fall tribute, so I add a little bit of caramel syrup.

If you stir this drink, it will get too diluted. Instead, make it as a bottled cocktail (see the Alternative Chilling section, page 140). If you fill the bottles with the mix, purge the headspace of oxygen using liquid nitrogen and cap them; they should last a long time. I store them at exactly 22°F (−5.5°C)—service temperature—in my Randell FX freezer. If all you have is a home freezer, let the bottles freeze! Just don't overfill them, or they'll explode when they freeze. When it comes time to serve, run the bottles under water till they just thaw out. This isn't the procedure I give for making bottled cocktails in the Alternative Chilling section; it is a slightly different technique based on the principles outlined in that chapter. I present it this way to show that you needn't follow any technique dogmatically. I don't. You can take advantage of any of the techniques in this book using whatever equipment you have at hand, once you grasp the principles.

This drink could also be bottled and chilled for service in a salt-ice bath.

Bottled Caramel Appletini

My Wicksons were 15 Brix. If yours are higher (many reported measurements are over 20), you will have to adjust your recipe or it will become toothachingly sweet. You don't need a refractometer to figure this out: if the drink is too sweet, adjust! This drink is not supposed to be a sugar bomb. Caramel syrup isn't as sweet at the sugar it contains, because some breaks down during caramelization.

**MAKES ONE 5⅕-OUNCE (155-ML) DRINK AT 16.5%
ALCOHOL BY VOLUME, 7.2 G/100 ML SUGAR, 0.45% ACID**

INGREDIENTS

2 ounces (60 ml) vodka (40% alcohol by volume)

¼ ounce (7.5 ml) Dolin Blanc vermouth (16.5% alcohol by volume)

1 ounce (30 ml) filtered water

1¾ ounces (52.5 ml) clarified Wickson crabapple juice

1 bar spoon (4 ml) 70-Brix caramel syrup (see Note)

NOTE: For 70-Brix caramel, add a small amount of water—around 1 ounce (30 ml)—to the bottom of a pan. On top of the water pour 400 grams of granulated sugar and heat till the mixture forms a rich, dark caramel, almost but not quite burned. Immediately add 400 ml of water. It will boil violently. Stir the mixture with a spoon to dissolve everything. After the syrup cools, measure the Brix. It should be somewhere between 66 and 70. If it is higher, add water. If lower, boil some water off.

Dash of orange bitters (preferably the recipe on page 211)

2 drops saline solution or a dash of salt

PROCEDURE

Mix everything together, bottle, chill, and serve.

This Auto-Justino was made from Ashmead's Kernel and aged for several months.

I promised you an appletini two ways, and for the second way I developed a new technique. At the beginning of this section I mentioned that clarifying an apple really removes some of the flavor from the juice. I wondered if I could get some of the unclarified pulpy flavor in a clear drink. I decided to add alcohol to the cloudy juice in hopes that the liquor would pull some of the flavor out of the pulp that would otherwise be lost when I clarified. I also decided to use very high-proof liquor, because I wanted the resulting liquid to be shelf-stable, similar to the Justinos I describe in the Clarification section. I added 400 ml of pure ethanol (the lab-grade stuff I have mentioned) to 600 ml of cloudy juice, with no added clarification enzymes. Something fantastic happened. The ethanol did in fact pull some good flavor out of the pulp. But this surprised me: the high-proof booze instantly caused the pectin in the apple juice to aggregate and auto-clarify. No waiting, no centrifuge, just crystal clear—and delicious! I call this the Auto-Justino. The Auto-Justino is shelf-stable and a bit over 40% alcohol by volume. Just add Dolin Blanc and bitters and stir it into an appletini. You can also use the Wickson Auto-Justino in shaken drinks. Although most of the pectin in the Auto-Justino aggregates and is strained out, there is enough residual pectin in the booze to make a nice layer of foam on top of a shaken drink.

The real problem with this technique is that good-quality, 96% or higher alcohol by volume liquor is hard to get, and the bad stuff smells like a hospital and tastes like poison even after it has been diluted. To see if the technique would work on lower-proof stuff, I tested a fifty-fifty mix of Bacardi 151-proof aged rum (75.5 % alcohol by volume). It worked fantastically well, and the resulting liquor was still above a respectable 37% alcohol by volume. I was surprised at how good the Bacardi tasted with the apple. I'm pretty sure the 151 is a stunt liquor built to fuel frat-party nonsense—it comes with flammability warnings and a metal antiflame pouring screen—but it is surprisingly well crafted.

The 151 at 75.5% alcohol by volume is near the lower limit of alcohol content for success with this technique; 57% alcohol by volume was a failure. If you can find even small amounts of tolerable 96% alcohol-by-volume booze, you can use it to fortify high-quality flavorful liquors up above 70% alcohol by volume for this technique. One such mix: 25% gin (47.5% alcohol by volume or above), 25 percent of the 96 percent alcohol by volume, and 50 percent juice.

When you make an Auto-Justino, you should use a spoon to gently mix the liquid around, coaxing the pectin globs closer to each other and allowing them to swab up stray cloudy bits in the liquid, polishing the liquor.

The Auto-Justino has lots of advantages: it is fast, it requires no equipment (except the juicer to make the apple juice), and it extracts good flavor from the pulp. This technique relies on the fact that apple juice contains pectin but is still rather thin. I tried it on thicker purees like strawberry without success.

FUTURE EXPLORATION

Plenty of mid- and late-season apples make great cocktails—dozens at least that any reader of this book in a temperate climate zone can get his or her hands on. Much more difficult is making a great cocktail with very early-season apples, like Yellow Transparent and Lodi. They are tough customers. They have very little flavor. They are called salt apples by some old-timers, who like to sprinkle them with salt and eat them less as a dessert-style snack and more as a savory, salty, refreshing-style snack. I have not yet successfully made a great cocktail with these apples. I think I haven't yet fully listened to what they are trying to tell me. I haven't figured out what they want to be yet, but I get closer every year. My guess is light, salty, and fleeting, maybe using agave for its here-and-gone fructose hit, and gin. Maybe I'll crack the code next year. I've had early-season apples that would be great in cocktails, but I haven't been able to source them regularly. I once got a shipment of Carolina Red June and Chenango Strawberry apples from Virginia. Blended together, they were the best early-season juice I've ever made, but I've never had enough to really test.

I have yet to explore the possibility of using highly tannic cider apples as a component of a cocktail—a ridiculous oversight you can be sure I will rectify next season. When you taste one of these apples in an orchard, you spit it out straightaway. In fact they are called spitters. My brain never got over the prejudice of not liking them in the orchard so I could see how they may be useful in a cocktail, or how they might taste after I remove some of the tannin with milk/egg/chitosan washing.

Last, I would love to travel to Kazakhstan, the home of the apple, to the Tien Shien fruit forest. By all accounts, this forest, stretching from Kazakhstan to western China, is an amazing place. Most fruits in their wild state are not nearly as good as their bred and domesticated counterparts, and many people think this is true of the apple—but no. Phil Forsline, the curator of the U.S. apple collection when McGee and I visited in 2007, had made a pet project of gathering wild apples from the Tien Shien forest. McGee and I tasted some of those wild apples and they were quite good, worthy of being named. Perhaps I could wander in that forest till I found a wild apple tree whose fruit had the rich taste, high acid, and high sugar necessary for good cocktail work. Then I'd have a variety I could literally call my own. That would be fun cocktailing.

Coffee

I have spent most of my life hating coffee and referring to it as an execrable, thin, bitter liquid. When I was growing up, my mom's favorite nighttime tipple was Kahlua and milk; even in that highly sugared state I couldn't stand it. I satisfied my college and grad-school caffeine needs with tea and Diet Coke.

In my late twenties, I decided things were going to change. I was going to like coffee, and I was going to like it strong. I forced myself to down shots of espresso. After a couple of weeks I learned to like it—then love it. I soon decided I needed pro espresso at home, but I couldn't afford a good machine. I started trolling restaurant auctions. I struck gold at a place that, after a DEA raid, had had the door padlocked for weeks with rotting food inside. Since I was one of the few prospective buyers willing to stand the stench during the auction, I left with a sweet eighties-vintage two-group Rancilio for just a hundred bucks. Thus began my many years' journey into the world of espresso.

Pulling an espresso shot.

That was in the late 1990s. The state of espresso art has come a long, long way since then, and many smart people have committed many years of study to that little brown shot. This is a book on cocktails, not coffee, so I'll avoid the nitty-gritty here, but I do want to walk you through my approach to a challenge I set for myself: to make a coffee cocktail that captures what I like about espresso in cocktail form. These cocktails don't necessarily contain espresso; they just embody what is great about it.

A quick note on other coffee forms: I have not been able to get excited about drip coffee, iced coffee, or coffee with milk. I don't like coffee-flavored things, even coffee ice cream. I'm not telling you this because I'm proud of it—I just want you to know my proclivities so you can better judge the rest of this section.

ESPRESSO CHARACTERISTICS: WHAT WE ARE AIMING FOR

For the purposes of this discussion, I define espresso as 1½ ounces (45 ml) of coffee brewed from 15 grams of compacted freshly ground coffee in 22 seconds using 92°C (198°F) water at a pressure of 135 psi (9.3 bars). I use more coffee grounds and less water than a traditional northern Italian would, and fewer coffee grounds and more water than a modern American barista would. Feel free to disagree with my ratio.

Espresso should be strong and pleasantly bitter, but not acrid. It shouldn't require sugar. So I want my coffee cocktail to taste strongly of coffee, and to avoid acridity and sweetness.

The high pressure under which espresso is brewed causes a foam, called crema, to form on top. The bubbles that form this foam are actually present throughout the shot. The high pressure also causes coffee oils to emulsify into the liquid (espresso, unlike drip coffee, is an emulsion). Foaming and emulsification give espresso its characteristic opacity, body, and texture. The texture of espresso dies fairly quickly, just as the texture of a shaken cocktail does. So to recreate the body of espresso in a cocktail, we definitely won't be stirring—stirring is for limpid drinks, and we are aiming for exactly the opposite. We will need to use shaking and carbonating to get the texture I'm after.

ESPRESSO DRINKS THAT ACTUALLY INCLUDE ESPRESSO

Espresso is an ideal cocktail ingredient, because it delivers its substantial flavor in small, cocktail-sized doses. Drip coffee cannot compete with espresso in cocktails; it carries too much water with it. If you try to make drip coffee strong enough to stand up to dilution, it becomes acrid. You might think that cold-brew coffee concentrate, which has become popular recently, is a good substitute: like espresso, it is bitter without being acrid. But the taste just isn't the same.

Using espresso instead of drip coffee takes care of our first coffee-cocktail criteria: strong but not acrid. To help us tackle the tougher problem, texture, let's take a quick look at iced coffee.

ICED ESPRESSO

Iced coffee is one of the few things I make a lot of, even though I loathe it. My wife loves iced coffee and cannot understand my contempt. In an attempt to make an iced coffee we both would like, I focused on improving its texture. Simple shaking was the answer. Espresso shaken with ice and a little sugar—a *caffè shakerato* in Italy—textures very nicely but must be consumed very quickly. Add milk and your texture problem is gone completely. Milk is a foam-making machine. A shaken iced espresso with milk is a pretty good drink, even for haters. If you don't try any other recipe in this book but you like an iced coffee, please try this one. Your life will be better for it!

Shakerato with Milk

MAKES ONE 6⅔-OUNCE (197-ML) DRINK 0% ALCOHOL BY VOLUME, 4.7 G/100 ML SUGAR, 0.34% ACID

INGREDIENTS

1½ ounces (45 ml) freshly made espresso cooled down to at least 60°C (140°F)

3 ounces (90 ml) whole milk

½ ounce (15 ml) simple syrup

2 drops saline solution or a pinch of salt

Copious ice

PROCEDURE

Combine the ingredients and shake like the devil. Either strain and serve in a chilled glass or serve it in a tall glass with some ice and a long straw. If you prefer a no-milk version, omit the milk, shake a bit longer, and serve in a chilled coupe glass.

ALCOHOLIC ICED ESPRESSO

Adding liquor to a shakerato throws off the dilution and body. To get the right texture with liquor, I suggest substituting cream for milk and letting the espresso chill a bit more.

Boozy Shakerato

MAKES ONE 7⅘-OUNCE (234-ML) DRINK AT 10.2% ALCOHOL BY VOLUME, 3.9 G/100 ML SUGAR, 0.29% ACID

INGREDIENTS

1½ ounces (45 ml) freshly made espresso cooled down to 50°C (122°F)

2 ounces (60 ml) dark rum (40% alcohol by volume; nothing too funky)

1½ ounces (45 ml) heavy cream (you can use light cream or even half and half if you prefer, but the drink won't be as good)

½ ounce (15 ml) simple syrup

2 drops saline solution or a pinch of salt

Copious ice

PROCEDURE

Combine the ingredients and shake. Strain into a chilled double old-fashioned glass (the volume is too much for a coupe).

TOP: Boozy Shakerato with cream, shaken with regular ice. BOTTOM: Boozy Shakerato with milk—no cream.

Boozy Shakerato 2

Another option to keep the texture without adding cream: freeze milk into ice cubes and shake with them. This is a variant of the juice-shake method in the Alternative Chilling section, page 140.

MAKES ONE 7½-OUNCE (225-ML) DRINK AT 10.7% ALCOHOL BY VOLUME, 4.1% G/100 ML SUGAR, 0.3% ACID

INGREDIENTS

1½ ounces (45 ml) freshly made espresso cooled down to 50°C (122°F)

2 ounces (60 ml) dark rum (40% alcohol by volume; nothing too funky)

½ ounce (15 ml) simple syrup

2 drops saline solution or a pinch of salt

3½ ounces (105 ml) whole milk, frozen into ice cubes

PROCEDURE

Combine the liquid ingredients and shake with the frozen milk cubes until all the ice has broken into slush. You should be able to hear the slushiness in your cocktail shaker. Strain into a chilled double old-fashioned glass (the volume is too much for a coupe).

Boozy Shakerato 2 made with milk ice cubes.

TEXTURE WITHOUT SHAKING: BUBBLES

Espresso starts out as a drink full of bubbles—carbon dioxide bubbles, in fact. When coffee is roasted, carbon dioxide generates inside the beans. When you force pressurized water through the coffee grounds, the CO_2 dissolves into the brew water. When the hot water reaches atmospheric pressure, the dissolved CO_2 bubbles out like it does in a foaming bottle of warm soda, causing espresso to foam. As coffee beans age and go stale, their CO_2 depletes, making them taste bad and taking away their texturizing powers.

Since the bubbles in espresso are just CO_2, why not just carbonate a chilled espresso cocktail to get the texture I want? Because carbonated coffee tastes weird, that's why. (You can buy a coffee soda called Manhattan Special that appeals to some people—I'm not among them.) The amount of CO_2 in an espresso is small because the brew water is so hot; remember, the amount of CO_2 soluble in a liquid is inversely proportional to its temperature. You'd never describe espresso as tasting carbonated, right? To get bubbles without the prickly taste of CO_2, use nitrous oxide (N_2O), which creates bubbles much like CO_2 but tastes sweet instead of prickly.

I have a big nitrous oxide tank, so I can make drinks with N_2O the same way I would with CO_2. Most of you, except for the dentists, will not be able to buy it in large quantities. Luckily, N_2O is easy to purchase in cartridge form; refer back to the Rapid Infusion section, page 189. So let's make this drink in an iSi whipper.

Nitrous Espresso

Be aware that espresso foams like a demon when you add bubbles to it, so though you can make two drinks at a time, be careful not to overfill your whipper. Also note that nitrous is sweet, so as this drink sits around in the glass and the nitrous bubbles off, it will taste progressively less sweet. If you swish a nitrous drink around in your mouth before you swallow, it will give you a burst of sweetness as more nitrous is released from the liquid. Although for consistency I give the recipe for a single drink, this recipe is best doubled.

MAKES ONE 5-OUNCE (165-ML) DRINK AT 12.7% ALCOHOL BY VOLUME, 5.6% G/100 ML SUGAR, 0.41% ACID

INGREDIENTS

1½ ounces (45 ml) espresso

1¾ ounces (52.5 ml) vodka (40% alcohol by volume)

½ ounce (15 ml) simple syrup

1³⁄₇ ounces (52.5 ml) filtered water

2 drops saline solution or a pinch of salt

EQUIPMENT

Two 7.5-gram N_2O chargers

PROCEDURE

Combine all the ingredients in the iSi whipper and chill in the freezer till it is just about to freeze (this will also chill your whipper). Close the whipper and charge with an N_2O charger. Shake and then vent while holding a towel over the nozzle to deaerate the drink (try not to spray coffee everywhere; believe me, it is messy). Now charge with a second charger and shake for at least 12 seconds. Put the whipper down and let it remain undisturbed for around 90 seconds. Slowly—I mean slowly—vent the whipper. You are trying to preserve the bubbles here. Serve in a chilled flute glass.

ESPRESSO DRINKS THAT
DON'T ACTUALLY CONTAIN ESPRESSO

Early in my cocktail experimentation I became interested in distillation as a method for preserving aroma while removing bitterness. I attempted to make a coffee liquor with no bitterness and therefore no need for sugar. I had no success at all. None of the distillations I made, whether I used grounds or brewed coffee, tasted remotely like good, strong coffee. Turns out coffee flavor is not recognizable without all the bitter, heavy stuff that doesn't distill.

When distillation failed I turned to infusions, but nothing ever tasted quite right . . . until I developed rapid nitrous infusion. Without rapid infusion, my coffee concoctions were too weak or had a lingering, unpleasant, bitter aftertaste. Rapid infusion allowed me to make a coffee infusion that

COFFEE
ZACAPA
COCKTAIL

acted like espresso, with a pure, strong coffee taste, the pleasant bitterness of coffee, and no acridity. This liquor required only a modicum of sugar when used in a cocktail. The only problem with my coffee infusions now: they still required milk to achieve the right texture. You can guess how I felt about that. My second breakthrough came with the milk-washing technique (revisit the Milk Washing section, page 267). With milk washing you add milk to liquor, allow or force the milk to curdle, and then strain out the solids. The resulting liquors still contain whey proteins and make nice creamy, foamy drinks, but they don't taste milky.

Finally! A cold coffee drink I really enjoyed. The recipe below is a variation of the one in the Rapid Infusion section, in which I use djer, a West African spice, to make Café Touba (page 204). The liquor base is aged rum. First the infusion:

INGREDIENTS FOR COFFEE ZACAPA

750 ml Ron Zacapa 23 Solera rum, divided into a 500 ml portion and a 250 ml portion

100 ml filtered water

100 grams whole fresh coffee beans roasted on the dark side

185 ml whole milk

Citric acid or lemon juice, if needed

PROCEDURE

Grind the coffee in a spice grinder until it is slightly finer than drip grind. Combine 500 ml rum and the coffee in a half-liter iSi whipper, charge with one charger, shake, and then add a second charger. Shake for another 30 seconds. Total infusion time should be 1 minute 15 seconds. Vent. Unlike most infusions, don't wait for the bubbling to stop, or the liquor will be overinfused. Instead rest for just 1 minute, then pour through a fine-mesh filter into a coffee filter. If you pour the mixture directly into a coffee filter it will probably clog very quickly. The mixture should filter within 2 minutes. If not, your grind was too fine. Combine and stir the drained grounds in the coffee filter, then add the water evenly over the grounds and let it drip through (this is called sparging). That water will replace some of the rum that was trapped in the grounds during the infusion. The liquid that comes out of the grounds from sparging should be about 50 percent water and 50 percent rum.

At this point you should have lost roughly 100 ml of liquid to the grounds. About half of that lost liquid is water and the other half rum, so your final product has a slightly lower alcohol by volume than you started with.

Taste the infusion. If it is strong (which is good), add the additional 250 ml of rum to the liquor. If the infusion can't bear to be toned down without losing the flavor of the coffee, your grind size was too coarse; don't add any more rum, and reduce the amount of milk you use for milk washing to 122 ml.

While stirring, add the coffee rum to the milk and not the other way around, lest the milk instantly curdle. Stop stirring and allow the mixture to curdle, which it should do within about 30 seconds. If it doesn't curdle, add a little 15% citric acid solution or lemon juice bit by bit until the mixture curdles, and don't stir when it's curdling. Once the milk curdles, gently use a spoon to move the curds around without breaking them up. This step will help capture more of the casein from the milk and produce a clearer product. Allow the mix to stand overnight in the fridge in a round container; the curds will settle to the bottom and you can pour the clear liquor off the top. Strain the curds through a coffee filter to get the last of the liquor yield. Alternatively, spin the liquor in a centrifuge at 4000 g's for 10 minutes right after it curdles. That's what I do.

APPROXIMATE FINAL ALCOHOL BY VOLUME: 35%
MAKES ONE 3⁹⁄₁₀-OUNCE (117-ML) DRINK AT 15.8% ALCOHOL
BY VOLUME, AND 7.9% G/100 ML SUGAR, 0.38% ACID

INGREDIENTS FOR COFFEE ZACAPA COCKTAIL

2 ounces (60 ml) Coffee Zacapa

½ ounce (15 ml) simple syrup

2 drops saline solution or a pinch of salt

Ice

PROCEDURE

Shake all the ingredients in a cocktail shaker and strain into a chilled coupe glass. The drink should be creamy and frothy.

FUTURE STEPS

I'd like to revisit coffee distillation; I have not tried since my first ill-fated attempts. It might be possible to create a supercoffee liquor by first distilling liquor with coffee and then infusing that same liquor with coffee grounds—a double coffee effect.

I'd like to experiment with altering the espresso-brewing procedure specifically for cocktails. The amount of foam in an espresso is dependent, as we discussed, on the amount of CO_2 in the roasted beans. The darker the roast of the coffee, the more CO_2 is present, hence more foam. The foam in espresso has greater stability, however, if you use medium-roasted coffee, so I'd like to use a medium-roasted coffee and supplement the CO_2 during the brewing process. If I'm lucky, this will let me brew a 1½-ounce shot of espresso directly into an ounce shot of liquor to produce hot espresso shots with good texture. At home, my espresso machine is plumbed to my filtered water supply, as is my carbonator. It would be fairly simple to hook the output of my carbonator to the input of my espresso machine. If I brew with carbonated water, I should get more foam—I hope. My espresso machine uses a heat exchanger to heat the brewing water, which means that the water used for brewing goes from room temperature to hot very quickly and is under the full 135 psi (9.3 bars) of pressure the entire time it is being heated and delivered. It just might work.

I'd also like to experiment with brewing hot cocktails directly, using booze mixed with water in an espresso machine. It might be good or it might be terrible. I know I'll learn a lot trying.

The Gin and Tonic

I am ending this book where my own cocktail journey began: with the gin and tonic. It's the first drink I remember my dad making. He would fix one for himself and a tonic with lime for me. The first cocktail I analyzed closely, the G&T inspired me to develop many of my cocktail techniques. I still think about it daily. The gin and tonic repays deep thought.

The gin and tonic is seemingly so simple: gin, tonic water, and a squeeze of lime. The promise of the G&T is so great: crisp and refreshing, on the dry side, a bit tart, slightly bitter, aromatic, crystal clear with lots of bubbles. But G&Ts almost always disappoint. Sometimes there's too much gin, and therefore too little carbonation. Sometimes there's too little gin, and therefore too few aromatics and too much sweetness. All too often, warm gin and tepid tonic water are poured over copious quantities of watery ice, producing a drink tasting mainly of water. How could something so simple in concept be so difficult in practice? The answer is simple. It is *impossible* to make a good gin and tonic using traditional techniques. Yes, impossible. There is *no* ratio of gin to tonic with the right balance of flavors and enough carbonation. You may postulate that I am an outlier, that I was ruined by years of drinking straight tonic water with my dad. But I think if you search your heart, you will agree that you too are unsatisfied with the bubbles in a traditional gin and tonic, even if—or especially if—the G&T is one of your favorite drinks.

THE BEST G&T YOU CAN MUSTER IF YOU CAN'T MUSTER MUCH

The best gin and tonic you can make with traditional techniques uses gin that you've stored in your freezer and tonic water poured from a fresh bottle that you've kept in ice water (if you really must, use tonic water that has merely been refrigerated, but keep it in the coldest part of your fridge—the part that accidently freezes the lettuce from time to time). By "fresh bottle" I don't just mean unopened, I mean recently purchased. Plastic bottles lose their carbonation at an alarming rate, and smaller bottles lose carbonation faster than larger ones. Twenty-ounce bottles can lose an appreciable amount of carbonation in a month at room temperature. If you buy tonic water in glass bottles or cans—both of which are gas-impermeable—storage time isn't important.

Before you make the drink, you must decide what glassware to serve in. For my G&Ts I typically choose a champagne flute (and no ice), but I am always using force carbonation. In this nonforce scenario, the champagne glass feels wrong, and it's best just to serve the drink on the rocks in a standard highball glass. You will add 1¾ ounces (52.5 ml) of gin and 3¼ ounces (97.5 ml) of tonic water to the glass to make a 5-ounce (150-ml) drink. You can measure the gin with a jigger, but don't measure the tonic water that way—jiggering will cause too much carbonation loss. Instead, before you make your drink measure 5 ounces (150-ml) of water into the glass and note where the water level, or wash line, is. Try free-pouring water into the glass to that same level by eye and then measure afterward how accurate you were. After a couple attempts you will likely be getting to within a quarter-ounce or better every time. You can also try to learn where the wash line for 1¾ ounces is in your glass so you can free-pour the gin, but I'd just use a jigger instead. Now you are ready to make the drink.

Several minutes before drink time, make sure your glass is in the freezer getting cold. Cut your lime into quarters, of which you will need one per drink. At drink time, pull the glass and gin out of

the freezer and pour 1¾ ounces (52.5 ml) of gin into the glass *before you pour in the tonic*. Next, tilt your glass to a 45-degree angle and slowly pour the ice-cold tonic water into the glass. As you pour, slowly raise the glass to vertical and stop pouring when you have reached the 5-ounce (150-ml) wash line you memorized earlier. The order of operations is important. You want the two ingredients to mix thoroughly without any bubble-liberating activities like stirring. Pouring the tonic into the gin mixes better than pouring the gin into the tonic. Tonic water is denser than gin (even gin at freezer temperature),

5

HARD WORK FOR THE LAZY DRINKER—THE NO-TECH GIN AND TONIC: *1) Pre-freeze your glass and gin and pour the gin into the glass. 2) Tilt the glass and pour in fresh, ice-cold tonic water. 3) Now squeeze in some fresh lime and 4) gently place in un-tempered freezer ice. 5) Place the lime on top and drink it, lazy head.*

surfactants that would wreak havoc with the tonic's carbonation if it were added earlier.

Then add freezer-cold ice—not tempered ice. Don't drop the ice into the drink like a Neanderthal. Gently slide it in using a bar spoon. It's important that you add the ice last; if it goes into the glass before the liquids, it will promote foaming as the tonic is poured and will present a barrier to mixing. Added at the end, the ice *promotes* mixing. If you use ice directly out of your freezer, it will add very little additional dilution. The ice cubes will crack from thermal shock, but that's okay in this application. Drop the lime quarter into the top of the glass if you like that sort of thing. If you dropped the lime directly into the liquid, it would create constant bubble nucleation, but in our on-the-rocks scenario the lime will sit just above the drink, leaving bubbles unharmed and lending a nice aroma as the glass is raised to the lips.

If you have a Sodastream or some other force-carbonation apparatus, you can take this recipe—1¾ ounces (52.5 ml) of gin and 3¼ ounces (97.5 ml) of tonic water—and put the mix in your freezer till crystals just begin to form, then force-carbonate per the instructions in the Carbonation section, page 288 (for a Sodastream you'll have to double the recipe). In this case I'd serve the drink in a chilled champagne flute without ice—you've earned it. Squeeze the lime into the drink after you have poured it and do not put the lime into the flute unless you want to spoil your carbonation work. Clarify the lime juice first for an even better result.

so the tonic will sink through the gin. Also, there is more tonic in the recipe than gin, and when mixing two liquids you'll mix more efficiently if you add the larger volume of liquid to the smaller one. As a bonus, adding gin to the glass first will melt any errant ice crystals on the inside of the glass—crystals that would become bubble nucleation sites and cause copious foaming if the tonic hit them.

Next, squeeze as much juice as you'd like into the drink from a quarter of a lime. Adding lime before tonic would help the drink mix better, but lime juice contains bubble nucleation sites and bubble-stabilizing

THE WAY OF THE G&T

In 2005, I realized that I would never be satisfied with a traditional G&T. It was a profound moment. I learned that carbonation is an ingredient, and I knew I had to master carbonation so that I could separate the volume of tonic from the quantity of bubbles in the cocktail. I felt compelled to break down the entire gin and tonic and rebuild it from first principles. I won't go into detail on my carbonation travails here, because I discuss them ad nauseum in the Carbonation section. Let's talk about the other ingredients—the tonic water and the gin.

TONIC 101

Tonic water is a mixture of water, sweetener (usually high-fructose corn syrup in the United States), citric acid, quinine sulfate, and "flavors." Strangely, many people believe that tonic water is non-sweet and non-caloric, as seltzer water is. Not so. Tonic water is as sweet as soda, usually between 9.5 and 10 percent sugar by weight.

I like tonic water. I always have. I do not want to reinvent tonic or add new flavors to it. What I want is merely this: hyperfresh, hypercrisp, crystal-clear, and impeccably clean-tasting tonic—the second-most refreshing beverage in the world (behind seltzer).

THE QUININE

The ingredient that sets tonic water apart from other lemon-lime sodas is quinine, an intensely bitter plant alkaloid that fluoresces intensely under UV or black lights (a fact of which many a clubgoer is aware). Quinine comes from the bark of the South American cinchona tree and has been used as a medicinal herb in present-day Bolivia and Peru since before recorded history, and since the sixteenth century by Europeans as a malaria cure. Unlike many herbal remedies, which function primarily as placebos, quinine is a legit drug. It was immensely important to Europeans looking to subjugate malaria-ridden parts of the world in the 1800s. In the mid-nineteenth century people learned that taking small amounts of quinine weekly or daily protected against malaria, and tonic water was born. The amount of quinine in today's tonic water isn't enough to act as an effective prophylactic, but in some countries with malaria problems, as I witnessed firsthand in Senegal, people regularly consume tonic in hopes of avoiding the disease.

The papers I have read on malaria prophylaxis indicate that an effective daily dosage is around 0.3 grams of quinine sulfate. The U.S. legal limit for quinine sulfate in tonic water is 85 milligrams per liter, so you'd have to drink 3.5 liters for effective protection. That is a lot, and commercial tonic water usually has much less quinine than the legal limit would allow.

I wanted to get my hands on some quinine. I considered tracking down some cinchona bark, which is relatively easy to do, and steeping it, but decoctions of cinchona are brown and contain suspended detritus even after they are run through a coffee filter. Since my ideal G&T is crystal clear, this approach was not going to work. At the time of my early experiments I hadn't come up with any good clarification techniques, so I figured the bark would damage my carbonation. Quinine isn't the only substance in cinchona, so I was also pretty sure that I would introduce some unwanted nonquinine flavors. I decided I had to get the pure stuff.

Sourcing purified quinine wasn't easy. Quinine is sometimes used in the treatment of nighttime leg

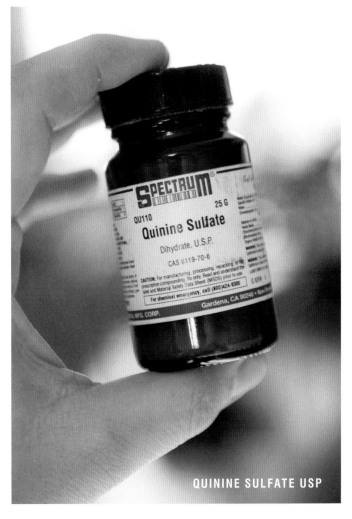

QUININE SULFATE USP

cramps, and until 1994 you could buy it over the counter for that purpose. At the time of my early experiments it was routinely prescribed by doctors. My mom is a doctor! No sweat! "No way," she said, without hesitation. Besides the obvious ethical violations, she said, there was no way she was writing me a prescription for a potentially harmful medicine that I planned to serve in a cocktail. Apparently consuming too much quinine causes a syndrome known as cinchonism, featuring nasty symptoms that range from simple nausea and dizziness, to the more frightening temporary hearing loss and blindness, all the way up to death from cardiac arrest or renal failure. Luckily, quinine is incredibly bitter and therefore almost impossible to overdose on accidentally—*if* it is used properly. You'd never willingly drink a cocktail with too much quinine in it, I pleaded. My mom (not surprisingly) turned a deaf ear. I ended up purchasing it from a chemical supply house. You must be extremely careful about what you purchase through a chemical supplier. Many chemicals come in different grades. If you are using a chemical for food or drink, it needs to be USP (United States Pharmacopeia)-grade, food-grade, or equivalent. Lower grades of chemicals can have dangerous impurities in them. This is a good across-the-board safety tip. Unfortunately, USP-grade quinine sulfate is expensive. As of this writing, 10 grams of the stuff will set you back almost a hundred bucks from a chemical supply house, and 100 grams will cost almost $500.

IMPORTANT QUININE SAFETY

Quinine is dangerous if used improperly. As little as one-third of a gram, the therapeutic dose for malaria prophylaxis back in the day, is enough to cause mild symptoms of cinchonism in some people. I will repeat that: one-third of 1 gram. **Never let anyone who isn't aware of the safety concerns work with quinine.** Quinine must be diluted before it's consumed. I predilute by making quinine simple syrup, and I use a scale accurate to one-hundredth of a gram. Unless you have a scale that is at least this accurate, you have no business working with quinine. After the quinine is diluted to a safe concentration, there is no real danger of overdose, unless some crazy person pounds a whole liter of your quinine simple syrup.

Once again: **do not attempt to use quinine in its undiluted powdered form**. The amount of powdered quinine you would want for a single drink is vanishingly small, nearly impossible to measure. Even if you could measure it properly, quinine tends to clump and is difficult to dissolve; the clumps can't be tasted if they don't hit your tongue, so they don't give warning of mismeasurement and potential overdose. You must check any quinine syrups you make to be sure all the quinine is completely dissolved, and pour them through a fine-mesh strainer as a final step.

USING QUININE AND CINCHONA

In my early experiments I made quinine water for cocktails so I could alter bitterness without altering any other components, such as sweetness and tartness. After years of honing my recipe, I now just add quinine directly to simple syrup. I find the syrup much easier to use, make, store, and dose. Here's my recipe:

Quinine Simple Syrup

MAKES 1 LITER

INGREDIENTS

0.5 gram quinine sulfate USP

1 liter simple syrup (615 grams of water and 615 grams sugar mixed till the sugar is totally dissolved; note that these ingredients are given by *weight* but will produce 1 liter of syrup by *volume*)

PROCEDURE

Carefully weigh the quinine in a small, bone-dry, non-stick, and nonstatic-producing container. You don't want any quinine sticking to your measuring container. Put the rest of the quinine away where no one will mess with it by mistake. Add the simple syrup to a blender and the quinine to the syrup while the blender is running. Let the blender blend for a minute or so on medium speed. Turn off the blender and wait for the bubbles to come out of the syrup solution. It should be clear, with no quinine specks left in it. If you still see white powder, blend some more. Strain the syrup through a fine strainer into a storage container.

Make the recipe as I have written it and experiment with it awhile before you change it. If you find my recipe too bitter for your taste, add regular simple syrup to it rather than making the recipe with less quinine. Measuring and dispensing less than 0.5 gram accurately can be tricky in bar environments. If you find that this syrup is not bitter enough, you have a more complicated problem. Half a gram grams of quinine sulfate per liter of simple syrup is right at the solubility limit of the quinine. You'd be hard-pressed to dissolve more in. You would have to make syrup that has less sugar in it than 1:1 syrup. Unfortunately, weaker syrups won't have as long a shelf life.

The solubility limit of quinine is actually one of the beauties of this recipe. If you take all the precautions—visual checks, straining—it is very hard to overdose on the quinine using this recipe. The standard sweetness of tonic water is 10 percent sugar by weight, so a liter of tonic made with our syrup will have 170 ml (208 grams) of quinine simple syrup and therefore 0.069 grams of quinine—much less than the legal limit of .083 grams per liter. Even to approach the limit with this syrup, you'd have to make tonic water that was almost 13 percent sugar by weight, which would be unpalatably sweet.

If you'd rather not deal with quinine, you can work directly with cinchona bark instead, like this:

Cinchona Syrup

MAKES 1.2 LITERS

INGREDIENTS

20 grams (around 3 tablespoons) powdered cinchona bark (available online or at herbal shops; if you can't find powder, grind bark chips in a spice grinder)

750 ml filtered water

750 grams granulated sugar

PROCEDURE

Add the powdered bark to the water in a saucepan and bring to a simmer over medium-high heat. Lower the heat and simmer for 5 minutes, and then allow to cool. Strain through a fine strainer and then through a coffee filter. Press on the bark to extract liquid (or, as an alternative, spin the cinchona water in a centrifuge). Redilute the cinchona water to 750 ml (you will have lost some water in the infusion process) and add the sugar. Blend to dissolve.

CINCHONA BARK

THE LIME AND THE GIN

In addition to quinine and sugar, tonic water contains citric acid and "flavors." To my palate, those flavors are merely lime, or possibly lemon and lime. One of my big gripes with commercial tonic water is the lameness of the citrus flavor. After I tackled quinine, a more difficult problem came into focus: the lime. Back in 2006 I had no good way to carbonate liquids containing juice. I hadn't yet developed any of my clarification techniques, so I could only use freeze-thaw gelatin clarification, which took several days to complete, or old-school consommé-style clarification, which involved boiling. Neither was okay. Lime juice must be used the day it is juiced, and it should never be heated.

I was experimenting with low-temperature vacuum distillation in a cobbled-together rotary evaporator, so I tried to distill the flavor of fresh limes at room temperature. I distilled straight lime juice, lime juice with peels, and both of those mixed with gin. I liked the distillates better without the peels, but the real discovery was that lime distilled with gin was infinitely better than lime distilled on its own, because ethanol is much better at holding on to volatiles than water is. (Years later I found the secret to distilling top-notch flavors without ethanol: a liquid nitrogen condenser. LN freezes all the volatiles on the condenser and captures them. But that discovery was far in the future.) My ability to preserve all the flavors that I was boiling off in my homemade vacuum was pretty poor, so I soon upgraded to a 1980s-vintage rotary evaporator, sourced on eBay for a couple hundred bucks.

That rotovap furthered my lime and gin distillation efforts immensely. When it arrived, it was filthy and smelled of carbon tetrachloride. (At least, it smelled like my memory of carbon tetrachloride. My high school chemistry teacher, Mrs. Zook, kept a stash of it for her favorite students' experiments with nonpolar solvents.) I washed the bejeezus out of that thing. It was old and required a lot of care, and as I logged hundreds of hours flying it, I learned a lot about what makes a rotary evaporator tick.

For starters, I learned what *doesn't* distill. Acids from lime juice, for instance, don't distill. Neither do sugars. In order to make my lime and gin distillations taste like lime juice, I had to add back lime acids. Lime juice contains a blend of citric and malic acids in a 2:1 ratio, with a pinch of succinic acid thrown in. This was my first clue as to why commercial tonic water wasn't as good as it could be: the makers use only citric acid. Citric acid on its own tastes only of lemon. It isn't till you add the malic acid (which, on its own, tastes like a green-apple Warhead candy) that the combination starts tasting like lime. Both citric and malic acid are easy to get, so it is criminal that the tonic makers don't add the malic. The real acid secret, however, was the tiny, tiny amount of succinic acid I added. On its own, succinic acid tastes terrible: bitter/salt/acid/unpleasant. Strangely, however, in minute amounts (a couple hundredths of a percent) it makes the flavor of the whole much better.

About this time I also began experimenting with distilling my own gin. I would add whatever flavors I liked—usually some combination of Thai basil, cilantro leaf, roasted oranges, and cucumber, with some juniper thrown in so I could call it gin. Some of those distillations were really good, but none of them were really gin. Learning to distill gave me a real respect for professional distillers. I decided to leave the gin to the pros.

My dad's gin of choice was Bombay. Not Bombay Sapphire, but old-school Bombay London dry gin, with the green label featuring Queen Victoria's sour mug. It is a fine product that I like better than the Sapphire. But when I make a G&T, I reach for Tanqueray. Here is my gin and tonic procedure, circa 2007:

Make a two-to-one mixture of citric and malic acid and dissolve the acid in water. Make some 1:1 simple syrup (I hadn't yet started making quinine simple). Make some diluted quinine water. Juice a bunch of limes. Add the lime juice to Tanqueray and distill it in a rotary evaporator at room temperature with a condenser chilled to at least −20°C (−4°F) so that 700 ml of liquid is distilled off for each liter of gin used. Get ice. Pour the gin into some ice and stir it down till cold and partially diluted (be careful not to overdilute at this stage). Add the citric-malic acid blend to taste, then simple syrup, then quinine water, then a pinch of salt and a pinch of succinic acid, then taste and adjust (I would go through four or five rounds of adding this or that till I thought it tasted right). Chill and carbonate (I didn't yet have liquid nitrogen, so I chilled the mix in the same chiller that I used to chill the condenser of my rotary evaporator).

THE BOTTLE-STRENGTH GIN AND TONIC

My distillation experiments with the G&T led to some bad ideas. The bottle-strength gin and tonic was an experiment in how far I could push the flavors of a gin and tonic. Could I make a gin-and-tonic shot that was at the same alcoholic strength as the gin that also tasted good? Of course I could. I had a rotary evaporator. I could easily pull some water out of the gin and replace it with tonic flavors, but there were issues.

For several reasons, I knew that a G&T shot would have to be served very cold. First, chilling blunts the impact of alcohol on the nose and tongue so other flavors won't be overpowered. Second, chilling increases the amount of CO_2 I could push into the booze at any given pressure, and highly alcoholic mixtures need a lot of CO_2 to taste really carbonated. I knew from tests with straight vodka that chilled shots taste best between −16°C (3°F) and −20°C (−4°F). Any temperature much below −20°C (−4°F) starts to be painful. I set my goals for the lowest guaranteed nonpainful temperature, −20°C (−4°F). Serving a shot that cold brings its own problems. The balance between sugar and acid is temperature-dependent. Your tongue's sensation of sugar is blunted by cold much more than its sensation of acid is. You need to add more sugar to a −20°C shot than one at −7°C (like one of my regular carbonated drinks) to get the same sweetness. Thus, bottle-strength gin and tonics are drinkable only in a very restricted temperature range. If you overchill one even a couple of degrees, you'll be cold-burning people's tongues. If you allow them to warm above −16°C, the shots start to taste sickly sweet and overly alcoholic. The bottle-strength gin and tonic was good *only* between −16°C (3°F) and −20°C (−4°F)—a mere 4°C of leeway! I knew I could serve the shots in perfect shape, but if the people I served it to did silly things like have a conversation instead of just drinking when I served them, the drink would warm up. I could see people standing around talking while the drink turned to crap in their hands, and it made me sweat. I realized that I couldn't force people to drink a shot immediately and in its entirety.

This experiment turned me off highly alcoholic carbonated drinks. I now serve carbonated cocktails with lower alcohol-by-volume content. They are more temperature-resilient, less cloying, and less stupefying, and I am better for it. The bottle-strength G&T is a stunt that I won't repeat again, but here's the technique if you're curious:

Mix a liter of Tanqueray with a half-liter of fresh lime juice and distill to 700 ml in a rotary evaporator at room temperature with a condenser set to at least

–20°C. (Tanqueray starts off at 47.3% alcohol by volume here in the United States. Very little alcohol is left behind in the rotovap, so the 700 ml of distillate has an alcohol by volume of about 67%.) You now have a little less than 300 ml of room to add flavors to the redistilled Tank and still maintain a bottle strength of 47.3% alcohol by volume. Add concentrated lime acid and simple syrup, quinine, and salt to taste, then chill the mix down to –20°C (–4°F). Test for balance, adjust, then dilute back to 1 liter. Rechill the batch and carbonate at 50 psi. Store the carbonated G&T in a chiller at –20°C till it is ready to serve. Serve in very chilled shot glasses.

CLARIFICATION MAKES MY LIFE EASIER

Distilled lime essence doesn't keep any longer than fresh lime juice, so I always found myself slogging through a lot of distillation right before an event. This was a real hassle, because I could distill only about 1 liter per hour, during which time I was tied to the rotovap and couldn't do anything else.

When I finally figured out how to clarify lime juice, my life got a whole lot easier. I didn't have to use the rotovap to make a gin and tonic anymore! I could clarify *liters* of lime juice, and suddenly I could make huge volumes of gin and tonics for events. Life was good. The drinks were good. The only fly in the buttermilk: I was still chilling the drinks in large batches by hand, using liquid nitrogen prior to carbonation. This process made it difficult to serve the G&T in top condition at a *bar,* where chilling individual drinks with liquid nitrogen is problematic (see the Alternative Chilling section, page 140, for why). You can't stick the batch in a fridge—not cold enough—or in a freezer—too cold. To solve this problem I purchased a Randell FX fridge/freezer. It can maintain any temperature within a couple of degrees Fahrenheit. I just set it to –7°C (about 20°F) and let my carbonated drink batches chill for several hours prior to carbonation. Later the Randell keeps the drink in perfect condition for service.

If you are going to make your gin and tonics and drink them all in one day, go ahead and add the clarified lime juice before you carbonate. If you are going to keep your gin and tonics for more than a day (which makes sense for bar service), you should clarify fresh lime juice every day (never compromise on this) and add it to the precarbonated chilled drink at service time. The small amount of noncarbonated clarified lime juice doesn't mess with the carbonation. Here is my current recipe:

INGREDIENTS

Full 1¾ ounces (53.5 ml) Tanqueray gin
(47% alcohol by volume)

Short ½ ounce (12.5 ml) Quinine Simple Syrup
(page 367) or Cinchona Syrup (page 368)

Short 3 ounces (87 ml) filtered water

1–2 drops saline solution or a pinch of salt

⅜ ounce (11.25 ml) clarified lime juice (6%
acid)

PROCEDURE

Combine all the ingredients except the lime juice
and chill between –5° and –10°C (14°–23°F). Carbonate at 42 psi. Add the lime juice as the drink is
poured into a chilled flute. If you carbonate with the
lime juice, serve the drink that day. If you carbonate
without the juice and add it later, your G&T will keep
indefinitely.

So I don't leave you with a sense of incompleteness, here is a recipe for
tonic water with options for the acidulant. The fresh lime juice option must
be used fresh; the others will keep indefinitely but aren't quite as good. The
procedure is the same either way. This tonic is on the dry side.

Tonic Water Two Ways

MAKES 34 OUNCES (1021 ML) AT 8.8 G/100 ML SUGAR, 0.75% ACID

INGREDIENTS

4¾ ounces (142.5 ml) Quinine Simple Syrup
(page 367) or Cinchona Syrup (page 368)

4¼ ounces (127.5 ml) clarified lime juice
(6% acid), or premade lime acid (page 60)
(6% acid), or, if you don't have premade lime
acid, 5.1 grams of citric acid and 2.6 grams
malic acid and the tiniest pinch of succinic
acid dissolved in 4 ounces (120 ml) of water
(6% acid)

20 drops (1 ml) saline solution or a couple
pinches of salt

25 ounces (750 ml) filtered water

PROCEDURE

Combine all the ingredients and chill and carbonate
to between 40 and 45 psi.

GIN AND TONIC

FUTURE STEPS

Right now I'm interested in finding a drink that has the same feeling as a gin and tonic but contains no tonic, meaning no lime and no quinine. Why? I just want to. It is a challenge. I've come close. My two best candidates for tonic replacers thus far are schisandra berries and camu camu, a South American fruit. But, big drawback: the folks who sell them here in the United States think of them as medicines and superfoods, and they don't really care how they taste. That attitude drives me nuts.

SCHISANDRA

Schisandra berries (from the plant *Schisandra chinensis*) hail from China, where they are known as five-flavor berries. They are *almost* true to their name. They are tart and bitter (which is why they are good in tonic-style preparations) but a bit sweet and peppery as well. Four flavors. Supposedly they are also salty, but I don't really get that. I find the flavor intriguing.

Schisandra is used in China as a traditional medicinal herb. Here in the United States it is available as a dried berry of wildly varying quality. Avoid ones that are so desiccated as to resemble peppercorns. Look for ones that have a nice red color.

I have tried schisandra in water-based teas, directly steeped in gin, Justinoed into gin, and iSi-infused into gin. So far, direct infusion is winning out. Some of my tests have yielded very refreshing G&T-plus-pepper drinks that I like a lot. But I have not been able to get consistent results, probably owing to the variability of the product, so I have no recipe to share. I leave it to you to experiment!

CAMU CAMU

In 2012, I attended a lecture in Bogotá on the use of rare indigenous Colombian rainforest products. Typically I relish the opportunity to learn about any new ingredient. Unfortunately, the lecture was in Spanish and, stupidly, I can't speak Spanish. The speaker began the presentation with a fruit he claimed had more vitamin C than any on earth, an antioxidant powerhouse named camu camu (*Myrciaria dubia*). The fruit, I gathered from the little Spanish I could understand, contained 1.5 percent pure vitamin C. I could not have cared less. I get so much vitamin C on a daily basis that I make Linus Pauling look like he was about to get scurvy.

There was a tasting after the lecture. Out came some camu camu puree. I wasn't excited at first, since I had hoped to try fresh fruits. But apparently camu camu is harvested only by canoe-rowing collectors during the rainy season, from half-submerged wild plants. Fresh fruits would never survive the transport to the city and are converted to puree almost on the spot. The puree was bright red. I tasted it. Wow. I instantly knew I had found a perfect gin and tonic replacer! Imagine a fruit with the bitterness of tonic, the acidity of a lime, and an extra spiciness I can only describe as faintly Christmasy. I loved it. I pleaded, borrowed, and begged to get a single jar of the stuff. Because I was in Bogotá for a cocktail demonstration, I happened to have some Pectinex Ultra SP-L on hand, and a tiny $200 desktop centrifuge. I clarified the camu camu into a clear juice, mixed it with gin and sugar and salt, and carbonated it. Damn if that wasn't a fantastic drink. I was elated.

When I got home, I began researching camu camu. It *is* available in the United States, but mainly as a horrid powder. As I've said, superfood types don't tend to fret about taste. I finally found a source of the puree and ordered it. It was from Peru, not Colombia, but how different could it be? When it arrived, I anxiously shredded the package to get at it. When I saw the puree, I was crestfallen: it was yellow, not red, which meant one of three things: it was a different kind of fruit (or at least a different cultivar), it was harvested in an unripe state, or the skins had not remained in contact with the fruit pulp during processing. When I tasted it, my fears were confirmed. It didn't have the spiciness or the bitter bite. Damn. Someone out there in Colombia has access to this wonderful ingredient. Do the world a favor and use it!

SOURCES

EQUIPMENT AND BOOKS

Cocktail Kingdom
www.cocktailkingdom.com
These folks sell all the good cocktail gear, and they have the best reprints of classic cocktail books that I have ever seen. Check out the Bad Ass Muddler, the fine-spring strainer, and the bitters bottle dasher tops.

J. B. Prince
www.jbprince.com
This is an awesome chef's supply store that sells only high-quality stuff. It has a wide array of ice carving/handling equipment. If you find yourself in New York City, visit the showroom at 36 East 31st Street, eleventh floor. You won't be disappointed.

Mark Powers and Company
http://www.markpowers-and -company.com
I get all my soda and carbonation supplies here except the carbonator caps, including cold plates, carbonators, Corny kegs, tubing, and clamps. The staff is friendly and the prices are good.

Liquid Bread
http://www.liquidbread.com
Makers of the carbonator cap.

Katom
www.katom.com
This is a good inexpensive website for restaurant supplies like bar mats if you don't have access to real live kitchen-supply joints.

McMaster-Carr
www.mcmaster.com
Need some weird industrial doo-dad? Need it tomorrow? Are you willing to pay 30 percent more than you should just so you don't have to scour the Internet for another source? McMaster's got you covered. I use it constantly. Similar but less complete industrial suppliers are *www.grainger.com* and *www .mscdirect.com*.

Amazon
www.amazon.com
Yeah, you already knew to check here. I include it because many suppliers of scientific gear have started selling on Amazon, including purveyors of the $200 centrifuge and all the lab glassware you will ever need.

If you are looking for books, skip Amazon and buy from the folks at Kitchen Arts and Letters in New York, *www.kitchenartsandletters .com*. I support them. The knowledge they can offer up on individual books and authors is worth paying the extra nickels for.

For old books I used to use the aggregator *www.bookfinder.com* excusively, but it has become polluted with crappy reprints and I now rely on *www.ebay.com* and *www.abebooks.com*.

INGREDIENTS

Real Live Stores
If you live in New York you have access to two great ingredient and spice shops that sell everything you need to make any type of bitters. Every major metropolis probably has similar shops. If you can't find them, the Internet is your friend.

Kalustyans, 123 Lexington Ave., New York, NY 10016
www.kalustyans.com

Dual Specialty Store, 91 First Ave., New York, NY 10003
The website is too horrid for me to direct you there.

Wherever you live (unless Manhattan is home) you probably have access to a home-brew shop that carries a lot of useful ingredients for cocktails, including malic, citric, tartaric, and lactic acid, as well as equipment like carbonator caps and other fun stuff. There are also scads of home-brew shops online.

INTERNET SOURCES

Modernist Pantry
www.modernistpantry.com
A one-stop shop for modern ingredients such as Pectinex SP-L, hydrocolloids, and so on.

Terra Spice
www.terraspice.com
A good place for quality spices and extracts.

WillPowder
www.willpowder.net
Supplies modern ingredients from pastry chef Will Goldfarb.

Le Sanctuaire
www.le-sanctuaire.com
A supplier of high-end cooking equipment and ingredients. This site has the best selection of hydrocolloids from CP Kelco, the folks who do gellan and really good pectins.

TIC Gums
www.ticgums.com
The folks who make Ticaloid 210S and Ticaloid 310S, the mixture of gum arabic and xanthan gum that I use to make orgeats and oil syrups. They will sell to normal folks like us.

FURTHER READING

GENERAL BOOKS ON COCKTAILS AND SPIRITS

Here is a short list of cocktail books and authors I have found useful or entertaining over the years.

Baker, Charles H., Jr. *Jigger, Beaker and Glass: Drinking Around the World* Derrydale, 2001 (1939). A raucous romp through the decades-long, worldwide cocktail peregrinations of Charles H. Baker, Jr. A gem.

Berry, Jeff. *Beach Bum Berry Remixed*. 2nd ed. SLG, 2009. If you like Tiki, Berry is the man to trust. This volume is a compilation of two of his books.

Conigliaro, Tony. *The Cocktail Lab: Unraveling the Mysteries of Flavor and Aroma in Drink, with Recipes.* Ten Speed, 2013. My good friend Tony Conigliaro is the finest practitioner of the modern cocktail that I know. Here is a look at how he thinks.

Craddock, Harry. *The Savoy Cocktail Book.* Martino, 2013 (1930). This is *the* old-school recipe book that every self-respecting cocktail dork should have in his or her library. One of the few manuals from which the cocktail revolution was launched. I own it, although I don't really use it.

Curtis, Wayne. *And a Bottle of Rum: A History of the New World in Ten Cocktails.* Broadway Books, 2007. If you enjoy viewing history through the lens of products you like, you'll dig Curtis's historical account of rum through the ages.

DeGroff, Dale. *The Craft of the Cocktail.* Clarkson Potter, 2002. A must-own cocktail book. Dale is the king of the cocktail, and this is his great cocktail tome.

Embury, David A. *The Fine Art of Mixing Drinks.* Mud Puddle, 2008 (1948). This was the first cocktail book I owned. I found an old copy at a used-book store and picked it up on a lark. Lucky choice. Embury was a great writer. He was a lawyer by trade and a cocktail thinker by avocation. I once met an old-time bartender who had served Embury; he said the guy was a lousy tipper, which kind of ruined him for me. Why must heroes be jerks?

Haigh, Ted. *Vintage Spirits and Forgotten Cocktails: From the Alamagoozlum to the Zombie—100 Rediscovered Recipes and the Stories Behind Them.* Quarry, 2009. Good book.

Hess, Robert. *The Essential Bartender's Guide.* Mud Puddle, 2008. Hess is a longtime fixture of the cocktail intelligentsia. This is his intro-to-bartending guide. Many people, once they reach a certain level of proficiency, dismiss introductory books. I don't. The best thing about this kind of book is the direct view it can provide into the writer's way of thinking.

Liu, Kevin K. *Craft Cocktails at Home: Offbeat Techniques, Contemporary Crowd-Pleasers, and Classics Hacked with Science.* Kevin Liu, 2013. Liu's book on the science of cocktails is worth the read. One of the few books on the subject as of this writing.

Meehan, Jim. *The PDT Cocktail Book: The Complete Bartender's Guide from the Celebrated Speakeasy.* Sterling Epicure, 2011. PDT is one of the most renowned bars in the United States. Meehan has done a great service with *The PDT Cocktail Book*, giving an actual account of the recipes used there.

O'Neil, Darcy S. *Fix the Pumps.* Art of Drink, 2010. *Fix the Pumps* has been tremendously influential in the cocktail world. With it, O'Neil ushered in the resurgence of interest in the artistry of the old-school soda jerk. Bartenders mine this book for ideas on both alcoholic and nonalcoholic drinks.

Pacult, F. Paul. *Kindred Spirits 2.* Spirit Journal, 2008. While this edition is getting a little long in the tooth (let's hope he makes a third edition soon), no one does spirits reviews like Pacult.

Regan, Gary. *The Joy of Mixology.* Clarkson Potter, 2003. Fantastic book. I don't agree with everything Regan says, but his landmark book is a must-read. His drink systematics are particularly interesting.

Stewart, Amy. *The Drunken Botanist.* Algonquin, 2013. This book ran like wildfire through the cocktail community when it was published. It is delightful. It simplifies and glosses over some subjects, but that doesn't spoil the enjoyment. You will pick up some interesting facts from this book and they will spark ideas for the future.

Wondrich, David. *Imbibe!* Perigree, 2007. Really, you can't go wrong with anything Wondrich writes. Read this one and his other book, *Punch.* Wondrich is a born storyteller with a great feel for history in general and for the history of the common drinking person in particular. Ostensibly *Imbibe!* is the story of Jerry Thomas, the author of the first cocktail book and a famous American barman of the mid-nineteenth century, but Wondrich casts a much wider net.

BOOKS ON FOOD AND SCIENCE YOU MAY FIND HELPFUL

Damodaran, Srinivasan, Kirk L. Parkin, and Owen R. Fennema. *Fennema's Food Chemistry, Fourth*

Edition (Food Science and Technology). CRC, 2007. The standard textbook in food chemistry.

McGee, Harold. *On Food and Cooking: The Science and Lore of the Kitchen.* Scribner, 2004. If you don't already own *On Food and Cooking,* go out and buy it right now. It is *the* reference for science as applied to making things delicious. How many other food books are recognized simply by their acronym? *OFAC* is hugely influential, and hugely useful. The current edition, from 2004, is a vastly different book from the original, groundbreaking 1984 edition. McGee was forced to remove a lot of the historical vignettes and stories that peppered the first edition to make way for all-new information he added in the second. You will want both.

———. *The Curious Cook: More Kitchen Science and Lore.* Wiley, 1992. McGee's second book, now sadly out of print, is entirely different from *OFAC.* In *The Curious Cook,* McGee shows how to think like a scientist in the kitchen— which is entirely different from, and more important than, knowing a lot of science.

Myhrvold, Nathan, Chris Young, and Maxime Bilet. *Modernist Cuisine: The Art and Science of Cooking.* Cooking Lab, 2011. The greatest cookbook achievement of all time. If you want to investigate modern techniques, this is the place to start.

What follows is a nonexhaustive list of books and articles related to specific sections of this book.

PART 1: PRELIMINARIES
MEASUREMENT, UNITS, EQUIPMENT
If you are interested in home distilling, these two books are good places to start.

Smiley, Ian. *Making Pure Corn Whiskey: A Professional Guide for Amateur and Micro Distillers.* Ian Smiley 2003.

Nixon, Michael. *The Compleat Distiller.* Amphora Society, 2004.

INGREDIENTS
Nielsen, S. Suzanne, ed. *Food Analysis.* Springer 2010. Pages 232–37 describe how to calculate and convert various measures of titratable acidity.

Saunt, James. *Citrus Varieties of the World.* Sinclair International Business Resources, 2000.

PART 2: TRADITIONAL COCKTAILS
Weightman, Gavin. *The Frozen Water Trade: A True Story.* Hyperion, 2004. This book is the story of how ice became a commercial commodity in the nineteenth century prior to the advent of mechanical refrigeration. It explains the marketing genius that made iced drinks popular in the nineteenth century—they were not always ubiquitous—and it whet many a bartender's desire for beautiful clear ice like the ice of yore that was harvested from lakes.

Oxtoby, David W. *Principles of Modern Chemistry. 7th ed.* Cengage Learning, 2011. Oxtoby is my go-to reference for entropy, enthalpy, and other general chemistry information.

Garlough, Robert, et al. *Ice Sculpting the Modern Way.* Cengage, 2003. If you want to know more about ice carving.

PART 3: NEW TECHNIQUES AND IDEAS
ALTERNATIVE CHILLING
U.S. Chemical Safety and Hazard Investigation Board. "Hazards of Nitrogen Asphyxiation." Safety Bulletin No. 2003-10-B. June, 2003.

RED-HOT POKERS
Brown, John Hull. *Early American Beverages.* C. E. Tuttle, 1966. Old-time recipes, including those that use the flip dog.

RAPID INFUSIONS, SHIFTING PRESSURE
Parsons, Brad Thomas. *Bitters: A Spirited History of a Classic Cure-All, with Cocktails, Recipes, and Formulas.* Ten Speed, 2011. The book to buy if you want to learn about making traditional bitters.

CLARIFICATION
Moreno, Juan, and Rafael Peinado. *Enological Chemistry.* Academic, 2012. Chapter 19, Wine Colloids, deals with the subject of wine fining and clarification.

WASHING
BOOKS
Phillips, G. O. et al. *Handbook of Hydrocolloids. 2nd ed.* Woodhead, 2009. Really pricey, but it's the standard reference for hydrocolloids. Has good chapters on gellan, chitosan, milk proteins, and egg proteins, which are used in this book. It also has chapters on agar, gum arabic, pectin, and gelatin that might be helpful in other areas of cocktail science.

ARTICLES
Luck, Genevieve, et al. "Polyphenols, Astringency and Proline-rich Proteins." *Phytochemistry* 37. no. 2 (1994): 357–71. The title pretty much tells it all.

Ozdal, Tugba, et al. "A Review on Protein–Phenolic Interactions and Associated Changes." *Food*

Research International 51 (2013): 954–70. While the focus of this paper is the possibility that protein-phenolic interactions may reduce the number of "health effects" that phenolic compounds may confer—a subject I couldn't care less about, because today's health-giver is almost always tomorrow's enemy—it does give a good overview of the interactions.

Hasni, Imed, et al. "Interaction of Milk a- and b-Caseins with Tea Polyphenols." *Food Chemistry* 126 (2011): 630–39. This article deals with the specific reaction that makes Tea Time work.

Lee, Catherine A., and Zata M. Vickers. "Astringency of Foods May Not Be Directly Related to Salivary Lubricity." *Journal of Food Science* 77, no. 9 (2012). An article that goes against the commonly held view that astringency is solely related to the tongue's loss of ability to lubricate itself.

de Freitas, Victor, and Nuno Mateus. "Protein/Polyphenol Interactions: Past and Present Contributions. Mechanisms of Astringency Perception." *Current Organic Chemistry* 16 (2012): 724–46. Is what it says it is.

CARBONATION
BOOKS

Liger-Belair, Gérard. *Uncorked: The Science of Champagne.* Rev. ed. (Princeton University Press, 2013. Liger-Belair, the champagne physicist, has a sweet, sweet job. This is a revised edition of his book on champagne science for lay audiences.

Steen, David P., and Philip Ashurst. *Carbonated Soft Drinks: Formulation and Manufacture.* Wiley-Blackwell, 2006. Technical, dry, and expensive. Unless you need to know the nitty-gritty of commercial soda production, you can skip this book, but I often use it for reference.

Woodroff, Jasper Guy, and G. Frank Phillips. *Beverages: Carbonated and Noncarbonated.* AVI, 1974. Out of print, out of date, and dull, but I need it from time to time. It's one of a series of books on food technology published by AVI Press in the 1970s, standard-setting in its day.

ARTICLES

Liger-Belair, Gérard, et al. "Unraveling the Evolving Nature of Gaseous and Dissolved Carbon Dioxide in Champagne Wines: A State-of-the-Art Review, from the Bottle to the Tasting Glass." *Analytica Chimica Acta* 732 (2012): 1–15. Awesome article. Title is fairly explanatory.

————— "Champagne Cork Popping Revisited Through High-Speed Infrared Imaging: The Role of Temperature." *Journal of Food Engineering* 116 (2013): 78–85. Another fun read from Liger-Belair.

—————. "Carbon Dioxide and Ethanol Release from Champagne Glasses, Under Standard Tasting Conditions." *Advances in Food and Nutrition Research* 67 (2012): 289–340. Liger-Belair goes into depth on the effects that glass shape and pouring style have on carbon dioxide levels in champagne, and how these effects affect the drinking experience.

—————. "CO_2 Volume Fluxes Outgassing from Champagne Glasses: The Impact of Champagne Ageing." *Analytica Chimica Acta* 660 (2010): 29–34. The champagne master explains how aging champagne affects bubble size.

Cuomo, R., et al. "Carbonated Beverages and Gastrointestinal System: Between Myth and Reality." *Nutrition, Metabolism and Cardiovascular Diseases* 19 (2009): 683–89. Examines whether or not bubbles are bad for you. Short answer: no.

Roberts, C., et al. "Alcohol Concentration and Carbonation of Drinks: The Effect on Blood Alcohol Levels." *Journal of Forensic and Legal Medicine* 14 (2007): 398–405. Interesting study finding that two-thirds of the individuals tested got drunk faster on bubbly than on flat booze.

Bisperink, Chris G. J., et al. "Bubble Growth in Carbonated Liquids." *Colloids and Surfaces A: Physicochemical and Engineering Aspects* 85 (1994): 231–53. This article demonstrates that carbonation levels would need to be hundreds of times higher than normal to create spontaneous bubble nucleation.

Profaizer, M. "Shelf Life of PET Bottles Estimated Via a Finite Elements Method Simulation of Carbon Dioxide and Oxygen Permeability." *Italian Food and Beverage Technology* 48 (April 2007).

PART 4: LITTLE JOURNEYS

Hubbard, Elbert. *Little Journeys to the Homes of the Great.* Memorial edition in 14 volumes Roycrofters, 1916. Hubbard wrote a series of short, rambling, opinionated biographies called Little Journeys on famous people throughout history and from all walks of life. I read them as a child, and they really stuck with me. Though not much remembered today, Hubbard and his American Arts and Crafts community, the Roycrofters, were important and influential in the early 1900s. Some of the hand-printed Roycrofters editions, especially the illuminated ones done in green suede-covered boards, are treasures. I named my section after these books.

APPLES

Beach, Spenser Ambrose. *Apples of New York.* 2 vols. (J. B Lyon, 1905). Almost 110 years old and still the definitive book on apples, *Apples of New York* was the first of a series of mammoth books to come out of the Agricultural Experiment Station in Geneva, New York, where McGee and I went apple-tasting. The rest of the *Fruits of New York* series—grapes, pears, cherries, peaches, small fruits, and plums—were written by the inimitable U. P. Hedrick, but he is a subject for a different book. All the *Fruits of New York* books are worth looking at, and they are all available to read online.

Traverso, Amy. *The Apple Lover's Cookbook.* W. W. Norton, 2011. One of the few good apple books that is also a cookbook. Beautifully shot.

Hanson, Beth. *The Best Apples to Buy and Grow.* Brooklyn Botanic Garden, 2005. Great little inexpensive book. I love all the books in the Brooklyn Botanic Garden's horticultural series. Also check out *Buried Treasures: Tasty Tubers of the World*.

Morgan, Joan, et al. *The New Book of Apples: The Definitive Guide to Over 2,000 Varieties.* Ebury, 2003. This is a book that was written with the Brogdale collection of apples in Kent, England, as its tasting home base. The Brogdale is the U.K. equivalent of our apple collection in Geneva, N.Y., but also does pears, small fruits, nuts, and more. This book is great, but it definitely skews toward varieties that do well in the U.K. The history section is worth looking at.

COFFEE

BOOKS

Illy, Andrea, and Rinantonio Viani. *Espresso Coffee, Second Edition: The Science of Quality.* Academic Press, 2005. Great little technical book on espresso written by the scientific leader of the current generation of the Illy family. There is a similarly titled coffee-table book that isn't useful for technical info.

Rao, Scott. *Everything but Espresso.* Scott Rao, 2010. Scott Rao's books have been game-changers in the coffee world. You kind of need to check out this one and the one below.

———. *The Professional Barista's Handbook: An Expert Guide to Preparing Espresso, Coffee, and Tea.* Scott Rao, 2008.

Schomer, David C. *Espresso Coffee: Professional Techniques.* Espresso Vivace Roasteria, 1996. Perhaps a bit out of date now. Schomer was a revolutionary at the time (still is, I guess). It was a visit to his shop, Café Vivace, in Seattle in 2001 that showed me I didn't know diddly-squat about espresso. I bought this book on the spot.

ARTICLES

Illy, Ernesto, and Luciano Navarini. "Neglected Food Bubbles: The Espresso Coffee Foam." *Food Biophysics* 6 (2011): 335–48. Good paper on the working of foam (crema) in espresso.

GIN AND TONIC

CFR: Code of Federal Regulations Title 21—Food and Drugs §172.575. These are the U.S. government regulations on quinine.

Carter, H. R. "Quinine Prophylaxis for Malaria." *Public Health Reports (1896–1970)* 29, no. 13 (March 27, 1914): 741–49. Interesting paper on the history of quinine use and the doses necessary to prevent and/or treat malaria.

Recipe List

PAGE	RECIPE	INGREDIENTS	STATS
51	Simple Syrup	Equal parts sugar and water by weight. Mix to dissolve.	
53	Honey Syrup	64 grams of water to every 100 grams of honey.	
54	Butter Syrup	10 allspice berries, crushed 200 grams water 3 grams TIC Gums Pretested Ticaloid 210S 150 grams melted butter 200 grams granulated sugar Simmer allspice in water to infuse, strain, hydrate Ticaloid in water, emulsify in butter, then blend in sugar.	
55	Cold Buttered Rum	2 ounces (60 ml) spiced rum, such as Sailor Jerry Fat ounce (1⅛ ounces, 33.75 ml) Butter Syrup ½ ounce (15 ml) freshly strained lime juice Shake with ice and serve in a chilled single old-fashioned glass.	Makes one 5⅗-ounce (168-ml) drink at 16.4% alcohol by volume, 8.6 g/100 ml sugar, 0.54% acid
56	Any Nut Orgeat	FOR NUT MILK: 600 grams very hot water 200 grams nut of your choice FOR EVERY 500 GRAMS NUT MILK: 1.75 grams Ticaloid 210S 0.2 gram xanthan gum 500 grams granulated sugar Blend hot water and nuts, then pass nut milk through a fine filter. Hydrate Ticaloid into nut milk in a blender, then blend in sugar.	
60	Lime Acid	94 grams filtered water 4 grams citric acid 2 grams malic acid 0.04 gram succinic acid (optional)	
60	Lime Acid Orange	1 liter freshly squeezed orange juice 32 grams citric acid 20 grams malic acid	
61	Champagne Acid	94 grams warm water 3 grams tartaric acid 3 grams lactic acid (use the powder form)	
61	Saline Solution	20 grams salt 80 ml filtered water	
75	Martini	2 ounces (60 ml) gin or vodka, room at temperature ⅜–½ ounce (10–14 ml) Dolin dry vermouth (or your favorite), at room temperature 1 or 3 olives on a toothpick, or a lemon twist Stir with ice and serve in a chilled coupe glass.	

PAGE	RECIPE	INGREDIENTS	STATS
86	Manhattans for two	4 ounces (120 ml) Rittenhouse rye (50% alcohol by volume) Fat 1¾ ounces (53 ml) Carpano Antica Formula vermouth (16.5% alcohol by volume) 4 dashes Angostura bitters 2 brandied cherries or orange twists Stir with ice and serve in chilled coupe glasses.	Makes two 4⅓-ounce (129-ml) drinks at 27% alcohol by volume, 3.3 g/100 ml sugar, 0.12% acid
95	Whiskey Sour with Egg White	2 ounces (60 ml) bourbon or rye (50% alcohol by volume) Fat ½ ounce (17.5 ml) freshly strained lemon juice ¾ ounce (22.5 ml) simple syrup Pinch of salt 1 large egg white (1 ounce or 30 ml) Mix everything but the egg white in a tin, then add the egg white and shake without ice for at least 10 seconds. Add ice and shake an additional 10 seconds. Strain into a chilled coupe glass through a tea strainer.	Makes one 6.6-ounce (198-ml) drink at 15.0% alcohol by volume, 7.1 g/100 ml sugar, 0.53% acid
99	Classic Daiquiri	2 ounces (60 ml) light, clean rum (40% alcohol by volume) ¾ ounce (22.5 ml) simple syrup ¾ ounce (22.5 ml) freshly strained lime juice 2 drops saline solution (20%) or a pinch of salt Shake with ice and serve in a chilled coupe glass.	Makes one 5⅓-ounce (159-ml) drink at 15.0% alcohol by volume, 8.9 g/100 ml sugar, 0.85% acid
100	Hemingway Daiquiri	2 ounces (60 ml) light, clean rum (40% alcohol by volume) ¾ ounce (22.5 ml) freshly strained lime juice ½ ounce (15 ml) Luxardo Maraschino (32% alcohol by volume) ½ ounce (15 ml) freshly strained grapefruit juice 2 drops saline solution or a pinch of salt Shake with ice and serve in a chilled coupe glass.	Makes one 5⅘-ounce (174-ml) drink at 16.5% alcohol by volume, 4.2 g/100 ml sugar, 0.98% acid
104	Negroni for two	2 ounces (60 ml) gin 2 ounces (60 ml) Campari 2 ounces (60 ml) sweet vermouth 2 orange twists Stir with ice and serve in a chilled coupe glass or over a large rock.	Makes two 5.2-ounce (127-ml) drinks at 20.7% alcohol by volume, 9.4 g/100 ml sugar, 0.14% acid
107	Cliff Old-Fashioned	2 ounces (60 ml) Elijah Craig 12-year bourbon (47% alcohol by volume) ⅜ ounce (11 ml) Coriander Syrup 2 dashes Angostura bitters Orange twist One 2-inch-by-2-inch clear ice cube Build over a large rock in an old-fashioned glass.	Makes one 3-ounce (90-ml) drink at 32% alcohol by volume, 7.7 g/100 ml sugar, 0.0% acid
110	Coriander Syrup	125 grams coriander seeds, preferably with a fresh, citrusy aroma (for soda syrup, reduce to 100 grams) 550 grams filtered water 500 grams granulated sugar 5 grams salt 10 grams crushed red pepper	

PAGE	RECIPE	INGREDIENTS	STATS
115	The Shaken Margarita	2 ounces (60 m) tequila (40% alcohol by volume) ¾ ounce (22.5 ml) Cointreau ¾ ounce (22.5 ml) freshly strained lime juice ¼ ounce (7.5 ml) simple syrup 5 drops saline solution or a generous pinch of salt Shake with ice and serve in a chilled coupe glass.	Makes one 5⁹⁄₁₀-ounce (178-ml) drink at 18.5% alcohol by volume, 6.0 g/100 ml sugar, 0.76% acid
118	Blender Marg	1 ounce (30 ml) Cointreau ¾ ounce (22.5 ml) La Puritita mezcal ½ ounce (15 ml) Yellow Chartreuse ½ ounce (15 ml) freshly strained lime juice 10 drops Hellfire bitters or spicy nonacidic stuff of your choice 5 drops saline solution or a generous pinch of salt About 4 ounces (120 grams) ice Combine ingredients in a blender and blend briefly.	Makes one 5.3-ounce (158-ml) drink at 17.2% alcohol by volume, 7.9 g/100 ml sugar, 0.57% acid
119	Generic Blender Sour	2¼ ounces (67.5 ml) liquid that contains around 0.9 ounces (27 ml) pure ethanol and 12.75 grams sugar (see Sugared Booze recipe, below) ½ ounce (30 ml) freshly strained lemon, lime, or other sour juice 4 ounces (120 ml) ice 2–5 drops saline solution or a generous pinch of salt Combine ingredients in a blender and blend briefly.	
119	Sugared Booze for blended drinks	212 grams superfine sugar (regular granulated sugar is okay but will take longer to dissolve) 1 liter either 80 or 100 proof (40 or 50% alcohol by volume) spirits	Makes 1140 ml (38 ounces) liquor at either 44% or 35% alcohol by volume
120	Rittenhouse Blender Sour	2 ounces (60 ml) sugared Rittenhouse rye (44% alcohol by volume; see Sugared Booze recipe, above) ½ ounce (15 ml) freshly strained lemon juice ¼ ounce(7.5 ml) freshly strained orange juice 4 drops saline solution or a generous pinch of salt 4 ounces (120 grams) ice Combine ingredients in a blender and blend briefly.	Makes one 5¼-ounce (157-ml) drink at 16.7% alcohol by volume, 7.8 g/100 ml sugar, 0.61% acid
120	Blender Daiquiri	2¼ ounces (67.5 ml) sugared Flor de Caña white rum (35% alcohol by volume; see Sugared Booze recipe, above) ½ ounce (15 ml) freshly strained lime juice 4 drops saline solution or a generous pinch of salt 4 ounce (120 grams) ice Combine ingredients in a blender and blend briefly.	Makes one 5¼-ounce (157-ml) drink at 15% alcohol by volume, 8.1 g/100 ml sugar, 0.57% acid
142	Frozen Daiquiri for two	4 ounces (120 ml) white rum (40% alcohol by volume), preferably clean-tasting and cheap, like Flor de Caña 4 ounces (120 ml) filtered water 1½ ounces (45 ml) simple syrup 4 drops saline solution or 2 pinches of salt 1¾ ounces (52.5 ml) freshly strained lime juice Combine ingredients in a Ziploc bag and freeze. When frozen, blend for a couple seconds.	Makes two 5.6-ounce (169-ml) drinks at 14.2% alcohol by volume, 8.4 g/100 ml sugar, 0.87% acid

PAGE	RECIPE	INGREDIENTS	STATS
144	Ebony for two	5 ounces (150 ml) Carpano vermouth (16% alcohol by volume, roughly 16 grams sugar/dl, roughly 0.6% acid) 1½ ounces (45 ml) vodka (40% alcohol by volume) 2½ ounces (75 ml) filtered water 4 drops saline solution or 2 pinches of salt Flat ¾ ounce (21 ml) freshly strained lemon juice Combine ingredients in a Ziploc bag and freeze. When frozen, blend for a couple seconds.	Makes two 4⅘-ounce (145.5-ml) drinks at 14.4% alcohol by volume, 8.4 g/100 ml sugar, 0.74% acid
144	Ivory for two	5½ ounces (165ml) Dolin Blanc vermouth (16% alcohol by volume, roughly 13 grams sugar/dl, roughly 0.6% acid) 1 ounce (30 ml) vodka (40% alcohol by volume) 2 ounces (60 ml) filtered water 4 drops saline solution or 2 pinches of salt Flat ¾ ounce (21 ml) freshly strained lime juice Combine ingredients in a Ziploc bag and freeze. When frozen, blend for a couple seconds.	Makes two 4⅗-ounce (138-ml) drinks at 13.9% alcohol by volume, 7.9 g/100 ml sugar, 0.81% acid
147	Strawberry Bandito	2 ounces (60 ml) strawberry juice (8 g/dl sugar, 1.5% acid) (or you can use 2½ ounces-75 grams frozen strawberries and 15 grams ice) 2 ounces (60 ml) Jalapeño Tequila or regular blanco tequila (40% abv) ¼ ounce (7.5 ml) freshly strained lime juice Short ½ ounce (12.5) simple syrup 2 drops saline solution or a pinch of salt Shake liquid ingredients with juice ice and serve in a chilled glass.	Makes one 4⅗-ounce (140-ml) drink at 17.1% alcohol by volume, 9.0 g/100 ml sugar, 0.96% acid
149	Shaken Drake	1½ ounces (45 ml) Helbing Kümmel liqueur (35% alcohol by volume) ½ ounce (15 ml) vodka (40% alcohol by volume) 1 bar spoon (4 ml) grade B maple syrup (87.5 g/dl sugar) 5 drops saline solution or a generous pinch of salt 2 ounces (60 ml) freshly squeezed and strained grapefruit juice (10.4 g/dl sugar, 2.4% acid) frozen into 1-ounce (30-ml) ice cubes Shake liquid ingredients with juice ice and serve in a chilled glass.	Makes one 4.6-ounce (139-ml) drink at 15.6% alcohol by volume, 10.2 g/100 ml sugar, 1.03% acid
150	Scotch and Coconut	1½ ounces (45 ml) Ardbeg 10-year Scotch (46% alcohol by volume) ½ ounce (15 ml) Cointreau (40% alcohol by volume) ¼ ounce (7.5 ml) freshly strained lemon juice 2 drops saline solution or a pinch of salt 2½ ounces (75 ml) fresh coconut water (6.0 g/dl sugar) frozen into 1¼ ounce (37.5 ml) ice cubes 1 orange twist 1 star anise pod Shake liquid ingredients with juice ice and serve in a chilled glass. Express orange over the top and float the star anise.	Makes one 4⁷⁄₁₀-ounce (142-ml) drink at 18.6% alcohol by volume, 5.9 g/100 ml sugar, 0.32% acid

PAGE	RECIPE	INGREDIENTS	STATS
153	Manhattans by the Pitcher	14 ounces (420 ml) Rittenhouse rye (50% alcohol by volume) 6¼ ounces (187.5 ml) Carpano Antica Formula vermouth (16.5% alcohol by volume, roughly 16% sugar, 0.6% acid) ¼ ounce (7.5 ml) Angostura bitters 10½ ounces (315 ml) ice water (water that has been chilled with ice; don't add the ice) Garnish of your choice	Makes seven 4⅖-ounce (132-ml) drinks at 26% alcohol by volume, 3.2 g/100 ml sugar, 0.12% acid
155	Bottled Manhattans, Professional Style	Three 750 ml bottles Rittenhouse rye One 1 liter Bottle Carpano Antica Formula vermouth 1 ounce (30 ml) Angostura bitters 1700 ml filtered water	Makes thirty 4½-ounce (136-ml) drinks at 26% alcohol by volume, 3.2 g/100 ml sugar, 0.12% acid
172	TBD: Thai Basil Daiquiri	5 grams (7 large) Thai basil leaves 2 ounces (60 ml) Flor de Caña white rum or other clean white rum (40% alcohol by volume) ¾ ounce (22.5 ml) freshly strained lime juice Short ¾ ounce (20 ml) simple syrup 2 drops saline solution or a pinch of salt Nitro- or blender-muddle herbs with rum. Shake with ice and serve in a chilled coupe glass.	Makes one 5⅓-ounce (160-ml) drink at 15% alcohol by volume, 8.9 g/100 ml sugar, and 0.85% acid
174	Spanish Chris	3.5 grams (a small handful) fresh tarragon leaves 1½ ounces (45 ml) La Puritita mezcal or any fairly clean blanco mezcal (40% alcohol by volume) ½ ounce (15 ml) Luxardo Maraschino (32% alcohol by volume) ¾ ounce (22.5 ml) freshly strained lime juice ½ ounce (15 ml) simple syrup 3 drops saline solution or a generous pinch of salt Nitro- or blender-muddle herbs with spirits. Shake with ice and serve in chilled coupe glass.	Makes one 5-ounce (149-ml) drink at 15.3% alcohol by volume, 10.0 g/100 ml sugar, 0.91% acid
175	The Flat Leaf	6 grams fresh parsley leaves or 4 grams fresh lovage leaves (a small handful) 2 ounces (60 ml) gin (47.3% alcohol by volume) 1 ounce (30 ml) freshly strained bitter/sour orange juice or ¾ ounce (27.5 ml) freshly strained lime juice ½ ounce (15 ml) simple syrup 3 drops saline solution or a generous pinch of salt Nitro- or blender-muddle herb with gin. Shake with ice and serve in chilled coupe.	Makes one 5½-ounce (164-ml) drink at 17.7% alcohol by volume, 7.9 g/100 ml sugar, 0.82% acid
176	The Carvone	6 grams (a good handful) fresh mint leaves 2 ounces (60 ml) Linie aquavit (40% alcohol by volume) Short ½ ounce (13 ml) simple syrup 3 drops saline solution or a generous pinch of salt 1 lemon twist Nitro- or blender-muddle herbs with aquavit. Shake with ice and serve in a chilled coupe glass with the twist of lemon.	Makes one 3⁹⁄₁₀-ounce (117-ml) drink at 20.4% alcohol by volume, 6.8 g/100 ml sugar, 0% acid

PAGE	RECIPE	INGREDIENTS	STATS
187	Red-Hot Ale, Poker Style	1 ounce (30 ml) Cognac 3 ounces (90 ml) malty but not hoppy abbey ale, such as Ommegang Abbey ale ¼ ounce (7.5 ml) simple syrup ¼ ounce (7.5 ml) freshly strained lemon juice 3 dashes Rapid Orange Bitters 2 drops saline solution or a pinch of salt 1 orange twist Mix ingredients and hit with a red hot poker.	Makes one 4⅗-ounce (138-ml) drink at 15.3% alcohol by volume, 3.5 g/100 ml sugar, 0.33% acid. Finished volume and alcohol by volume vary with poking time.
187	Red-Hot Ale, Pan Style	2½ teaspoons (12 grams) granulated sugar 1 ounce (30 ml) Cognac 3 ounces (90 ml) abbey ale 3 dashes Rapid Orange Bitters ¼ ounce (7.5 ml) freshly strained lemon juice 2 drops saline solution or a pinch of salt 1 orange twist Heat sugar in a pan till almost burned, add liquor (it will flame), add other ingredients to extinguish, and dissolve sugar. Serve with the twist.	Makes one 4⅗-ounce (138-ml) drink at 15.3% alcohol by volume, sugar incalculable, 0.33% acid. Finished volume and alcohol by volume vary with burn time.
188	Red-Hot Cider, Poker Style	1 ounce (30 ml) apple brandy, such as Lairds Bottled in Bond (50% alcohol by volume) 3 ounces (90 ml) hard apple cider (use decent stuff; I usually use a Norman-style cider) ½ ounce (15 ml) simple syrup ¼ ounce (7.5ml) freshly strained lemon juice 2 dashes Rapid Orange Bitters 2 drops saline solution or a pinch of salt 1 cinnamon stick Mix ingredients and hit with a red-hot poker.	Makes one 4⅗-ounce (138-ml) drink at 15.3% alcohol by volume, 6.5 g/100 ml sugar, 0.31% acid. Finished volume and alcohol by volume vary with poking time.
188	Red-Hot Cider, Pan Style	3 teaspoons (12.5 grams) granulated sugar 1 ounce (30 ml) apple brandy (50% alcohol by volume) 3 ounces (90 ml) hard apple cider 2 dashes Rapid Orange Bitters ¼ ounce (7.5 ml) freshly strained lemon juice 2 drops saline solution or a pinch of salt 1 cinnamon stick Heat sugar in a pan till almost burned, add liquor (it will flame), add other ingredients to extinguish, and dissolve sugar. Serve with the cinnamon stick.	Makes one 4⅗-ounce (138-ml) drink at 15.3% alcohol by volume, sugar incalculable, 0.31% acid. Finished volume and alcohol by volume vary with burn time.
200	Turmeric Gin	500 ml Plymouth gin 100 grams fresh turmeric thinly sliced into 1.6-mm (¹⁄₁₆-inch) disks Infuse with 2 chargers for 2½ minutes.	Yield: 94% (470 ml)

PAGE	RECIPE	INGREDIENTS	STATS
200	Glo-Sour	2 ounces (60 ml) Turmeric Gin ¾ ounce (22.5 ml) freshly strained lime juice Flat ¾ ounce (20 ml) simple syrup 3 drops saline solution or a generous pinch of salt 1–2 dashes Rapid Orange Bitters Shake with ice and serve in a chilled coupe glass.	Makes one 5⅓-ounce (160-ml) shaken drink at 15.9% alcohol by volume, 8.0% g/100 ml sugar, 0.84% acid
203	Lemongrass Vodka	300 ml vodka (40% alcohol by volume) 180 grams fresh lemongrass, sliced into disks Infuse with 2 chargers for 2 minutes.	Yield: 90% (270 ml)
203	Lemon Pepper Fizz	58.5 ml Lemongrass Vodka 12 ml clarified lemon juice 18.75 ml simple syrup Dash Rapid Black Pepper Tincture 2 drops saline solution or a pinch of salt 76 ml filtered water Mix, chill, carbonate, and serve in a chilled flute.	Makes one 5½-ounce (166-ml) carbonated drink at 14.3% alcohol by volume, 7.1 g/100 ml sugar, 0.43% acid
205	Coffee Zacapa	750 ml Ron Zacapa 23 Solera or other aged rum, divided into a 500 ml portion and a 250 ml portion 100 ml filtered water 100 grams whole fresh coffee beans roasted on the darker side 185 ml whole milk Infuse with 2 chargers for 1¼ minutes.	Approximate final alcohol by volume: 31% Yield: Approximately 94% (470 ml)
205	Djer Syrup	400 grams filtered water 400 grams granulated sugar 15 grams djer (grains of Selim), or 9 grams of green cardamom pods and 5 grams black pepper Blend ingredients together, then strain.	
206	Café Touba	2 ounces (60 ml) Coffee Zacapa ½ ounce (15 ml) Djer Syrup 3 drops saline solution ½ ounce (15 ml) cream (if you have not milk-washed the rum) Shake with ice and serve in a chilled coupe glass.	Makes one 3⁹⁄₁₀-ounce (115-ml) shaken drink at 16.1% alcohol by volume, 8.0 g/100 ml sugar, 0.39% acid
207	Jalapeño Tequila	45 grams green jalapeño peppers, seeded, deveined, and very thinly sliced 500 ml blanco tequila (40% alcohol by volume) Infuse with 2 chargers 1½ minutes.	Yield: over 90%
208	Chocolate Vodka	500 ml neutral vodka 40% alcohol by volume neutral vodka 75 grams Valrhona cocoa nibs Infuse with 2 chargers 1½ minutes.	Yield: over 84% (425 ml)

PAGE	RECIPE	INGREDIENTS	STATS
209	Schokozitrone	2 ounces (60 ml) Chocolate Vodka ½ ounce (15 ml) freshly strained lemon juice ½ ounce (15 ml) simple syrup 1:1 2 dashes Rapid Chocolate Bitters 2 drops saline solution or a pinch of salt Candied ginger Stir with ice and serve in a chilled coupe glass with candied ginger garnish.	Makes one 4⅓-ounce (128-ml) stirred drink at 19.2% alcohol by volume, 7.4 g/100 ml sugar, 0.70% acid
211	Rapid Orange Bitters	0.2 grams whole cloves (3 cloves) 2.5 grams green cardamom seeds, removed from pod 2 grams caraway seeds 25 grams dried orange peel (Sevilles preferable) 25 grams dried lemon peel 25 grams dried grapefruit peel 5 grams dried gentian root 2.5 grams quassia bark 350 ml neutral vodka (40% alcohol by volume) 25 grams fresh orange peel (no pith, orange only) Infuse with 1 charger in simmering water for 20 minutes, then cool before venting.	Yield: 52% (185 ml)
213	Rapid Chocolate Bitters	3.0 grams mace (3 whole) 350 ml neutral vodka (40% alcohol by volume) 100 grams Valrhona cocoa nibs 1.5 grams dried gentian 1.5 grams quassia bark Infuse with 2 chargers for 60 minutes.	Yield 85% (298 ml)
214	Rapid Hot Pepper Tincture	8 grams red habanero peppers, seeded, deveined, and very finely sliced 52 grams red serrano peppers, seeded, deveined, and very finely sliced 140 grams green jalapeño peppers, seeded, deveined, and very finely sliced 250 ml pure ethanol (200 proof; 195 is fine) 100 ml filtered water Infuse with 2 chargers for 5 minutes.	Yield: over 90% (315 ml)
214	Rapid Black Pepper Tincture	15 grams Malabar black peppercorns 10 grams Tellicherry black peppercorns 5 grams green peppercorns 3 grams grains of paradise 2 grams cubebs 200 ml neutral vodka (40% alcohol by volume) Infuse with 2 chargers for 5 minutes.	Yield: 80% (160 ml)
216	Rapid Hops Tincture	250 ml neutral vodka (40% alcohol by volume) 15 grams fresh Simcoe hops Infuse with 2 chargers. If hot, simmer for 30 minutes then chill. If cold, infuse for 30 minutes.	Yield: 85% (212 ml)

PAGE	RECIPE	INGREDIENTS	STATS
217	Dual Hot and Cold Hops Tincture	30 grams fresh Simcoe hops, divided into two 15 gram piles 300 ml neutral vodka (40% alcohol by volume) Infuse 15 grams hops with one charger, simmer for 30 minutes, cool, and vent. Add additional 15 grams hops, then infuse with 2 chargers an additional 30 minutes.	Yield: 85% (212 ml)
228	Cucumber Martini	6⅔ ounces (200 ml) cold gin 1⅔ ounces (50 ml) cold Dolin Blanc vermouth ⅓ ounce (10 ml) cold simple syrup Dash of cold saline solution 2 chilled cucumbers (577 grams) 1 lime Maldon salt Celery seeds	
232	All-Purpose Sweet-and-Sour	400 ml simple syrup (or 250 grams granulated sugar and 250 grams filtered water) 400 ml freshly strained lime juice, lemon juice, or lime acid 200 ml filtered water Healthy pinch of salt	Makes 1 liter
233	Tomato Infusion Liquid	100 grams granulated sugar 20 grams salt 5 grams coriander seeds 5 grams yellow mustard seeds 5 grams allspice berries 3 grams crushed red pepper (omit if garnishing Jalapeño Tequila) 100 grams filtered water 500 grams white vinegar	
260	Bananas Justino	3 peeled ripe bananas (250 grams) per 750 ml liquor 2 ml Pectinex Ultra SP-L	Approximately 32% alcohol by volume if starting with 40% alcohol by volume liquor
260	Dates Justino	187 grams Medjool dates per 750 ml liquor 2 ml Pectinex Ultra SP-L Additional 250 ml liquor	Alcohol content roughly unchanged
260	Red Cabbage Justino	400 grams red cabbage dehydrated to 100 grams 500 ml Plymouth gin 1–2 ml Pectinex Ultra SP-L	Alcohol content roughly unchanged
260	Apricots Justino	200 grams dehydrated Blenheim apricots 1 liter liquor 3–4 ml Pectinex Ultra SP-L 250 ml filtered water	Approximately 35% alcohol by volume if starting with 40% liquor
260	Pineapples Justino	200 grams dried pineapple 1 liter liquor 2 ml Pectinex Ultra SP-L	Alcohol content roughly unchanged

PAGE	RECIPE	INGREDIENTS	STATS
267	Benjamin Franklin's Milk Punch	Take 6 quarts of Brandy, and the Rinds of 44 Lemons pared very thin; Steep the Rinds in the Brandy 24 hours; then strain it off. Put to it 4 Quarts of Water, 4 large Nutmegs grated, 2 quarts of Lemon Juice, 2 pound of double refined Sugar. When the Sugar is dissolv'd, boil 3 Quarts of Milk and put to the rest hot as you take it off the Fire, and stir it about. Let it stand two Hours; then run it thro' a Jelly-bag till it is clear; then bottle it off.	
268	Tea Vodka	1 liter vodka (40% alcohol by volume) 32 grams Selimbong second-flush Darjeeling tea 250 ml whole milk 15 grams 15% citric acid solution or a fat 1 ounce (33 ml) freshly strained lemon juice Infuse tea into vodka till quite dark. Strain. While stirring, add vodka to milk and break with citric acid. Let settle and filter out curds.	
269	Tea Time	2 ounces (60 ml) milk-washed tea-infused vodka ½ ounce (15 ml) honey syrup ½ ounce (15 ml) freshly strained lemon juice 2 drops saline solution or a pinch of salt Shake with ice and serve in a chilled coupe glass.	Makes one 4⅗-ounce (137-ml) drink at 14.9% alcohol by volume, 6.9 g/100 ml sugar and 0.66% acid
271	Dr. J	2 ounces rum ¾ ounce lime-strength orange juice flat ¾ ounce simple syrup Pinch of salt Drop of vanilla extract Shake with ice and serve in a chilled coupe glass.	Makes one 5⅓-ounce (159-ml) drink at 15.0% alcohol by volume, 8 g/100 ml sugar, 0.14% acid
273	Egg-Washing Technique	750 ml liquor of your choice at 40% alcohol by volume or higher 1 extra-large egg white 1 ounce (30 ml) filtered water	
275	Cognac and Cabernet	1 ounce (30 ml) egg white (1 large) 2 ounces (60 ml) Cognac (41% alcohol by volume) 4 ounces (120 ml) cabernet sauvignon (14.5% alcohol by volume) 2 ounces (60 ml) filtered water ½ ounce (15 ml) clarified lemon juice or 6% citric acid solution ½ ounce (15 ml) simple syrup 4 drops saline solution or a generous pinch of salt	Makes two 4⅘-ounce (145-ml) drinks at 14.5% alcohol by volume, 3.4 g/100 ml sugar, 0.54% acid
280	Chitosan/Gellan Washing Technique	15 grams liquid chitosan solution (2% of the booze amount) 750 ml booze 15 grams Kelcogel F low-acyl gellan (2% of the booze amount)	
282	Carbonated Whiskey Sour	2⅝ ounces (79 ml) filtered water 1¾ ounces (52.5 ml) chitosan/gellan-washed bourbon at (47% alcohol by volume) ⅝ ounce (19 ml) simple syrup 2 drops saline solution or a pinch of salt Short ½ ounce (12 ml) clarified lemon juice (or add the same amount of unclarified lemon juice after you carbonate) Combine ingredients, chill, and carbonate. Serve in a chilled flute.	Makes one 5⅖-ounce (162.5-ml) drink at 15.2% alcohol by volume, 7.2 g/100 ml sugar, 0.44% acid

PAGE	RECIPE	INGREDIENTS	STATS
285	Peanut Butter and Jelly Vodka	25 ounces (750 ml) vodka (40% alcohol by volume) 120 grams creamy peanut butter 125–200 grams Concord grape jelly	Yield: Variable, between 60 and 85% Finished alcohol by volume: 32.5%
286	PB&J with a Baseball Bat	2½ ounces (75 ml) Peanut Butter and Jelly Vodka (32.5% alcohol by volume) ½ ounce (15 ml) freshly strained lime juice 2 drops saline solution of a pinch of salt Shake briefly with ice and serve in a chilled coupe glass.	Makes one 4⁷⁄₁₀-ounce (140-ml) drink at 17.3% alcohol by volume, 9.0 g/100 ml sugar, 0.77% acid
318	Simple Lime Soda	1 ounce (30 ml) simple syrup ¾ ounce (22.5 ml) clarified lime juice or lime acid base 4¼ ounces (127.5 ml) filtered water 2 drops saline solution or a pinch of salt	Makes 6 ounces (180 ml) at 10.5 g/100 ml sugar, 0.75% acid
319	Strawbunkle Soda	½ ounce (15 ml) simple syrup 3¾ ounces (112.5 ml) clarified strawberry juice 1¾ ounces (52.5 ml) filtered water. 2 drops saline solution or a pinch of salt Combine ingredients, chill, and carbonate.	Makes 6 ounces (180 ml) at 10.1 g/100 ml sugar, 0.94% acid
323	X, Y, or Z and Soda	Short 2 ounces (57 ml) liquor Fat 3½ ounces (108 ml) filtered water Combine ingredients, chill, and carbonate. Serve in a chilled flute.	
324	Carbonated Margarita	Short 2 ounces (58.5 ml) light-bodied, clean tequila like Espolón Blanco (40% alcohol by volume) Fat 2½ ounces (76 ml) filtered water Short ½ ounce (12 ml) clarified lime juice Short ¾ ounce (18.75 ml) simple syrup 2–5 drops saline solution or a generous pinch of salt 1 orange twist Combine ingredients, chill, and carbonate. Serve in a chilled flute with the twist of orange.	Makes one 5½-ounce (165-ml) drink at 14.2% alcohol by volume, 7.1 g/100 ml sugar, 0.44% acid
325	Carbonated Negroni	1 ounce (30 ml) gin 1 ounce (30 ml) Campari 1 ounce (30 ml) sweet vermouth ¼ ounce (7.5 ml) clarified lime juice 2¼ ounces (67.5 ml) filtered water 1–2 drops saline solution or a pinch of salt 1 grapefruit twist Combine ingredients, chill, and carbonate. Serve in a chilled flute with the twist of grapefruit.	Makes one 5½-ounce (165-ml) drink at 16% alcohol by volume, 7.3 g/100 ml sugar, 0.38% acid
326	Champari Spritz	Fat 1½ ounces (48 ml) Campari (24% alcohol by volume, 24% sugar) ⅜ ounce (11 ml) champagne acid (6% acid) 3⅛ ounces (94 ml) filtered water 1–2 drops saline solution or a pinch of salt Combine ingredients, chill, and carbonate. Serve in a chilled flute.	Makes one 5½-ounce (165-ml) drink at 7.2% alcohol by volume, 7.2 g/100 ml sugar, 0.44% acid

PAGE	RECIPE	INGREDIENTS	STATS
327	Gin and Tonic	Full 1¾ ounces (53.5 ml) Tanqueray gin (47% alcohol by volume) Short ½ ounce (12.5 ml) Quinine Simple Syrup or Cinchona Syrup Short 3 ounces (87 ml) filtered water 1–2 drops saline solution or a pinch of salt ⅜ ounce (11.25 ml) clarified lime juice Combine ingredients, chill, and carbonate. Serve in a chilled flute.	Makes one 5½-ounce (165-ml) drink at 15.4% alcohol by volume, 4.9 g/100 ml sugar, 0.41% acid
328	Chartruth	Fat 1¾ ounces (54 ml) Green Chartreuse (55% alcohol by volume, 25% sugar) Short 3¼ ounces (97 ml) filtered water 1–2 drops saline solution or a pinch of salt Short ½ ounce (14 ml) clarified lime juice Combine ingredients, chill, and carbonate. Serve in a chilled flute.	Makes one 5½-ounce (165-ml) drink at 18.0% alcohol by volume, 8.3 g/100 ml sugar, 0.51% acid
331	Gin and Juice: Agar-Clarified	Scant 2 ounces (59 ml) Tanqueray gin (47% alcohol by volume) Short 2¾ ounces (80 ml) agar-clarified grapefruit juice Fat ¾ ounce (26 ml) filtered water (If a slightly sweeter drink is desired, replace a bar spoon (4 ml) of the water with simple syrup, that will make the drink 6.3% sugar, 1.10% acid) 1–2 drops saline solution or a pinch of salt Combine ingredients, chill, and carbonate. Serve in a chilled flute.	Makes one 5½-ounce (165-ml) drink at 16.9% alcohol by volume, 5.0 g/100 ml sugar, 1.16% acid
331	Gin and Juice: Centrifugally Clarified	Fat 1¾ ounces (55 ml) Tanqueray gin (47% alcohol by volume) Fat 1¾ ounces (55 ml) centrifuge-clarified grapefruit juice Short 1½ ounces (42 ml) filtered water Fat ¼ ounce (10 ml) simple syrup Scant bar spoon (3 ml) champagne acid (30 grams lactic acid and 30 grams tartaric acid in 940 grams water) 1–2 drops saline solution or a pinch of salt Combine ingredients, chill, and carbonate. Serve in a chilled flute.	Makes one 5½-ounce (165-ml) drink at 15.8% alcohol by volume, 7.2 g/100 ml sugar, 0.91% acid
339	Granny Smith Soda	5 ounces (150 ml) clarified Granny Smith juice 1 ounce (30 ml) filtered water 2 drops saline solution or a pinch of salt Combine ingredients, chill, and carbonate.	Makes 6 ounces (180 ml) at 10.8 g/100 ml sugar, 0.77% acid
340	Honeycrisp Rum Shake	2 ounces (60 ml) clean-tasting white rum 40% alcohol by volume Either short ½ ounce (12 ml) lime juice, or 0.7 gram malic acid dissolved in 10 ml water and 2 drops saline solution or a pinch of salt 3 ounces (90 ml) unclarified Honeycrisp apple juice frozen into three 1-ounce (30-ml) cubes Shake the ingredients with juice ice and serve in a chilled coupe glass.	Makes one 5⅖-ounce (162-ml) drink at 14.8% alcohol by volume, 7.8 g/100 ml sugar, 0.81% acid
345	Kentucky Kernel	1¾ ounces (52.5 ml) chitosan/gellan-washed Makers Mark bourbon 45% alcohol by volume 2½ ounces (75 ml) clarified Ashmead's Kernel juice 1 ounce (30 ml) filtered water 2 drops saline solution or a pinch of salt Mix, chill, carbonate, serve in a chilled flute.	Makes one 5¼-ounce (157.5-ml) carbonated drink at 15% alcohol by volume, 8.6 g/100 ml sugar, approximately 0.6% acid

PAGE	RECIPE	INGREDIENTS	STATS
347	Bottled Caramel Appletini	2 ounces (60 ml) vodka (40% alcohol by volume) ¼ ounce (7.5 ml) Dolin Blanc vermouth 1 ounce (30 ml) filtered water 1 ¾ ounces clarified Wickson crabapple juice 1 bar spoon (4 ml) 70-Brix caramel syrup Dash of orange bitters 2 drops saline solution or a dash of salt Mix, add to bottle, chill, serve in a chilled coupe glass.	Makes one 5⅕-ounce (155-ml) drink at 16.5% alcohol by volume, 7.2 g/100 ml sugar, 0.45% acid
347	70-Brix Caramel	About 1 ounce (30 ml) filtered water 400 grams granulated sugar Heat in pan till very dark but not burned. Add 400 ml water to hot pan, remove from heat, and stir till everything is dissolved.	
353	Shakerato with Milk	1½ ounces (45 ml) freshly made espresso cooled down to at least 60°C (140°F) 3 ounces (90 ml) whole milk ½ ounce (15ml) simple syrup 2 drops saline solution or a pinch of salt Shake with ice and serve in a chilled glass.	Makes one 6⅔-ounce (197-ml) drink at 0.0% alcohol by volume, 4.7g/100 ml sugar, 0.34% acid
354	Boozy Shakerato	1½ ounces (45 ml) freshly made espresso cooled down to 50°C (122°F) 2 ounces (60 ml) dark rum (40% alcohol by volume) 1½ ounces (45 ml) heavy cream ½ ounce (15 ml) simple syrup 2 drops saline solution or a pinch of salt Shake with ice and serve in a chilled glass.	Makes one 7⅘-ounce (234-ml) drink at 10.2% alcohol by volume, 3.9 g/100 ml sugar, 0.29% acid
355	Boozy Shakerato 2	1½ ounces (45 ml) freshly made espresso cooled down to 50°C (122°F) 2 ounces (60 ml) dark rum (40% alcohol by volume) ½ ounce (15 ml) simple syrup 2 drops saline solution or a pinch of salt 3½ ounces (105 ml) whole milk, frozen into ice cubes Shake liquid ingredients with milk ice and serve in a chilled glass.	Makes one 7½-ounce (225-ml) drink at 10.7% alcohol by volume, 4.1 g/100 ml sugar, 0.30% acid
357	Nitrous Espresso	1½ ounces (45 ml) espresso 1¾ ounces (52.5 ml) vodka at 40% alcohol by volume ½ ounce (15 ml) simple syrup 1½ ounces (52.5 ml) filtered water 2 drops saline solution or a pinch of salt Mix, chill, carbonate with N_2O, and serve in a chilled flute.	Makes one 5½-ounce (165-ml) drink at 12.7% alcohol by volume, 5.6% g/100 ml sugar, 0.41% acid
359	Coffee Zacapa Cocktail	2 ounces (60 ml) Coffee Zacapa ½ ounce (15 ml) simple syrup 2 drops saline solution or a pinch of salt Shake with ice and serve in a chilled coupe glass.	Makes one 3⁹⁄₁₀-ounce (117-ml) drink at 15.8% alcohol by volume, 7.9% g/100 ml sugar, 0.38% acid

PAGE	RECIPE	INGREDIENTS	STATS
362	The Best G&T You Can Muster If You Can't Muster Much	Glass from freezer 1¾ ounces (52.5 ml) gin from freezer 3¼ ounces (97.5 ml) ice-cold fresh tonic water Squeeze of lime Ice from freezer	
367	Quinine Simple Syrup	1 liter (1230 grams) of simple syrup and 0.5 gram of quinine sulfate USP. Mix and strain to remove possible undissolved quinine.	
368	Cinchona Syrup	20 grams powdered cinchona bark (around 3 tablespoons) 750 ml filtered water 750 grams granulated sugar Simmer bark in water 5 minutes, cool, strain through coffee filter, redilute to 750 ml, then dissolve in sugar.	Makes 1.2 liters
373	Tonic Water Two Ways	4¾ ounces (142.5 ml) Quinine Simple Syrup or Cinchona Syrup 4¼ ounces (127.5 ml) clarified lime juice, or premade lime acid, or 5.1 grams citric acid and 2.6 grams malic acid and the tiniest pinch succinic acid dissolved in 120 ml water 20 drops (1 ml) saline solution or a couple pinches of salt 25 ounces (750 ml) filtered water Mix, chill, carbonate.	Makes 34 ounces (1021 ml) at 8.8 g/100 ml sugar, 0.75% acid

INDEX

Page numbers in *italics* refer to photographs, charts, and tables. Page numbers in **boldface** refer to recipes.

ethanol (*continued*)

in Generic Blender Sour, 119

percentage in cocktail ingredients of, 136–37

and pressure when carbonating, 196

in Rapid Hot Pepper Tincture, 214

SP-L and, 244

eutectic freezing, 163

evaporation, evaporative cooling, 45–47, 224

exothermic reactions, 290

eyedroppers, 29, 30, 48, 49

Fahrenheit (F), 20

fats, 54, 283

fat washing, 264, 282, 283–86

pioneers of, 283, 285

technique for, 285, 285

Fernet Branca, 183

filter clarification, 236, 237

filtration, 73, 314

fining, *see* wine-fining

fireplace pokers, 177

Fitzgerald, Ryan, juicing prowess, 33

fixed rotor centrifuges, 37

Flat Leaf, The, **175**

flavored oils, infusion with, 233

flavor extraction in muddling, 166

flavors:

of apples, 334–35, 341–43, *342*

clarification and, 248, 338

concentration of, 124

control through rapid nitrous infusion, 189–97

fat washing for adding, 282, 283–86

infused into porous solids, 222

perception of, 47

in red-hot poker drinks, 177, 183

selective stripping of, *see* washing

flips, 177

flocculation in clarification, 247

Flor de Caña white rum:

in Blender Daiquiri, 120

in Frozen Daiquiri, 142, *142*

in TBD, 172

fluffing (carbonation), 305, 309, 320–21, *321*

fluid gels, 256, *256*

fluorescence and quinine, 364

foam, foaming, 91–94, 97, 267, 268, 273, 274, 305, *311*

clogging from, 308, 311, *311*

CO_2 loss through, 296–302, *302*, 305, 316

in espresso, 352, 357

iSi whipper and, 191–92, 357

in making apple juice, *336*

milk as agent for, 315, 353

fobbing (carbonation term), 297–98

FoodSaver, 219

Forsline, Phil, 350

Franklin, Benjamin, milk punch recipe of, **267**

Freeman, Eben, 73, 283

free-pour technique, 18–29

freezers:

alternative chilling techniques using, 140–51

in fat washing, 283

ice making in, 66, 68–69, *68*, 73

in making juice shakes, 145–50

in making slushies, 141–44

in making stirred drinks en masse, 152–56

freeze-thaw clarification, *240*

agar, 239–40, 241, 250, 253–54, *253*

gelatin, 238–39, *239*, 240, 369

inconsistent processing in, 240, *240*

releasing and thawing techniques for, 253–54

tips for, *253*

freezing point:

of beverages, 299

temperatures below, 74–79

of water, 67, 163

French Culinary Institute, New York, 219

Fresh Lime Gimlet, **132**

Frostbite and liquid nitrogen, 159

Frozen Daiquiri, **142**, *142*

frozen drinks, 141–50, 340

fructose, 50, 51, 183

fruit, frozen, 146

fruit juices, 32

frozen, 340

straining of, 36

sugar content of, 57

as sweeteners, 50

techniques for making, 33–35, *33*

fruit peels:

garnishing technique, *109*, *110*

in Rapid Orange Bitters, *210*, 211

removing pith with enzymes, 248

to drop or not to drop in drink, 110

fruit salad, alcoholic, 232, 233